S0-BFS-874

Water Renovation and Reuse

Water Renovation and Reuse

Edited by

HILLEL I. SHUVAL

Environmental Health Laboratories
The Hebrew University
Jerusalem, Israel

ACADEMIC PRESS New York San Francisco London 1977

A Subsidiary of Harcourt Brace Jovanovich, Publishers

ACADEMIC PRESS, INC.
111 Fifth Avenue, New York, New York 10003

United Kingdom Edition published by
ACADEMIC PRESS, INC. (LONDON) LTD.
24/28 Oval Road, London NW1

Library of Congress Cataloging in Publication Data

Main entry under title:

Water renovation and reuse.

(Water pollution series)
Includes bibliographies.
1. Water reuse. 2. Sewage–Purification.
I. Shuval, Hillel I., Date
TD429.W37 628′.3 76-12421
ISBN 0–12–641250–2

PRINTED IN THE UNITED STATES OF AMERICA

Contents

PART II EXPERIENCE AND PRACTICE AROUND THE WORLD

List of Contributors

Numbers in parentheses indicate the pages on which the authors' contributions begin.

SORAB J. ARCEIVALA (275), World Health Organization, Yenisehir, Ankara, Turkey

D. A. BAILEY (395), Yorkshire Water Authority, Yorkshire, England

GEORGES BELFORT (129), Human Environmental Sciences Program, School of Applied Science and Technology, The Hebrew University, Jerusalem, Israel

DOLLOFF F. BISHOP (427), Municipal Environmental Research Laboratory, U.S. Environmental Protection Agency, Cincinnati, Ohio

LAWRENCE K. CECIL (93), Consulting Chemical Engineer, Champaign, Illinois

LUCIEN DUCKSTEIN (191), Departments of Systems and Industrial Engineering and of Hydrology and Water Resources, University of Arizona, Tucson, Arizona

G. E. EDEN (395), Water Research Centre, Stevenage Laboratory, Stevenage, Herts, England

AKIVA FEINMESSER (73), Ministry of Agriculture, Jerusalem, Israel

OLIVER O. HART (353), National Institute for Water Research, Council for Scientific and Industrial Research, Pretoria, Republic of South Africa

K. JONES (395), Water Pollution Research Laboratory, Stevenage, Herts, England

William F. Jopling (217), Water Sanitation Section, California Department of Health, Berkeley, California

Chester C. Kisiel[1] (191), Departments of Systems and Industrial Engineering and of Hydrology and Water Resources, University of Arizona, Tucson, Arizona

Takeshi Kubo (331), Japan Sewage Works Agency, Minato-ku, Tokyo, Japan

F. M. Middleton (3), Municipal Environmental Research Laboratory, U.S. Environmental Protection Agency, Cincinnati, Ohio

W. J. Müller (255), Lehrstuhl für Wasserversorgung, Abwasserbeseitigung und Stadtbauwesen, Technische Hochschule, Darmstadt, Germany

Josef Noy[2] (73), Soil and Irrigation Field Service, Ministry of Agriculture, Jerusalem, Israel

Henry J. Ongerth (217), Water Sanitation Section, California Department of Health, Berkeley, California

Gedaliah Shelef (309), Department of Environmental Engineering, Technion-Israel Institute of Technology, Haifa, Israel

Hillel I. Shuval (33), The Hebrew University, Jerusalem, Israel

G. J. Stander (117), Water Research Commission, Pretoria, Republic of South Africa

Akinori Sugiki (331), Japan Sewage Works Agency, Toda City, Saitama-ken, Japan

Lucas R. J. van Vuuren (353), National Institute for Water Research, Council for Scientific and Industrial Research, Pretoria, Republic of South Africa

[1]Deceased.
[2]*Present address:* At Ruppin Institute of Agriculture, Emek, Hefer, Israel.

Preface

You never know the worth of water till the well runs dry.

BENJAMIN FRANKLIN

It is the goal of this book to present a detailed and up-to-date review of the general principles and technological developments of water renovation and reuse and to provide documented case studies of reuse experience and practice throughout the world.

Part I, "General and Technological Aspects," includes eight chapters dealing with different aspects of the problem, each chapter written by an authority in the field. Part II, "Experience and Practice around the World," also includes eight chapters which report on water reuse practice in developed as well as developing countries, including arid zones and those endowed with abundant water resources.

There is a growing understanding throughout the world of the urgent need to conserve, recycle, and reuse our limited water resources. The science, technology, and practice of water renovation and reuse have gone through a number of phases in the past one-hundred years. The initial phase was motivated by two quite different thrusts: one based on the conservationists' concept that society's wastes should be conserved and utilized to preserve the fertility of the soil, while the other, more pragmatic approach, was directed toward eliminating river pollution.

The conservation approach is expressed in Victor Hugo's "Les Miserables," published in 1868, in which the author eloquently deplored the dumping of the sewage of Paris into the river, running wastefully to the sea. He wrote: "All the human and animal manure which the world loses, if returned to the land, instead of being thrown into the sea, would suffice to nourish the world."

The pollution control approach is exemplified by the report of the first Royal Commission on Sewage Disposal in England in 1865. The Commission recommended that "The right way to dispose of town sewage is to apply it continuously to land, and it is by such application that the pollution of rivers can be avoided." At the end of the last century, the concept of land treatment of sewage by grass filtration and broad irrigation was initiated in the United Kingdom, Germany, and the United States primarily as a method of sewage treatment to reduce river pollution rather than as a rational method of conserving water or returning nutrients to the soil. A number of early projects, however, stressed the conservation aspect of water reuse. Most of these early land disposal projects were eventually abandoned as the cities grew because the extensive land areas required were no longer available and because of esthetic and public health considerations.

In the second phase, which continues to this day, the driving force has mainly been the need to conserve and reuse water in arid areas. At first, we see the main efforts of water renovation and reuse for agriculture developing in the water-short areas of the United States such as California and Texas and in countries such as South Africa, Israel, and India. In Israel, for example, wastewater reuse became a declared national policy in 1955. The National Water Plan included reuse of all major sources of municipal wastewater in the program for the development of the country's limited potential water resources.

Today, due to the ever increasing demands for more and more water, plans for water renovation and reuse are spreading to many areas of the world not normally considered arid. Such programs have broadened to industrial, recreational, and even municipal use in addition to the already well-established forms of reuse in agriculture.

The third phase overlaps the second and is based once again on the urgent need to reduce river and lake pollution. In areas in which expensive high levels of advanced waste treatment are required to protect waters, it has become apparent to planners that once so much effort has been devoted to treating the wastewater it might be more logical to reuse it directly rather than dump it back into the rivers.

The most extreme expression of this phase has resulted from the policy of "zero pollution" which has been promulgated recently in the United States. This policy may lead to programs of land disposal or other forms of reuse solely as pollution control measures whether or not there is an objective need for water reuse.

Simultaneously with this current phase of planned, *direct* reuse, we are witnessing massive *indirect* or *covert* reuse of wastewater as a result of the almost universal withdrawal of water supplies for urban, industrial, and agricultural purposes from heavily polluted rivers. The down-stream sections of the world's major rivers carry significant loads of wastewater, much of it only partially treated, if at all. During periods of minimal base flow, rivers such as the Rhine, Thames, and Ohio may carry anywhere between 20 and 50% urban and industrial wastewater. Water withdrawn from such sources is without doubt one of the most common forms of wastewater reuse. It has been estimated that some 100 million people throughout the world are being supplied today with drinking water by this form of *indirect* wastewater reuse.

There is increasing evidence that conventional water treatment plants are not fully capable of removing the hundreds of potentially harmful organic and inorganic pollutants that appear in such water sources. Nor can they be fully depended upon to remove or inactivate all harmful microorganisms of sewage origin. Viruses have been shown to be particularly resistant to conventional treatment methods of heavily polluted water with high concentrations of organic matter. Advanced wastewater treatment technology now being developed is needed even more urgently to meet the problems arising from *indirect* or *covert* reuse than for any future plans that may eventually develop for *direct* reuse for municipal purposes.

Wastewater renovation and reuse technology has today become a major area of interest to engineers, biologists, chemists, agronomists, public health officers, and water resources authorities. Their concern may vary from the need to prevent surface water pollution, the desire to conserve and recycle soil nutrients, and the development of additional water resources for agriculture, industry, or urban uses as well as the protection of public health.

We are living in a world that is rapidly despoiling and exhausting its limited water resources. As time goes on, the rational conservation, renovation, and reuse of water will play a major role in protecting our precious water sources, recycling them in a rational way for the better use of man.

Over the past twenty-five years, the various aspects of this problem have been extensively researched. In addition, vast practical experience has been gained in many parts of the world in actual water reuse practice. Water reuse today is rapidly developing a sound, scientific base and can draw on many new and important technological developments. This volume will provide designers and scientists, as well as policy makers, with a better understanding of the complex nature of this vital and growing area of water resources management which is so closely related to the protection of human health and well-being.

Hillel I. Shuval

Part I

General and Technological Aspects

1

Advanced Wastewater Treatment Technology in Water Reuse

F. M. Middleton

I. INTRODUCTION

It is clear that in the United States and many other parts of the world recycling and reuse of all our resources will have to become a way of life. The President of the United States stated the case well when presenting to the Congress the first report on Environmental Quality (1970). He said, "We can no longer afford the indiscriminate waste of our natural resources; neither should we accept as inevitable the mounting costs of waste removal. We must move increasingly toward closed systems that cycle what are now considered wastes back into useful and productive purposes."

Water has always been used and reused by man. The natural water cycle, evaporation and precipitation, is one of reuse. Cities and industries draw water from surface streams and discharge wastes into the same streams, which, in turn, become the water supplies for downstream users. In the past, dilution and natural purification were usually sufficient for such a system to perform satisfactorily, but, in recent years, population and industrial growth have made it evident that wastewater must be treated before discharge to maintain the quality of the stream. More often than not, treatment has become inadequate or nonexistent.

Conservation measures would save much water that is now wasted. Manufacturing processes can often be altered to cause less pollution and water in an industrial plant can be recycled. Nevertheless, cities and industries will still require large amounts of water. Pollution control measures require the treatment of wastewater to restore it to good quality so it may be reused. Wastewater so treated can be considered an additional water resource, and its planned reuse for purposes other than drinking can result in large savings of clean water supplies. From the health point of view, the direct reuse of wastewater may be different only in degree—or perhaps not at all—from the indirect or unintentional reuse resulting form the withdrawal of polluted water from rivers. The good management of all of our water resources is the key of optimum use.

Pollution control has now become a necessity in most countries and, backed by the force of law, large-scale abatement projects are in progress. Most wastewaters contain only small amounts of contaminants. Municipal wastewaters are often only 0.1% contaminant. It is obvious that a huge reuseable water resource exists in the wastewaters from cities and industries. The potentials for the various types of wastewater reuse are discussed elsewhere in this book.

The purpose of this chapter is to describe wastewater treatment processes, particularly advanced systems, and give examples of combinations of processes to achieve a variety of water qualities for reuse.

II. SPECIFIC CONSIDERATIONS GOVERNING REUSE

The reuse of treated effluents is most applicable where large volumes of water are used and the wastes are not too contaminated. Industrial wastes may be heavily contaminated and therefore may not offer much potential for the recovery of clean water. The location of the treatment plant and the possible transport of the renovated water are important considerations. A wastewater renovation plant need not always be located at the same place as the municipal wastewater disposal plant, nor should the renovation process be dependent upon treating the total flow. Treatment processes work most efficiently and economically when dealing with a steady flow of wastewater, rather than with the irregular flow normally experienced from urban sources. This condition can be obtained by withdrawing only a part of the urban wastewater, as depicted in Fig. 1, which shows how water renovation and reuse can be planned to best advantage in the community.

One very important question is whether the wastewater will be reused only once or whether it will be recycled many times; multiple recycling results in a buildup of certain dissolved materials, especially inorganic ions, that may make demineralization necessary. Most reuses do not lead to a high degree of recycling. Irrigation, which is an increasingly common reuse of wastewater is highly variable, but it also may not normally offer an opportunity for multiple recycling unless it is serving as a step in a treatment system for producing water for domestic use.

Domestic reuse offers the best recycle opportunity, but, even then, the amount of water recycled falls short of the total amount of water used. The wastewater arriving at the treatment plant is generally found to be less than the amount originally supplied to the municipal water system. Losses occur, and they may be quite large in warm dry areas, where domestic reuse is most likely to be practiced. In the United States, it is estimated that these losses range from less than 20% in humid areas to about 60% in arid areas. Losses of this magnitude call for a substantial amount of make-up water, which, in turn, keeps the mineral concentration from building up excessively. The degree of demineralization needed is thus substantially less than it would be in the absence of losses and make-up water. It can be achieved by demineralizing either the renovated wastewater or the supplementary water source. In certain circumstances, the latter may be more effective.

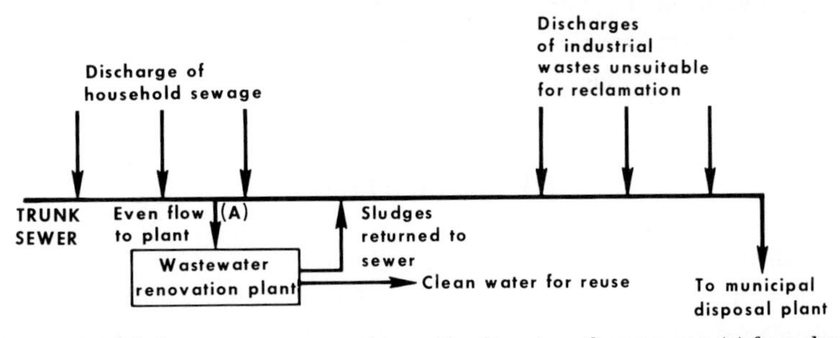

Fig. 1. Simplified wastewater reuse scheme. The diversion of wastewater (a) from the trunk sewer to the wastewater renovation plant should be chosen at a point where it is known that the trunk sewer contains only household sewage.

Another consideration is the character of the wastewater entering the renovation plant, especially if this waste includes some industrial pollutants. Care should be taken to exclude materials that would be detrimental to the application for which the reclaimed water is to be used. This is especially true for the domestic reuse. Such materials may not be only those usually considered toxic. Ordinary salt brines, for example, are undesirable if the renovated wastewater is to be demineralized. A survey of the sewer system will determine how much of the available wastewater could be reused. Water highly contaminated with metals or containing a high total concentration of dissolved solids may be unacceptable.

Another point to be considered is the distribution of the renovated water. A multiplicity of piping systems, each one containing renovated water of a different quality, would be scarcely practical and would multiply any potential hazards. However, if there are, in the vicinity of the treatment plant, a few large users of reclaimed water for nondomestic purposes, distribution is simple and inexpensive. If the users are widely dispersed, however, one piping system in addition to the existing municipal water supply system is almost certain to be the most that is economically realistic.

Treated wastewater may be deliberately used in a planned way for a variety of purposes some of which are shown in Fig. 2. Intentional reuse is not new, but there is a growing recognition of the need for it.

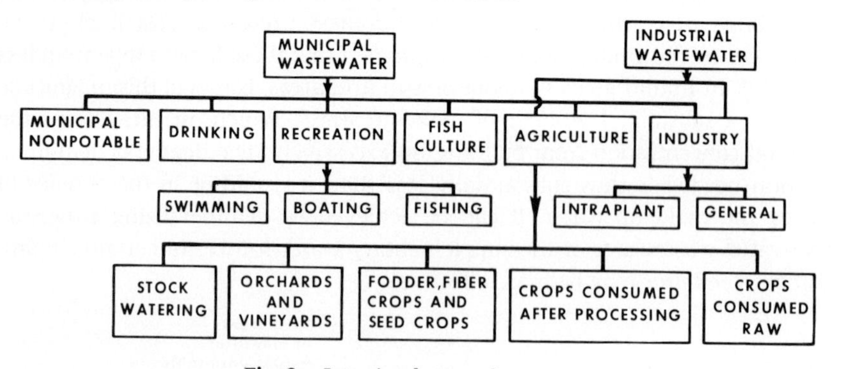

Fig. 2. Intentional reuse of wastewater.

III. DEFINITIONS

The following definitions apply to this chapter.

Municipal wastewater. The spent water of a community: it consists of water that carries wastes from residences, commercial buildings, and industrial plants and surface- or groundwaters that enter the sewerage system.

Indirect reuse. Indirect reuse of wastewater occurs when water already used one or more times for domestic or industrial puposes is discharged into fresh surface or underground waters and is used again in its diluted form.

Direct reuse. The planned and deliberate use of treated wastewater for some beneficial purpose, such as irrigation, recreation, industry, the recharging of underground aquifers, and drinking.

In-plant water recycling. The reuse of water within industrial plants for conservation and pollution control purposes.

Industrial wastewater. The spent water from industrial operations, which may be treated and reused at the plant, discharged to the municipal sewer, or discharged partially treated or untreated directly to surface waters.

Advanced waste treatment. Treatment systems that go beyond the conventional primary and secondary processes. Advanced waste treatment processes usually involve the addition of chemicals (biological nitrification–denitrification, the use of activated carbon), filtration, or separation by use of membranes.

IV. COMPOSITION OF WASTEWATERS

Unpolluted surface- and groundwaters contain various minerals and gases depending upon the geology and surface terrain. Use of water by a city adds a variety of materials such as grit, dirt, oil, bacteria, fertilizer, pesticides and miscellaneous organic matter from streets or land erosion; human waste (organic matter, bacteria, viruses, salts); laundry waste (inorganic salts, phosphates, salts, surfactants); industrial waste (heat, inorganic salts, color, metals, organics, toxic materials, oils, and the product itself). Even with the myriad materials in wastewaters, municipal wastes are 99.9% water.

Because of the many materials in wastewater at very low concentrations, it is impracticable to measure all of them. Measurement of classes of contaminants has become the rule; thus, the measure of the organic matter in sewage relies upon the biochemical oxygen demand (BOD) and the chemical oxygen demand (COD) and the total organic carbon (TOC). The BOD measures the oxygen required by organisms, the COD measures the oxygen required to chemically oxidize the organics, and the TOC directly measures the organic carbon. Determinations of solids are relatively easy. The microbiological characteristic principally depends upon measuring the coliform group of bacteria.

As wastewater reuse increases, it will be necessary to measure and monitor the water quality to a much greater degree. Good progress is being made in this field of work.

V. TREATMENT PROCESSES

For much of the world the accepted method of wastewater treatment is that illustrated in Fig. 3. This system is referred to by a variety of titles, such as conventional treatment, biological treatment, and primary plus secondary treatment. This treatment train has been specifically developed to remove suspended solids, biodegradable organics, and microorganisms from wastewater. In the past, it was only necessary to substantially remove these three classes of pollutants from wastewater prior to discharge to avoid adverse environmental effects. This is no longer generally true because of expansion in population, expansion of industry, and much greater demand for recreational and esthetic use of water resources.

Although, in many locations, the conventional system no longer is sufficient of itself, it is the base upon which some newer, more effective, treatment systems are constructed. Therefore, a short description of this system and the degree of treatment which it can provide is given below. Standard textbooks should be consulted for details of treatment processes.

A. Screening and Settling

As illustrated in Fig. 3, wastewater is first passed through preliminary treatment of screening and grit removal. Preliminary treatment is utilized to protect downstream pumps and pipes from harm from large articles and abrasives which are often found in sewage. Next, primary sedimentation is provided to remove relatively large organic solids. Such treatment is referred to as primary. In the case of municipal wastewaters, only about one-third of the oxygen demanding materials are removed by primary treatment. Bacteria and viruses are partially removed by settling.

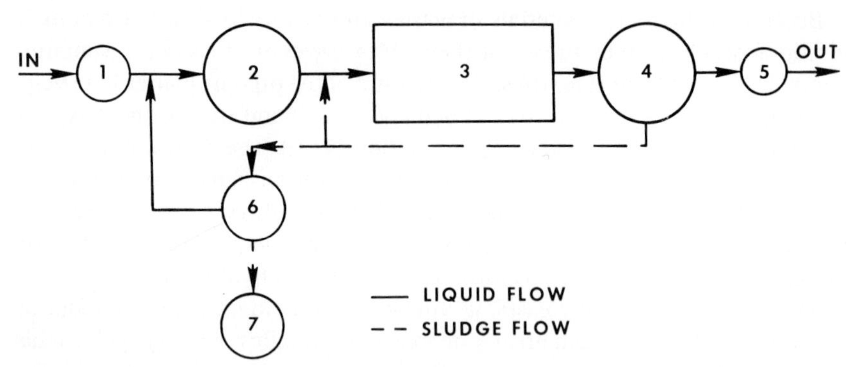

Fig. 3. Biological treatment system. 1, Preliminary treatment; 2, primary sedimentation; 3, biological oxidation, 4, secondary sedimentation; 5, disinfection; 6, sludge dewatering; 7, ultimate sludge disposal.

Wastewater effluents from municipal plants using primary treatment can be used for irrigation of crops, such as cotton, but not for direct human consumption. About one-half of all reused municipal wastewater is used for irrigation.

B. Biological Processes

The next step is biological oxidation in which a large mass of micro-organisms is contacted with the sewage in an aerobic environment. In this step, the microbes consume the soluble and colloidal organics producing more microbes and carbon dioxide and water. Two systems that are used for biological oxidation are the trickling filter and activated sludge. Details of these will not be discussed here. The microbes active in biological oxidation are separated from the flow in the secondary sedimentation tank. Being organic matter, they cannot be discharged with the effluent. Some are recycled to the head end of the biological oxidation process in order to maintain an adequate population in this unit, the remainder are sent to the sludge-handling section. After passage through the secondary sedimentation tank, the flow is often disinfected, usually by the application of chlorine and discharged.

Well-treated secondary effluents can be applied widely to reuse, including cooling for industries, irrigation of most crops and grass watering.

Table I illustrates the performance which can be expected of a typical well-operated conventional treatment plant. This type of plant cannot provide significant removal of phosphorus, nitrogen, or salts. Suspended solids, organic removal, and microorganism removal are significant.

C. Advanced Processes

Advanced processes can be applied to wastewaters to meet stringent pollution control requirements and to provide high-quality water for many reuses. The advanced processes are sometimes combined with the conventional

Table I
Typical Performance—Conventional Secondary Treatment

Pollutant	Effluent (mg/liter)	% Removal
Suspended solids	20–30	80–90
BOD	15–25	80–90
COD	30–60	70–80
Ammonia–N	15–25	0–10
Phosphorus–P	6–10	0–40
Coliform	1/ml	99.999

processes or may be used in a tertiary manner following conventional treatment. Principal processes include chemical treatment, adsorption with activated carbon, filtration, reverse osmosis, electrodialysis, microscreening, ion exchange, chlorination, and ozonation. The principles of these processes and the design basis for their use are described in Weber's book (1972).

1. COAGULATION AND FLOCCULATION

Coagulation and flocculation by the use of chemicals are often needed to enhance the removal of solids that do not normally settle. The terms coagulation and flocculation are sometimes used interchangeably, but two distinct processes occur. Colloidal particles carry charges and coagulating chemicals neutralize the charge and allow the particles to come together. Then, the neutralized particles bond together into larger particles or flocs. This constitutes flocculation. The larger particles aided by the weight of the chemicals will now settle. Coagulation occurs almost the instant the chemicals are mixed with the solution: flocculation requires some time for the agglomeration and growth of flocs.

Certain chemicals also serve to remove metal contaminants and phosphorus from wastewaters. The formation of flocs and the subsequent settling effectively removes large numbers of bacteria and viruses.

Lime, alum, iron salts, and organic synthetic polymers (sometimes called polyelectrolytes) are the chemicals commonly used for coagulation. Chemicals can be applied at several alternative locations within the plant. However, the selection of chemicals and their application must be tailored to the individual water and quality desired.

Lime is a common, cheap chemical available in most places. When lime reacts with municipal wastewater, three principal reactions occur:

Lime reacts with the bicarbonate alkalinity.

$$Ca(OH)_2 + Ca(HCO_3)_2 \longrightarrow 2\ CaCO_3 + 2\ H_2O \quad (1)$$

Lime reacts with orthophosphate to precipitate hydroxyapatite.

$$5\ Ca^{2+} + 4\ OH + 3\ HPO_4^{2-} \longrightarrow Ca_5OH\ (PO_4)_3 + 3\ H_2O \quad (2)$$

Lime raises the pH of the water, and above pH 10.5 magnesium, which is common to most waters, precipitates.

$$Mg^{2+} + Ca(OH)_2 \longrightarrow Mg(OH)_2 + Ca^{2+} \quad (3)$$

Lime can be added at the primary step in treatment plants, or it can be used as tertiary treatment. Lime should not be added to the aeration step in activated sludge plants. Small amounts of ferric chloride are sometimes added with the lime to assist in flocculating the colloidal phosphate and to weight the floc for better settling. The use of lime to treat municipal wastewaters is further explained by Culp and Culp (1971) and in a design manual by Black and Veatch (1976). Proper handling and staging of lime treatment permits recovery of relatively pure calcium carbonate ($CaCO_3$), which can then be reclaimed by recalcining according to Eq. (4).

$$CaCO_3 \xrightarrow{\text{heat}} CaO + CO_2 \tag{4}$$

The CO_2 formed can be used in the recarbonation step in the two-stage lime system.

Lime treatment forms more sludge on a dry basis than does alum or iron, but not more on a wet basis. Lime sludges usually dewater better than other chemical sludges. Also, lime may help to thicken alum sludges and aid in the dewatering of these sludges.

The aluminum ion is a good coagulant. Alum [$Al_2(SO_4)_3$] is the form most commonly used in wastewaters. Alum reacts with the bicarbonate (HCO_3^-) in the water, and a hydrolyzing action takes place as follows:

$$Al(SO_4)_3 + 6\ HCO_3^- \longrightarrow 2\ Al(OH_3) + 3\ SO_4^{2-} + 6\ CO_2 \tag{5}$$

The gelatinous aluminum hydroxide floc adsorbs colloidal particles and settles. Ferric chloride is sometimes added to aid in settling the floc.

Phosphates are present in most wastewaters and alum reacts with the phosphates:

$$Al_2(SO_4)_3 + 2\ PO_4^{2-} \longrightarrow AlPO_4 + 3\ SO_4^{2-} \tag{6}$$

Sodium aluminate ($NaAlO_2$) can also serve as a source of the aluminum ion, but it does not coagulate well in soft waters. Both alum and sodium aluminate are available in the wet and dry form. Alum reduces the pH, and sodium aluminate raises the pH. The ratio of the dose of aluminum to phosphorus is higher than indicated by the theoretical amounts shown in Eq. (6). Alum sludges are difficult to dewater.

The iron salts, ferrous chloride ($FeCl_2$), ferric chloride ($FeCl_3$), ferrous sulfate [$Fe(SO_4)_3$], and ferric sulfate [$Fe_2(SO_4)_3$], all react with water to form the flocculant ferric hydroxide [$Fe(OH)_3$]. Ferrous salts are not useful unless they first oxidize to the ferric state. Hence, ferrous salts can be used ahead of aeration basins, but may not work well after such basins. Above pH 7.0, iron reacts with phosphorus to form $FePO_4$. Waste pickle liquors from processing steel contain large amounts of $FeCl_2$ and are used to control phosphorus in municipal wastes. Some residual iron may pass through the process into the effluent if filtration is not provided. Also, when iron sludges are digested, some of the phosphate may be released into the supernatant.

Synthetic organic polymers consist of repeating chemical units that make up the molecule. A polyacrylamide is an example. Some of the polymers possess ionizable functional groups and thus may be cationic (positively charged) or anionic (negatively charged). If no ionizable groups are present, the compounds are nonionic.

The chemistry and action of these materials in waters are little understood. They exhibit remarkable abilities in aiding coagulation and flocculation of other chemicals and often serve alone as flocculants. A theoretical discussion of these materials is found in a report (Committee on Coagulation, 1971).

2. SOLIDS REMOVAL

Removing the natural solids and those formed by chemical treatment is the next step in obtaining a clean water from wastewater. Sedimentation by gravity is the simplest form of removal. Sludge-removal facilities must, however, be provided. Typical clarifier and sludge-removal equipment are shown in Fig. 4. Recently, shallow sedimentation devices consisting of a series of inclined tubes have come into use. The principle of the tube settlers is shown in Fig. 5. Culp and Culp (1971, pp 39–47) have discussed sedimentation devices in detail.

Even after the best settling devices have been used, fine materials still remain in wastewaters, and additional straining or filtration may be required. Filtration is simply a process to clarify a suspension by causing it to flow through

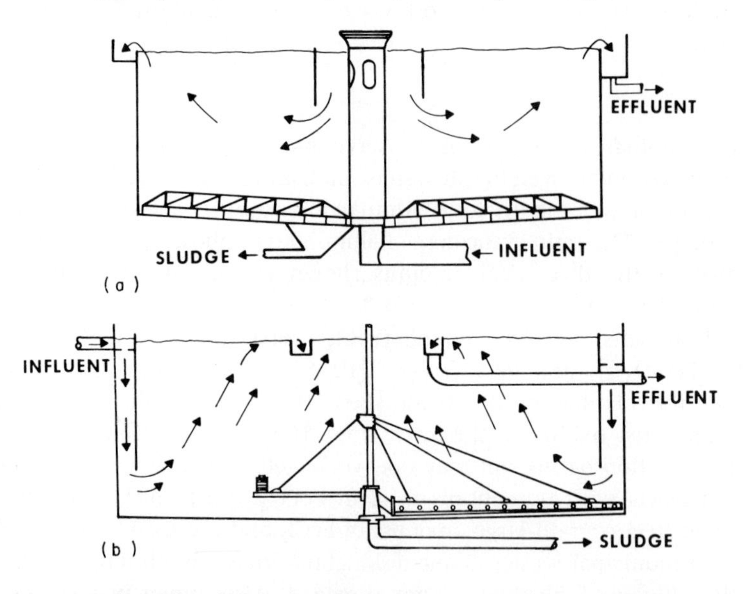

Fig. 4. Typical clarifier feed and sludge removal mechanisms. (a) Circular center feed clarifier with a scraper sludge removal system. (b) Circular peripheral feed clarifier with a hydraulic suction sludge removal system.

permeable media. Ives (1971) has compiled a comprehensive review of filtration.

Sand filters have been used for many years. More recently, mixed-media filters have been used. Anthracite coal on top of a sand bed is effective for many uses; other filter beds may contain three or more graded media of different densities and properties. Selection of the filtering process is governed by local conditions and by tests on the material to be filtered. Coagulant chemicals may be added just ahead of the filters for improved solids removal.

Microscreening devices that consist of a rotating drum covered with a stainless steel screen are available (Fig. 6). Influent water enters the drum internally and passes radially outward through the screen. The solids deposited on the inner surface of the screen are removed by jet streams of water. These solids are returned to the head of the plant. Typical screen openings are shown in Table II. Microscreens are best applied as polishing units to remove low concentrations of solids. These units and how they operate with wastewater have been described by Roy F. Weston, Inc. (1971).

Diatomaceous earth filters are frequently used for industrial applications, but less so in the municipal water field. In a Johns-Manville Company developed filtration system (the Moving Bed Filter), a constantly renewed fil-

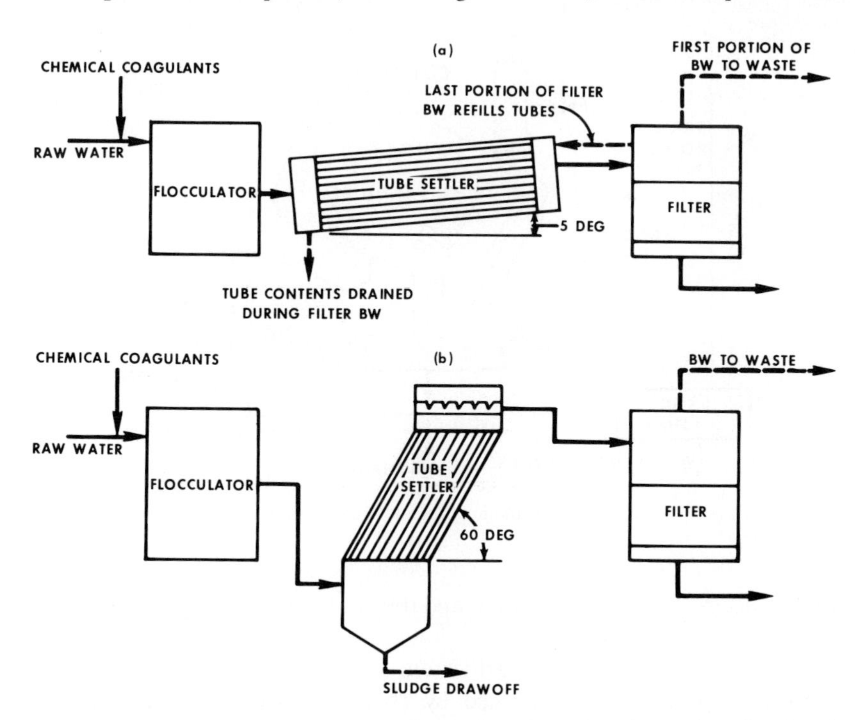

Fig. 5. Tube settler configurations: (a) horizontal tube settler, (b) steeply inclined settler. (Courtesy of Neptune Microfloc, Inc.)

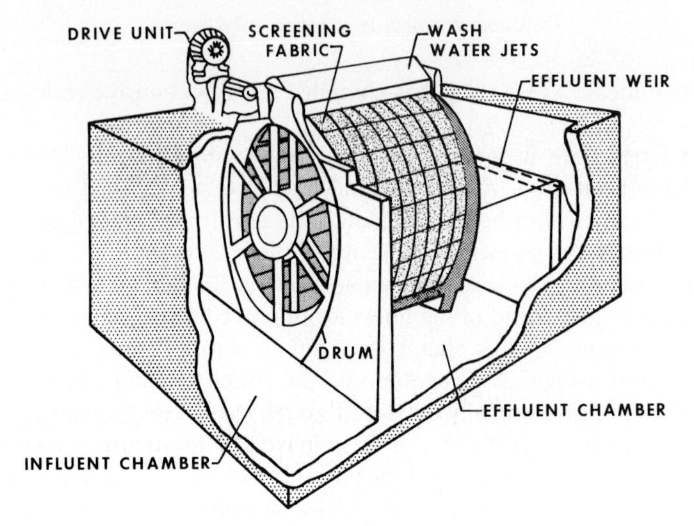

Fig. 6. Microscreen unit.

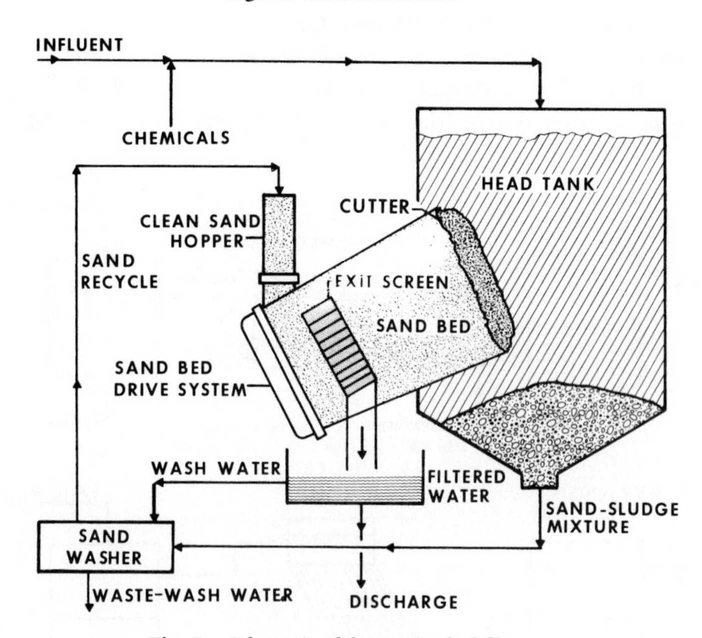

Fig. 7. Schematic of the moving-bed filter.

tering surface is exposed to the liquid and the sand is washed and recycled back into the system (Fig. 7).

Figure 8 shows a complete clarification system. Once the solid materials and those that can be precipitated by chemicals are removed, only the organic matter, dissolved solids, and various organisms that escaped prior removal remain.

Table II
Microscreen Fabric Sizes

Opening (μm)	Openings/in.^{2}
23	165,000
35	80,000
60	60,000

3. REMOVAL OF ORGANICS

As the reuse purpose approaches that of potable water, dissolved organics should be at the lowest possible concentration. After normal treatment processes have been applied, organic materials can be reduced by passing the water through a bed of activated carbon or by treating the water with powdered carbon.

Activated carbon is manufactured from coal, paper mill by-products, lignite, petroleum, and coke. Almost any organic material can be converted into activated carbon. The source material is heated in a controlled atmosphere to 750° to 950°C, and the carbon thus produced has a porous structure and a very large surface area per unit of weight (1000 m^2/gm is common).

Surface energy forces enable the carbon to exhibit remarkable capability for adsorbing dissolved organic materials from solution, although all organics are not adsorbed and rarely does 100% adsorption occur. Carbons differ in their characteristics, such as strength, density, fragility, and adsorptive capacities.

Carbon treatment is usually applied after wastewaters receive preliminary treatment by chemicals or by conventional settling and biological processes. Carbon is considered essential for treating wastewaters destined to become a part of a potable supply. Fortunately, carbon is effective in removing materials such as insecticides, a variety of hydrocarbons, and larger molecular weight

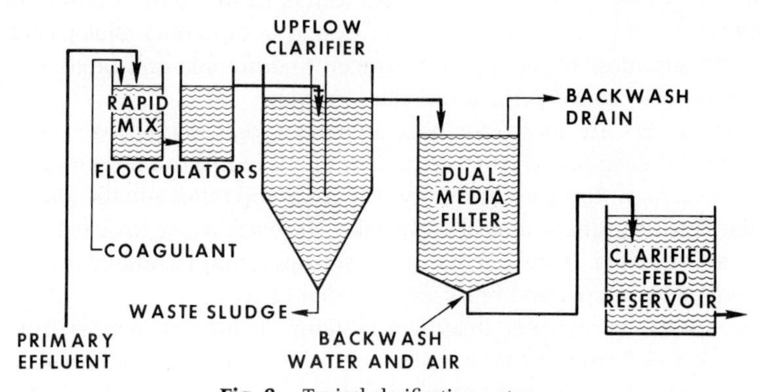

Fig. 8. Typical clarification system.

substances. Extremely water-soluble substances such as sugars are poorly removed.

The granular carbon derived from coal for treating wastewaters can be reactivated in a furnace and reused. A 5 to 10% loss occurs in the reactivation. Powdered carbon can also be used, but because of the difficulties of reactivation, it is usually discarded after use. Powdered carbon costs about 30 cents/lb and granular carbon costs about 50 cents/lb.

Granular carbon can be used in pressure vessels or in open beds. The water to be treated is passed through the bed at a rate to allow for a 30-min to 1-hr contact time. Carbon beds can be arranged in parallel or in series. Figure 9 shows one type of pressure carbon filter; Fig. 10 shows an activated carbon treatment system.

The efficiency of the carbon is generally based upon removal of organic matter as measured by the chemical oxygen demand (COD) or the total organic carbon (TOC). Table III presents the results of several such tests, while Table IV presents the costs associated with using granular carbon.

The effective removal of organic material from wastewater results, only partly, from adsorption. The carbon surface provides a suitable media for biological degradation and, thus, additional organic removal is obtained. Data on adsorption capacity vary, but COD removal ranges between 0.25 and 0.87 lb/lb of carbon.

Carbon is ready for reactivation when a preselected breakthrough of the material of interest occurs. For example, a 10 mg COD/liter in the carbon system effluent might call for reactivation.

Granular carbon columns also serve as filters. Filtering action is not their principal purpose, however, and high concentrations of solids should not be applied.

4. NITROGEN REMOVAL

Nitrogen is present in municipal wastewaters in the 15–25 mg/liter range. The nitrogen occurs in organic compounds and as ammonia. Biological treatment converts most of the organic nitrogen to ammonia and prolonged biological treatment converts the ammonia to nitrates.

If wastewaters are reused for agricultural purposes, the nitrogen serves as a nutrient and need not be removed. If the water is reused in a recreational lake, the nitrogen may stimulate overgrowth of algae and other aquatic plants and, thus, damage the esthetic appeal of the lake. Nitrogen is also toxic to fish when present as ammonia. Industrial reuse of wastewaters may be limited if ammonia is present, since copper and brass are corroded by it.

Conventional wastewater treatment systems do little to remove nitrogen, but once it is determined that nitrogen control is important, a removal system can be designed. Biological reactions can convert most of the nitrogen to

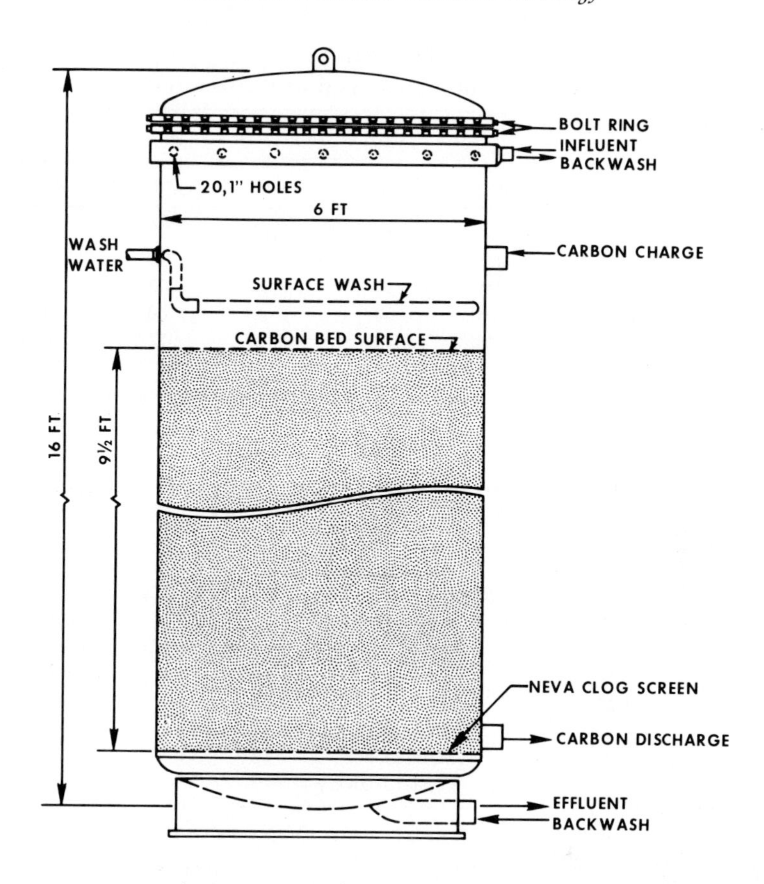

Fig. 9. Pressurized contactor for activated carbon.

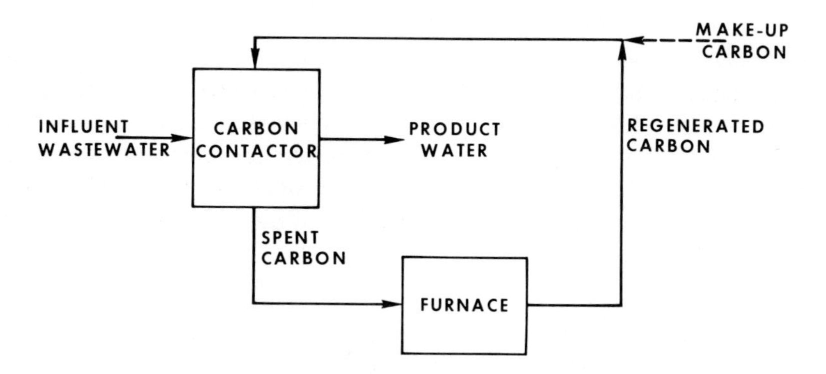

Fig. 10. Activated carbon treatment system.

Table III
Activated Carbon Treatment of Wastewater[a]

Location	Carbon	Organics (mg/liter) In	Out	% Removal	Organics measurement
Pomona	Granular	43	10	77	COD
Colorado Springs	Granular	43	13	70	TOC
S. Lake Tahoe	Granular	12	3	75	TOC
Lebanon	Powdered	20	4	80	TOC
Tucson	Powdered	27	7.5	72.5	COD

[a]Treatment of the wastewater by the activated sludge process preceded the carbon.

Table IV
Economics of Granular Carbon Treatment

Capital cost breakdown: Function	%	Operating cost breakdown: Function	%	Total cost, cents/1000 gal	
Carbon inventory	20	Make-up carbon	20	1 mgd	30
Contactors	41	Labor	12	10 mgd	10
Regeneration	12	Power	12	100 mgd	4
Pumps	27	Fuel	2		
		Backwash water	4		
		Maintenance	10		
		Capital recovery	40		

nitrates, and further bacterial action can convert the nitrates to gaseous nitrogen. Even after the best removal system is applied, a residual total nitrogen of 1 or 2 mg/liter remains. Little damage is likely from these levels; however ammonia can be removed by treating it with chlorine. A natural ion exchange material, clinoptilolite, is selective for ammonia, and this may also be used.

a. Biological Removal of Nitrogen. Most wastewater treatment systems employ biological processes. The activated sludge process is recognized as a good process and is widely used around the world. A biological nitrification process can be incorporated into such a system to convert the ammonia to nitrates. Specialized groups of bacteria perform the conversions. *Nitrosomas* oxidizes the ammonia to nitrite, the *Nitrobacter* oxidizes the nitrite to nitrate. The nitrification reactions are

$$2\,NH_4^+ + 3\,O_2 \xrightarrow{Nitrosomas} 2\,NO_2^- + 4\,H^+ + 2\,H_2O \quad (7)$$

$$2\,NO_2 + O_2 \xrightarrow{Nitrobacter} 2\,NO_3^- \quad (8)$$

Many varieties of bacteria can utilize the oxygen in nitrates and reduce them to nitrogen gas. The bacteria require organic food, and in the system discussed here, methanol is added to provide a carbon source for the bacteria.

The denitrification reaction is

$$6\ NO_3^- + 5\ CH_3OH \xrightarrow{\text{bacteria}} 3\ N_2 + 5\ HCO_3^- + 7\ H_2O + OH^- \quad (9)$$

A treatment system for removal of organic material and phosphorus that incorporates nitrification and denitrification is shown in Fig. 11.

b. Ammonia Removal by Air Stripping. At pH 7, only the ammonium ion NH_4^+ is present in wastewater. As the pH is raised from 7 to 12, a mixture of NH_4^+ ion and ammonia gas NH_3^+ occurs and the NH_3^+ increases as the pH goes up. Ammonia gas can be blown from the water by passing large volumes of air through the wastewater. A cooling tower is the device used. The factors involved and process descriptions for ammonia stripping are described by Culp and Culp (1971, pp. 52–67).

c. Ammonia Removal by Chlorination. Chlorine can convert ammonia to nitrogen gas. A series of reactions occur:

$$Cl_2 + H_2O \longrightarrow HOCl + HCl \quad (10)$$

$$NH_4 + HOCl \longrightarrow NH_2Cl + H_2O + H^+ \quad (11)$$

$$2\ NH_2Cl \longrightarrow N_2 + 3\ HCl + H_2O \quad (12)$$

Hydrochloric acid is formed and, unless neutralized by the natural alkalinity of water, an alkali must be added.

Thus, if sodium hydroxide is used, salt is formed.

$$NaOH + HCl \longrightarrow NaCl + H_2O \quad (13)$$

Mixing and pH control are critical to the successful operation of this process.

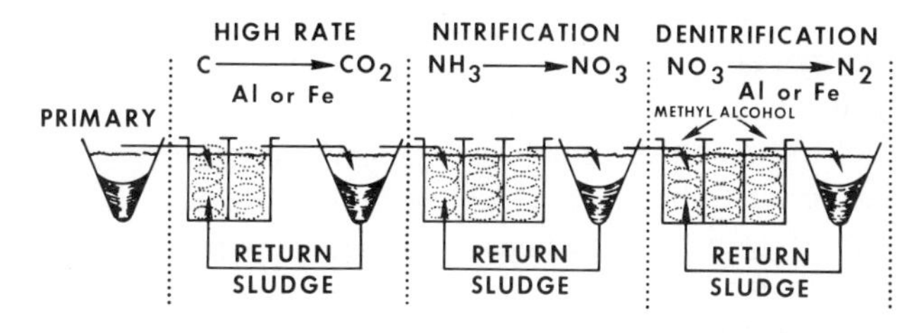

Fig. 11. Three sludge system for phosphorus and nitrogen removal.

5. DISINFECTION

Disinfection is usually the last defense against any remaining microbial contaminants that may have survived the various treatment processes. Drinking water from surface sources is almost universally disinfected, and disinfection is also applied to wastewater treatment plant effluents and to water intended for reuse.

Chlorine, although it has some shortcomings, is the most used disinfectant and has played a major role in the prevention of waterborne diseases. It is most effective as hypochlorous acid (HOCl), which is easily formed when chlorine is added to clean water that is free of organic matter and ammonia. Hypochlorous acid kills coliform bacteria in a few minutes, but takes an hour to kill some viruses. If organic materials are present, particularly nitrogenous materials, the chlorine will react with them and may not appear in the free form until a considerable amount is added.

In the presence of ammonia, mono- and dichloramines are formed when chlorine is added. These are also disinfectants, but they are much slower acting. In the presence of an excess of chlorine, the ammonia is destroyed, but each part of nitrogen present as ammonia requires nearly 10 parts of chlorine (break-point chlorination) for its destruction. It has been suggested that organic chlorine compounds originating from chlorinated effluents discharged into streams may be toxic to fish life and possibly detrimental to human health.

Ozone, a powerful oxidant, is also used as a disinfectant. The World Health Organization's (1971) International Standards for Drinking Water states that exposing water to 0.4 mg/liter of free ozone for 4 min is sufficient to inactivate viruses. In France, ozone has been used extensively for disinfecting drinking water. Ozone does not react with ammonia under most use conditions; and it decomposes to oxygen and, thus, leaves no objectionable residue.

At disinfection doses, ozone also lessens the color of effluents. The destruction of organic residues, however, calls for much greater doses, hence this method then becomes costly. Although ozone is more expensive than chlorine, its use is expected to grow, especially for wastewater disinfection.

Iodine, UV-radiation, and γ-radiation are good disinfectants, but have not come into large-scale use to treat wastewater.

Disinfection is most likely to be effective if the water is nearly free of organic matter and ammonia. Contact time is important; the recommended exposure per liter of water is at least 0.5 mg of free chlorine or the equivalent of other disinfectants of 1 hr. Substantial kills of organisms occur when any disinfectant is applied and sensitive organisms such as the *Vibrio cholera* are killed easily. Enteric viruses, in general, and hepatitis virus, in particular, require much more treatment than do most bacteria, but the recommended contact period of 1 hr will provide a margin of safety.

VI. RESIDUES RESULTING FROM TREATING WASTEWATERS

In the process of cleaning wastewaters for reuse, the various contaminants may be removed in the form of solid sludges containing both organic and inorganic materials. Salts that are soluble and cannot be precipitated may be removed by ion exchange or membrane processes. A concentrated stream of brine results and creates a disposal problem. These processes are discussed in Chapter 6.

In considering wastewater reuse systems, it is of vital importance to plan for use or disposal of the residues remaining after treatment. Their handling and disposal costs may be half the total cost of treating wastewaters.

In municipal wastewater treatment plants, trash is removed by a coarse screen and is usually burned or buried. Heavy inorganic materials, such as sand, are taken out in the grit chamber. These materials are often used as solid fill. Settling in the primary tank for 2 to 4 hr results in primary sludge consisting of fecal matter, food wastes, and miscellaneous organic matter. When the activated sludge process is used, bacteria and higher organisms that utilize the organic food material in the wastewater generate additional sludge. Some of this sludge is recycled to the process, and some is combined with primary sludge. Trickling filters produce attached biological growths on the rock and other media, and this material sloughs off from time to time to collect as sludge in the final settling tank.

The combined primary and waste-activated sludge most often goes to an an anaerobic digester where the materials decompose to a more stable material. Methane gas and carbon dioxide evolved by the bacteria in the process may be vented to the atmosphere or the methane can be burned and the heat used to advantage in the plant. The mixed sludges may also be incinerated instead of being digested.

When digested sludges are dried by mechanical filters or on sand beds, the dried sludge may be incinerated, dumped in a landfill, or spread onto the land as a soil supplement. The fertilizer value of sludge is about 1 cent/lb. When wastewaters are treated to remove phosphorus, the fertilizer value of sludges increases. Sludge need not always be dried before applying it to the land. It can be transported by truck, rail, or pipeline to disposal sites. The further degradation of the sludge on the land adds to the humus in soil, and the carbon dioxide liberated under the plants may stimulate growth. Sludges can improve worn-out soils, mine wastes, and mineral dumps.

If wet sludges are applied to the land, the possibility of disease transmission cannot be overlooked, although evidence is lacking that disease is transmitted by this practice. The sludge can be pasteurized economically, and some countries

Table V
Land Application of Wastewater, Sludge, and Sewage in Different U.S. Locations

Land application of substance	U.S.	Metric
Water		
One person uses per day	100 gal	400 liter
Or	12 ft^3	0.4 m^3
This will cover	1000 ft^2	100 m^2
And infiltrate per day	1/6 in	4 mm
Water		
100,000 people use	10 mgd	40,000 m^3/day
If spread	1/6 in.	4 mm
This would cover	2000 acres	1000 hectare
Or	3.6 mi^2	10 km^2
The total water per year will be	60 in.	1500 mm
Sludge		
One person produces per day	0.2 lb	0.1 kg
at 5% solids contained in	0.5 gal	2 liter
100,000 people produce solids per day	10 tons	9 tons
Land Assimilation of Sludge		
One acre can assimilate per year	20 tons	
One ha can assimilate per year		45 tons
100,000 people need	180 acres	70 hectare
Land Assimilation, Illinois		
Chicago Metropolitan Sanitary District produces		
Wastewater per day	1 billion gal	4×10^6 m^3
Sludge per day	1000 tons	900 tons
For irrigation of wastewater	360 $miles^2$	1000 km^2
Spreading of sludge	25 $miles^2$	70 km^2
Water Assimilation, U.S.		
200,000,000 people use per day	20 billion gal	80 million m^3
Irrigation would require	7200 $miles^2$	20,000 km^2
This is a little more than the area of New Jersey		
Sludge Spreading, U.S.		
200,000,000 people produce per day	20,000 tons	18,000 tons
Spreading would require	570 $miles^2$	1500 km^2
This is a little less than the area of Ocean County, New Jersey		
Decomposition of organic Matter (oil), Texas		
In Texas soil, per year	1 ft	30 cm
Per acre year	1000 tons	900 tons

require this during the grazing season. Lime treatment of sludges has disinfecting action, increases the agricultural value, and neturalizes toxic metals.

Table V gives some factors of interest relating to the quantities of wastewater sewage, and sludge produced by different numbers of people and the area needed for applying such quantities to the land.

Chemical sludges present some problems. Lime sludges, however, can be partially recalcined and the calcium oxide formed can be reused in the process. The chemical content of alum and iron sludges cannot be reclaimed.

VII. SYSTEMS AND COSTS (1974 DOLLARS) FOR WATER RENOVATION AND REUSE

Once local authorities have determined the types of wastewater reuse desired, the selection of treatment systems begins. Each situation is different, and the services of a consultant skilled in renovation technology are required. Some examples of treatment systems will be given. Costs given are examples and do not apply to any specific local situation.

Health protection is of paramount importance in all wastewater reuse applications. Table I in Chapter 2 shows suggested treatment processes that would meet health criteria for wastewater reuse as proposed by WHO.

A. Irrigation Reuse

Reusing of municipal wastewaters for irrigation is the oldest and largest reuse. Advanced treatment of wastewaters is not always required, but each potential reuse should be thoroughly analyzed to determine the quality required. The irrigational reuse of wastewater has been surveyed by Schmidt and Clements (1975) and by Pound and Crites (1973).

The City of Phoenix, Arizona, operates the largest wastewater reclamation and irrigation program in the United States. About 100 mgd (378.5×10^3 m^3/day) of secondary treated effluent is used to irrigate crops, add moisture to a 70-acre (28.3 ha) fish and game marsh, and supply a ground water aquifer with recharge. Stringent industrial discharge standards are enforced to ensure that toxicants are not in the effluents. The systems used at Phoenix are shown in Fig. 12.

B. Recreational Reuse

Reuse of wastewater in recreational lakes is becoming increasingly popular in the arid areas of the United States. The State of California has pioneered in recreational reuse, with several lakes in existence for a number of years. Lakes at Santee and Indian Creek Reservoir near Lake Tahoe are probably the best known. Tucson, Arizona, is installing recreational lakes derived from treated wastewaters.

Wastewater used in recreational lakes must satisfy both health standards and standards that will make the lake acceptable from the recreational standpoint.

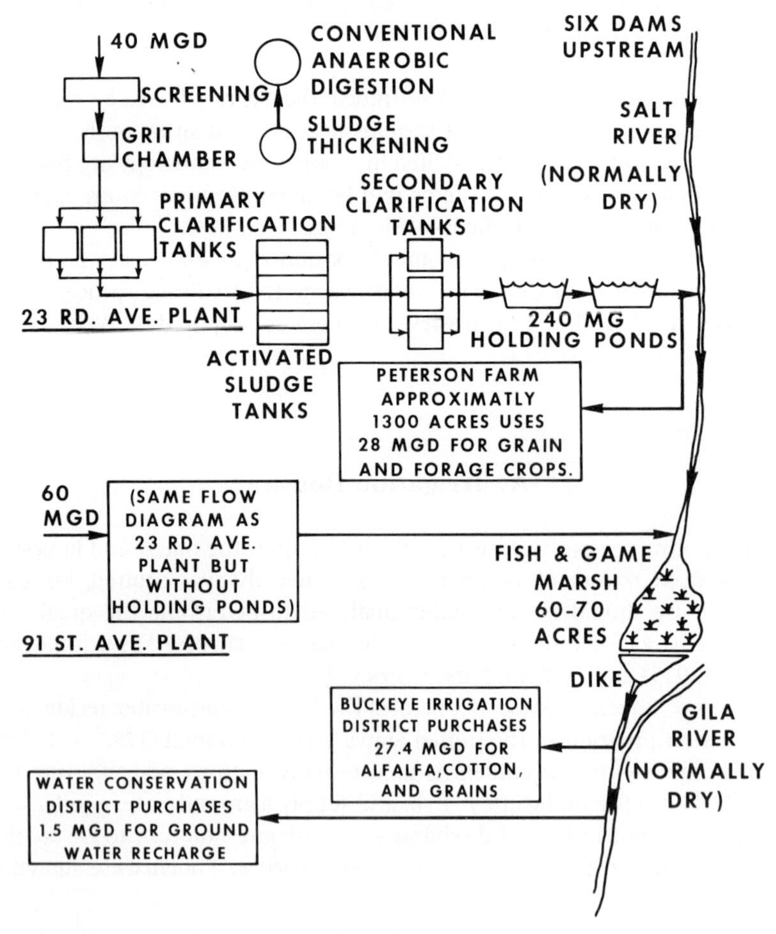

Fig. 12. Municipal water reclamation and irrigation reuse system, Phoenix, Arizona.

Although there are no national public health standards, at least one state, California, has written standards that can be used as an example. For unrestricted recreational use, wastewater in California must be biologically treated, chemically flocculated (or equally effective method), filtered to produce a turbidity of not more than 10 turbidity units, and adequately disinfected. Adequate disinfection is defined as 2.2 coliforms/100 ml (mean most probable number, MPN). The mean is determined over a 7-day period. These standards were arrived at after several years of careful observation, including monitoring for viruses, at Santee.

The public health standards cannot be used alone to define an adequate treatment system, since they do not deal in detail with problems of excessive algae growth in the lake or with problems related to ammonia in the water. Limiting the amount of the major nutrients (generally involves phosphorus or nitrogen, or both) can control excessive algae growth.

Not enough experience in this area has been gained to define an optimum treatment system for producing water for recreational reuse. Actually, there will probably never be just one optimum system. At Santee, for example, there is a natural percolation bed that forms a major part of the treatment.

For cost-estimating purposes, Fig. 13 shows a treatment system that should satisfactorily produce recreational lake water. The system includes conversion of nitrogen to nitrate and phosphorus removal down to the level of about 0.1 mg/liter, as P. Organic removal should be excellent because of the two-stage biological treatment and filtration. Suspended solids in the treated water should be almost zero. Costs for a 10-mgd (3.78×10^3 m^3/day) plant are shown in Table VI. This plant size was chosen because it is of a size that might be used

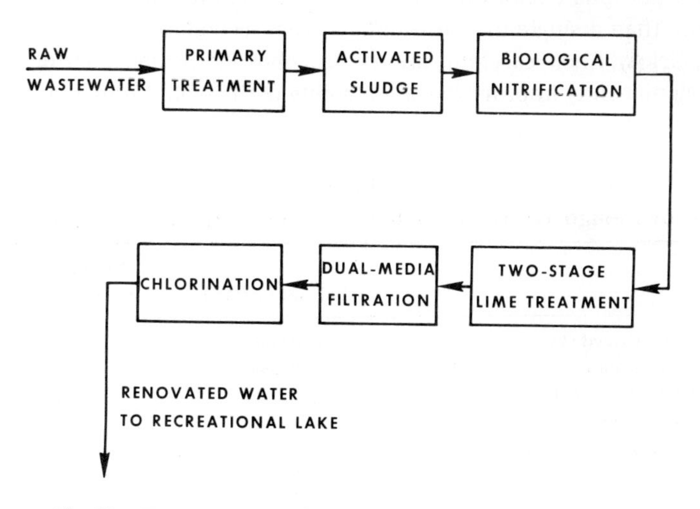

Fig. 13. Treatment system for producing recreational lake water.

for producing recreational water. For comparison, the total operating cost for a 1-mgd (3.8×10^3 m^3/day) plant would be \$1.00/1000 gal (3.8 m^3) and for a 100-mgd (378.5×10^3 m^3/day) plant, 22 cents/1000 gal (3.8 m^3).

C. Industrial Reuse

Industrial use of municipal wastewater has been practiced for some time. Early examples in the United States include the Bethlehem Steel Company at Baltimore, Maryland, and the Nevada Power Company at Las Vegas, Nevada. Until recently, nearly all of the reuse was for cooling water. Interest in using the wastewater for other industrial purposes is increasing, however. A number of industrial applications are reported by Schmidt and Clements (1975, pp. 42–72).

In the United States, it has been characteristic of industrial reuse that one plant only uses the effluent from a given sewage treatment plant. This occurs because the need for large volumes of cooling water and the one plant can use a significant fraction of the treatment-plant output. The result is a very simple distribution system. Because of the cost of distribution systems, this tendency toward a small number of users, probably located close to the treatment plant, is likely to continue.

Because there are many possible industrial uses for renovated wastewater, a wide variety of treatment systems might be employed. Secondary effluent with no further treatment has been used in some cases for cooling purposes. Calcium phosphate scale problems are likely, however, when no precautions are taken for phosphate removal. For most reuse situations, especially those involving more than a single user, it is unlikely that secondary treatment alone will be satisfactory. Chemical clarification with a phosphorus-precipitating chemical would significantly improve the acceptability of the water from both the stand-

Table VI
Costs for 10-mgd Treatment System to Produce Recreational Lake Water

Process	Capital cost (dollars)	Total operating cost (cents/1000 gal) (3.8 m^3)
Primary and activated sludge	4,902,000	17.6
Biological nitrification	1,404,000	5.1
Two-stage lime treatment	1,935,000	11.3
Dual-media pressure filtration	1,250,000	5.5
Chlorination (15 mg Cl_2/liter)	150,000	1.2
	9,641,000	40.7

point of phosphorus content and turbidity. Figure 14 shows a treatment system that should have widespread usefulness. Nitrification could be included if ammonia is a problem. Clarification could be carried out with lime, aluminum salts, or iron salts. Lime, however, would appear to have several advantages over the other materials: it does not add extraneous ions such as chloride or sulfate; it removes some heavy metals; and it produces a water that is less likely to be corrosive. If clarification were carried out with good solids control, filtration would not be required. For the relatively small additional cost, however, its inclusion significantly increases dependability. At Las Vegas, a system using lime treatment is followed by a holding pond rather than a filter. The pond is very effective for solids control.

The costs for the system shown in Fig. 14, a 10-mgd (37.8×10^3 m^3/day) system, are shown in Table VII. For a 100-mgd (378.5×10^3 m^3/day) plant the total operating cost is 20 cents/1000 gal (3.8 m^3).

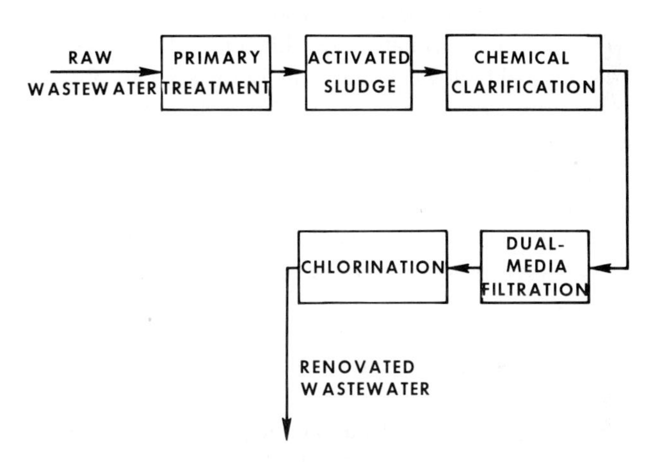

Fig. 14. Treatment system for producing industrial water.

Table VII
Costs for 10-mgd (37.8×10^3 m^3/day) Treatment System to Produce Water for Industrial Reuse

Process	Capital cost (dollars)	Total operating cost (cents/1000 gal) (3.8 m^3)
Primary and activated sludge	4,902,000	17.6
Two-stage lime treatment	1,935,000	11.3
Dual-media pressure filtration	1,250,000	5.5
Chlorination (15 mg/liter)	150,000	1.2
	8,237,000	35.6

D. Domestic Reuse of Nonpotable Water

Domestic use of wastewater must be divided into nonpotable and potable uses. Nonpotable uses are not particularly new. At the Grand Canyon in Arizona, wastewater has been reused for toilet flushing for many years. Treatment consists of biological oxidation followed by filtration and chlorination. On Pikes Peak in Colorado, a novel treatment system was recently installed to produce water for various nonpotable uses. It consists of a special pressurized, activated sludge system using a membrane for solids removal in place of the final settler.

Quality standards for nonpotable reuse have not been set on a national level. Local health authorities are responsible for approval of these systems. The major concern is for pathogenic organisms. Therefore, adequate disinfection is of great importance.

Since these systems have been small and for very specialized situations, making meaningful cost estimates are difficult. A large-scale system might consist of conventional biological treatment followed by filtration and chlorination. Total operating cost for a 10-mgd (37.8×10^3 m^3/day) plant would be about 25 cents/1000 gal (3.8 m^3).

Utilization of nonpotable domestic reuse systems is and will be greatly restricted by the need for a complete dual distribution system. In large cities, the cost would probably be prohibitive; in small towns, however, dual systems may be practical. Such systems have been proposed in the arid Southwest.

E. Domestic Reuse of Near-Potable Water

There has been a great deal of talk about potable reuse, but little activity. The only long-term operation has been that at Windhoek, Southwest Africa (Namibia), reported on by Cillie *et al.* (1967) and Stander and Funke (1967), where about one-third of the municipal water supply is renovated wastewater.

Because of the limited experience with direct reuse of renovated wastewater, there are no standards to apply to such waters. Standards for drinking water, such as those of the U.S. Public Health Service (1962), apply to water sources that are as free as possible from pollution. Many in the water field believe that renovated wastewater should meet additional standards beyond those written for largely unpolluted sources. The most important areas of concern are trace organic pollutants, metals, and pathogens, especially viruses. Analytical techniques to detect very low concentrations of organics and virus are in an early stage of development, are expensive, and are time-consuming. Analyses for metals are sensitive and generally adequate.

One answer to the problem is to insist that the water be overtreated to a point at which there could be absolutely no doubt of the water purity, even in the absence of adequate analytical techniques. Such procedure may be tolerated in some cases to obtain potable water, but this would not be possible for larger volume uses.

A system such as that shown in Fig. 15 might be used for the demineralization step, this is very much like the advanced waste-treatment plant at South Lake Tahoe, as it is presently operated. The product water from the system will meet the quantitative values of the U.S. Public Health Service Drinking Water Standards. Although there is no standard for ammonia, nitrification would be necessary to reduce the ammonia concentration to aesthetically acceptable levels. Limited virus determinations from the Tahoe plant showed no virus recovery from the product water. Whether demineralization would be required, or to what degree, would depend upon the fraction of recycle and the mineral content of the make-up water supply. A material balance would have to be made for each particular situation to determine demineralization requirements. A 40% total dissolved solids removal would probably cover most situations. Ion exchange is shown in the diagram as the method of demineralization because it presently appears to be as cheap as any other method and offers some degree of trace organic removal. Since ion exchange would probably be operated to give almost total removal of minerals, not all of the water would have to be treated. If demineralization were not included, there is a chance that the nitrate concentration would exceed the standard of 10 mg/liter, as nitrogen. In such cases, the necessary nitrate removal could be obtained by biological denitrification of part of the flow.

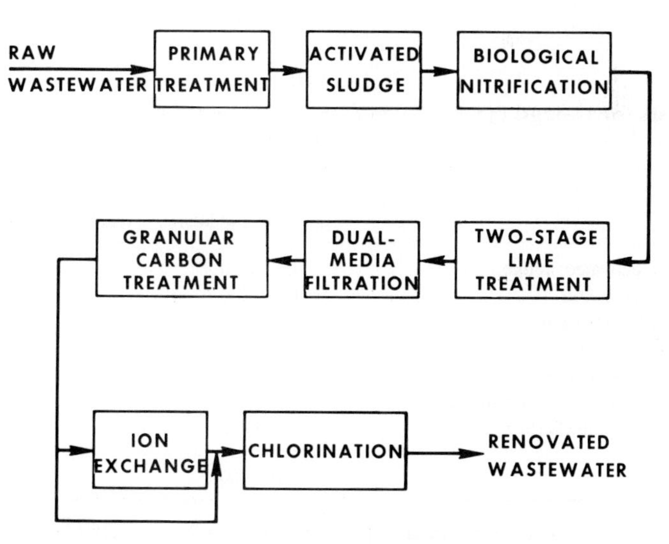

Fig. 15. Treatment system for producing near-potable water.

Costs for the system shown in Fig. 15, a 10-mgd (37.8×10^3 m^3/day) plant, are given in Table VIII. The operating cost, including amortization, of 68 cents/1000 gal (3.8 m^3) is reasonable when compared with costs from other sources in water-short areas. Total operating costs for a 100-mgd (378.5×10^3 m^3/day) plant would be about 43 cents/1000 gal (3.8 m^3).

Table IX shows the quality of water expected from the treatment system given in Fig. 15. Another treatment system that appears to have strong potential for producing potable water makes use of reverse osmosis. This system is not developed as far as the system shown in Fig. 15 and, therefore, cannot be considered as dependable as that system at present. The system would probably consist of conventional treatment, filtration, reverse osmosis, and disinfection. Approximate costs for this system are given in Table X. Considering cost reductions that are possible in reverse osmosis, the 62 cents/1000 gal (3.8 m^3) total operating cost compares very favorably with the costs for the system shown in Fig. 15.

The reverse osmosis system has the advantage of giving greater than 40% demineralization. Although not a necessity, this is a bonus. At plant sizes in the range of 1 mgd (3.8×10^3 m^3/day), the reverse osmosis system could have a significant cost advantage over the system in Fig. 15 because carbon treatment becomes quite expensive in that size range. The increase in carbon treatment cost is largely because of the cost of carbon reactivation for small plants.

A third treatment system that might be considered for producing potable water would involve distillation. Because of the high temperature to which the

Table VIII
Costs for 10-mgd (37.8×10^3 m^3/day) Treatment System to Produce Near-Potable Water

Process	Capital cost (dollars)	Total operating cost (cents/1000 gal) (3.8 m^3)
Primary and activated sludge	4,902,000	17.6
Biological nitrification	1,404,000	5.1
Two-stage lime treatment	1,935,000	5.5
Dual-media filtration	1,250,000	5.5
Granular carbon treatment	1,600,000	7.0
Ion exchange (40% mineral removal)	1,940,000[a]	20.2
Chlorination (15 mg Cl_2/liter)	150,000	1.2
	13,181,000	62.1

[a]Capital cost is for a 5-mgd (18.9 m^3/day) plant. It is assumed that one-half of the water will be treated for from 80 to 90% mineral removal.

water would be exposed, there would be 100% kill of all types of pathogens. The cost of the process, at present, is not competitive with other systems. Large plants using nuclear power may, however, make this an important process to consider in the future.

Table IX
Quality of Water from Renovation System for Producing Near-Potable Water

Parameter	Concentration in effluent[a] (mg/liter)	U.S. Public Health Service (1962) drinking-water standard (mg/liter)
Chemical oxygen demand	5	—
Coliform organisms (mean density)	Effluent breakpoint chlorinated	1/100 ml
Turbidity	< 1 unit	5 units
Color	< 2 units	15 units
Threshold odor number	No odor	3
Methylene blue active substances	< 0.5	0.5
Chloride	< 250	250
Carbon chloroform extract	< 0.05	0.2
Cyanide	0	0.01
Iron	< 0.03	0.3
Manganese	< 0.05	0.05
Nitrate (NO_3)	< 45	45
Phenols	0	0.001
Phosphorus (P)	< 0.5	—
Sulfate (SO_4)	< 250	250
Total dissolved solids	< 500	500

[a] The values given are a combination of results from a number of pilot studies (primarily the study at Pomona, California) and from the advanced waste-treatment plant as South Lake Tahoe, California.

Table X
Costs for 10-mgd (37.8×10^3 m^3/day) Treatment System to Produce Near-Potable Water Using Reverse Osmosis

Process	Capital cost (dollars)	Total operating cost (cents/1000 gal) (3.8 cm^3)
Primary and activated sludge	4,902,000	17.6
Dual-media pressure filtration	1,250,000	5.5
Reverse osmosis	5,230,000	38.0
Chlorination (15 mg Cl_2/liter)	150,000	1.2
	11,532,000	62.3

REFERENCES

Black and Veatch, Consulting Engineers. (1971). "Process Design Manual for Phosphorus Removal," U.S. Environmental Protection Agency, Washington, D.C.

Cillie, G. G., van Vuuren, L. R. J., Stander, G. J., and Kolbe, F. F. *Proc. Int. Conf. Water Pollut. Res. 3rd, 1966*, Water Pollution Control Federation, Washington D.C. pp. 1–35.

Culp, R. L., and Culp, G. L. (1971). "Advanced Wastewater Treatment." Van Nostrand-Reinhold, Princeton, New Jersey. pp. 18–26.

Environmental Quality. (1970). "The First Annual Report of the Council on Environmental Quality." Supt. of Documents, Washington, D.C.

Ives, K. J. (1971). *Crit. Rev. Environ. Contr.* **2**, (2), 293–335.

Pounds, C. E., and Crites, R. W. (1973). "Wastewater Treatment and Reuse by Land Application," Vols. I and II, Report No. 660/2-73-006a. U.S. Environmental Protection Agency, Washington, D.C.

Schmidt, C. J., and Clements, E. V. (1975). "Demonstrated Technology and Research Needs for Reuse of Municipal Wastewaters," EPA-670/2-75-038 pp 5—42 Environmental Protection Agency, Washington, D.C.

Stander, G. J., and Funke, J. W. (1967). Conservation of Water Reuse in South Africa. *Chem. Eng. Prog., Symp. Ser.* **63**, 1. Amer. Inst. Chem. Eng. New York.

State of the Art of Coagulation. (1971). *J. Am. Water Works Ass.* **63**, 99–108.

U.S. Public Health Service. (1962). "Drinking Water Standards." U.S. Dept. of Health, Education and Welfare, Washington, D.C.

Weber, W. J., Jr. (1972). "Physiochemical Processes for Water Quality Control." Wiley (Interscience), New York.

World Health Organization. (1971). "International Standards for Drinking Water." World Health Organ., Geneva, Switzerland.

Roy F. Weston, Inc. (1971). "Process Design Manual for Upgrading Wastewater Treatment Plants." U.S. Environmental Protection Agency, Washington, D.C.

2

Health Considerations in Water Renovation and Reuse

Hillel I. Shuval

I. INTRODUCTION

One of the primary considerations in the proper disposal of wastewater has traditionally been the protection of the health of the population directly served and of those that might be exposed to the wastes downstream. In earlier times, the main concern has been with the possible spread of communicable diseases by human body wastes, but as time went on, an awareness of the hazards to health associated with organic and inorganic chemicals from industrial and agricultural wastes has developed. This same basic concern with the protection of the public health must remain the sine qua non of any and all programs for water renovation and reuse.

Experience has shown that health requirements must be given very careful attention from the initial planning stages to ensure that adequate health criteria are established to meet various water objectives. Only when such health criteria have been set will it be possible to determine the design of treatment systems to meet such criteria on a sure and consistent basis.

II. TYPES OF CONTAMINANTS

The degree of concern about the various types of potential contaminants is of course dependent on the type of water reuse being considered. Such concern from the health point of view might be quite minimal for most microbial and chemical contaminants in the case of surface irrigation of industrial crops and would become extremely acute in the case of any planned reuse for domestic consumption. What follows is a brief review of the main microbial and chemical contaminants that may appear in wastewater which bear some relationship to various reuse strategies.

A. Microbial Contaminants

Since John Snow's classical investigations of cholera in London in 1854, it has been understood that water can serve as an efficient vector of human pathogenic microorganisms of sewage origin. We know today that the raw wastewater of a community usually carries the full spectrum of pathogenic bacteria, viruses, protozoans, and helminths, which are excreted by clinical cases and carriers associated with the enteric diseases endemic in the community. During periods of epidemics of enteric diseases, the concentration of pathogens can increase manyfold.

The agents of the following enteric diseases have often been detected in municipal wastewater: bacillary and amoebic dysenteries, cholera, typhoid and paratyphoid fevers, *Salmonella* gastroenteritis, tapeworm infections, shistosomiasis, ascariasis, and a number of virus diseases including poliomyelitis.

It is worth noting that the pathogenic agents of diseases not known to exist in the community can at times be detected in the wastewater, since the organisms may be excreted in quite high concentrations on a regular basis by undetected carriers. In a study of the appearance of *Salmonella* organisms in 96 samples of wastewater and polluted water in Tel Aviv, we were able to isolate 229 *Salmonella* strains which included 34 serotypes (Yoshpe-Purer and Shuval, 1970). The dominant type was *S. paratyphi B*. Many of the isolated organisms were rare in the community and a number had never been isolated from human cases in Tel Aviv. In another study carried out during the cholera outbreak in Jerusalem in 1970, we were able to detect *Cholera vibrios* in wastewater from neighborhoods where no clinical cases of the disease had as yet been reported.

Such findings provide further support to the premise that raw wastewater must be considered a most serious potential source of a wide variety of enteric pathogens, regardless of whether serious enteric disease epidemics are present in the community or not.

Kehr and Butterfield (1943) pointed out that while the level of coliforms found in community wastewater is fairly constant, the ratio of typhoid organisms to coliforms is a function of the endemic disease rate in the community. Thus, for any given concentration of coliforms found in wastewater the risk of pathogens being present would be ten- or one hundred-fold greater in certain Mediterranean, South American, or Asian countries than in the United States, since some such countries have 10 or even 100 times greater enteric disease rates. The risks associated with wastewater reuse using equivalent treatment processes would be correspondingly greater in such countries.

Enteroviruses present a particularly difficult problem since studies indicate that viruses are more resistant than coliforms to inactivation by natural factors in the water environment and to most water and wastewater treatment processes.

This means that under many circumstances a low coliform count in an effluent destined for reuse may not provide a clear assurance that the effluent is free of potentially infectious enteric viruses that survived the treatment processes.

Our studies (Shuval, 1970) and work of others indicate that domestic sewage carriers from 1 to 100 enteric viruses per ml. Enteroviruses found in wastewater may include more than 60 types—all of them considered pathogenic to man. These viruses include poliovirus, echoviruses and coxsackieviruses. Adenoviruses and reoviruses, although clinically considered to be respiratory, have been found in wastewater. Most important of all is probably the virus of infectious hepatitus, which has been shown by epidemiological studies to have caused over 50 waterborne epidemics. Although techniques for detecting this virus in water are now being developed, it had not been possible to do so in the past.

Methods have now been developed for detecting low levels of viruses in large volumes of water (Shuval and Katzenelson, 1972). It is likely that in the future routine virus assays of wastewater for some high levels of reuse will be required since, in certain cases, coliform tests appear inadequate.

B. Factors Affecting the Degree of Risk

The degree of risk of infection from sewage-borne pathogens in any reuse project depends on many factors, including the efficiency of wastewater treatment processes in removing or inactivating the pathogens; the survival of the pathogens in the wastewater effluent, in soil, and on crops in the case of agricultural reuse; and the infectivity or minimal infectious dose required to cause infection in man.

1. REMOVAL BY TREATMENT PROCESSES

While it is well accepted that conventional biological wastewater treatment processes provide only minimal removal of enteric bacteria, disinfection can often provide very high levels of bacterial inactivation. It is possible by optimal combinations of wastewater treatment and chemical disinfection to consistently achieve coliform counts in treated effluent lower than 100/100 ml. Enteric viruses, however, are usually manyfold more resistant to chlorination than are coliforms (Shuval, 1975a). With advanced wastewater treatment technology including physicochemical methods, it appears possible to effectively remove essentially all pathogens from an effluent stream. We have shown that ozone may prove to be particularly effective in such cases for the inactivation of viruses (Katzenelson *et al.*, 1974). Ozone is a most effective virucidal agent.

Our studies indicate that under controlled conditions a 99% kill of poliovirus can be achieved with 0.1 mg/liter of ozone residual in under 10 seconds while the same concentration of chlorine residual would require 10 minutes and iodine 100 minutes to achieve the same results. If effluent reuse for certain purposes calls for total effective removal of pathogens, it now appears to be within the limits of developing technology to achieve this goal.

The reliability of these processes remain to be evaluated under actual field conditions. Considering the health risks that may be involved in the case of a mechanical or human failure, such treatment processes should be fail-safe and monitoring procedures established to ascertain the microbial quality of the effluent before its distribution for reuse. This is now becoming technically feasible with the development of new procedures for the rapid detection of bacteria and viruses in water.

The question of microorganisms' survival in soil and on crops will be discussed under the heading of agricultural reuse.

2. MINIMAL INFECTIVE DOSE

With the possibility of obtaining significant reductions in the number of pathogens by active treatment processes or by die away in the soil or on crops, one must ascertain how many pathogens must be injested to cause infection or disease in man. It has been established that for certain *Salmonella* bacteria a person must injest many millions of viable organisms to become infected. For this reason, such *Salmonella* infections are most often associated with certain contaminated foods held at room temperature for periods of many hours which enables the massive multiplication of the initial inoculum of the pathogen. On the other hand, the ingestion of a few typhoid bacilli appears to be sufficient to cause infection in a certain percentage of susceptible humans who may have a low level of resistance. Very low levels of enteroviruses in water or on crops may present a potential health risk. It has been experimentally established that ingestion of as little as one tissue culture infectious dose of poliovirus (Plotkin and Katz, 1967) and other enteroviruses is sufficient to infect a percentage of susceptible persons. The minimal infective dose for infectious hepatitus has not been determined, but epidemiological evidence seems to indicated that the ingestion of but a few organisms might be sufficient to cause infection in some persons.

The ingestion of a relatively small number of cholera organisms may also lead to human infection. Infection with protozoan or helminthic pathogens may occur with a small number of ingested organisms as well.

With the above considerations in mind, it becomes apparent that very high removals of enteric pathogens are essential in any type of water reuse associated with human consumption of crops, body contact sports, or consumption as drinking water. The same goes for any form of reuse where effluent is sprayed into the air and aerosolized microorganisms can be dispersed over relatively wide areas, particularly in the vicinity of residential zones. It has been demonstrated that inhaled enteric bacteria can cause human infections in doses manyfold lower than when ingested (Sober and Guter, 1975). Inhaled salmonellae, for example are 1000 times more infective.

Since we cannot determine in advance the exact type of communicable disease organisms that may at times be present in the wastewater stream destined for reuse, it is reasonable to assume that it is a distinct possibility that highly infectious disease agents will indeed be present and that the ingestion of a very few of such organisms may cause human infection. Health criteria for different forms of water reuse must be based on this conservative assumption.

C. Chemical Contaminants

The unbridled increase in the use of hundreds of new and often structurally complex synthetic compounds in industry and agriculture has resulted in the appearance of many of these potentially toxic materials in municipal and industrial wastewater streams. Many of these chemicals which appear in wastewater are known not only for their acute toxic effects, but for their chronic effects which can be detected only after long periods of exposure. Materials having carcinogenic and mutagenic, as well as teratogenic, effects have been isolated in wastewater, polluted surface water, and drinking water from surface sources. Trace metals that may at times reach toxic concentrations have also been found on many occasions in wastewater streams, particularly those carrying a high percentage of industrial wastes (Shuval, 1962).

DETECTION OF MICROCHEMICALS IN WASTEWATER

Numerous efforts have been made to gain a better understanding of the toxic hazards of modern synthetic chemicals that find their way into wastewater and ultimately into drinking water sources. However, there are many difficult problems in concentrating, extracting, and identifying such compounds, many of which may be still unknown breakdown products of more complex chemicals that have undergone partial biodegradation.

The refractory organic components that remain after biological wastewater treatment and natural biodegradation processes in rivers often have been assayed as carbon chloroform extract (CCE) on the assumption that this test serves as a rough screening method of the presence of many of the potentially toxic synthetic organics. For example, the lightly polluted Columbia River at Bonneville Dam was found to contain 24 ppb of CCE, while the heavily polluted Detroit River near Wyandotte yielded 465 ppb (Middleton, 1960). Recycled wastewater at Chanute, Kansas, contained 992 ppb of CCE. The effluent of the Denver Wastewater Treatment Plant slated for eventual reuse for domestic purposes contains 2478 ppb of CCE as compared to 59 ppb found in the municipal water supply (Linstedt *et al.*, 1971).

More recently, a test for total organic carbon (TOC) has been introduced as an assay for residual organics. Effluent from well-operated biological sewage treatment plants may contain 40–60 ppm of TOC, while the Ohio River at Cincinnati which may at times carry as much as 15–20% sewage effluent has been reported to show a TOC of 20 ppm during certain periods (Cleary *et al.*, 1963).

The Rhine River in the Netherlands contains 10 ppm of TOC on the average, while the treated water after dune infiltration and the use of some activated carbon prior to rapid sand filtration still contains 4 ppm of TOC. In both the CCE and TOC test, little can be said of the exact nature of the organics present, particularly since it is known that part of the refractory organics may be humic acids or similar compounds of natural origin. However, many efforts have been made to identify the specific organic chemicals present in water and wastewater effluent.

Middleton (1960), using the carbon chloroform extract method (CCE) for concentrating the organics, identified from polluted riverwater, the presence of DDT, aldrin, o-chloronitrobenzene, tetralin, naphthalene, chloroethyl ether, acetophenone, diphenyl ether, pyridine, phenols, nitriles, acidic materials; miscellaneous hydrocarbons including substituted benzene compounds, kerosene, synthetic detergents, aldehydes, ketones, and alcohols. Some of these substances are known to be toxic. Many other compounds undoubtedly present, remained unidentified. Bunch *et al.* (1960), employing the best methods available to them at that time, were able to recover and identify less than 40% of the soluble organics remaining in biologically treated sewage and these were described only in general terms such as ether-extractable matter, protein, tannin, lignin, and alkyl benzene sulfonate.

The introduction of newer methods for sample concentration and analysis has led to vast improvements in identification and measurement of organic micropollutants in wastewater and polluted surface-water sources.

The development of concentration techniques for organics such as by means of macroretricular resins and reverse osmosis is rapidly replacing activated carbon adsorption used in the CCE method which has proven inadequate in many respects. High-resolution gas chromatography, in combination with mass spectrometry (G.C.–M.S.) aided by computer analysis, has provided much new information on the microorganics in water.

More than 1000 organic compounds have been identified in water by using such powerful techniques. In one such study drawing on data from numerous collaborating laboratories, 289 organic compounds detected in water are listed (World Health Organization, 1975b). Many of these compounds have known toxic effects. A few are known as carcinogens.

D. Presence of Carcinogens in Wastewater

Evidence is gathering from numerous sources that cancer causing chemicals are being released regularly into the water environment from municipal and industrial wastewater sources.

Heuper and Conway (1964) outlined the main sources of carcinogens that can appear in wastewater as follows.

1. *Petroleum products.* Petroleum refinery wastes containing polycyclic aromatic hydrocarbons, fuel oil, lubricating oils, and cutting oils are being introduced into lakes and rivers from garages, service stations, petrochemical plants,

metal-working plants, and ships. Contamination of public water supplies may also result from the use of kerosene, methylated naphthalenes, and similar petroleum products used as vehicles of insecticide sprays, or enter water from rain contaminated with air polutants or from tarred or asphalted roads.

2. *Coal tar.* Effluents from gas plants, coke oven operations, tar distilleries, tar-paper plants, and wood-pickling plants all contain carcinogens. Coal tar, pitch, creosote, and anthracene oil are known human carcinogens.

3. *Aromatic amino and nitro compounds.* Amino compounds such as β-naphthylamine, benzidine, and 4-aminodiphenyl are known to be human carcinogens from the incidence of cancer among workers in dye and rubber industries. These compounds along with their nitro analogues are released by dye and rubber manufacturing, pharmaceutical factories, textile dyeing plants, plastic production, and others.

4. *Pesticide, herbicide, and soil sterilants.* Compounds such as DDT, Dieldrin, Aramite, carbon tetrachloride, acetamide, thioacetamide, thiourea, thiouracil, aminotriazole, several urethane derivatives, isopropylchlorophenyl carbamate, and β-propiolactone are capable of eliciting benign and/or malignant tumors in various organs of experimental animals.

A new source of concern is the formation of certain halogenated hydrocarbons possessing toxic qualities as a result of the chlorination of wastewater effluent or treated water drawn from rivers heavily contaminated with organics. Jolly (1973) identified over 50 chlorinated hydrocarbons in chlorinated domestic wastewater effluent in the United States. While Rook (1974) obseved that chlorination of polluted river water in the Netherlands produced compounds such as chloroform, carbon tetrachloride, dichlorobromomethane, chlorodibromomethane, tribromomethane, and traces of other halomethanes and haloethanes. Bellar *et al.* (1974) have added further support to these findings and have shown that chloroform concentrations ranging from 37–152 ppb were present in chlorinated drinking water in five communities receiving water from either the Ohio or Mississippi River. This is estimated to be 10 times higher than the concentrations found in chlorinated drinking water from groundwater sources. Chloroform and carbon tetrachloride are definitely considered as carcinogens.

One must now ask whether the presence of cancer-causing chemicals in such low concentrations in water can actually cause cancer in man.

E. The Risk Associated with Low Levels of Carcinogens in Water

From Reports of the World Health Organization (1964) and the U.S. Public Health Service (1970), one must conclude that there is growing recognition by scientists that the majority of human cancers are due to chemical carcinogens in the environment. It has been estimated that somewhere between 60 and 90% of human cancers are environmental in origin and that the low levels of carcinogens to which the general public may be exposed could be the responsible causative agents (Epstein, 1974). For example, Dieldrin, a widely used chlorinated hydrocarbon insecticide, has been found to be carcinogenic in the lowest concentrations tested, 100 parts per billion (ppb) while the chemical aflatoxin, one of the most potent carcinogens, has been shown to produce liver tumors in trout when present in feed at concentrations as low as 0.4 ppb.

Although there still is much to be learned about environmental carcinogens and the risk associated with exposure to very low concentrations, there is a strong case in support of the position that "no level of exposure to a chemical carcinogen should be considered toxicologically insignificant for man" (Epstein, 1974).

Most information on the carcinogenic properties of chemicals has resulted from chronic animal experiments—the question of extrapolating the data to humans remains a problem. However, there appears to be increasingly strong evidence that indicates that materials found to produce cancer in animals will generally produce cancer in man. If anything, it is felt by science that animals may not be sensitive enough to certain materials which could produce cancer in man at very low exposure levels.

The possibility of long-term exposure to low levels of carcinogens by population groups consuming renovated water either directly or indirectly may be a prime factor in evaluating the safety of direct water reuse programs. Of particular concern is the report of Harris (1974) on the implications of cancer-causing substances in Mississippi River Water. The study presents presumptive epidemiological evidence that suggests a significant relationship between cancer mortality among white males and drinking water obtained from the Mississippi River in the New Orleans area. The report concludes that their analysis strongly suggests that drinking water from the Mississippi River is causally related to cancer mortality in the more than one million persons in Louisiana that depend upon that source for their drinking water supply.

A study by the U.S. Environmental Protection Agency (1972) of water in the lower Mississippi River reported the presence of heavy metals such as mercury, arsenic, lead, copper, chromium, zinc, and cadmium, as well as numerous organic chemicals, found in the finished water supplies. Chloroform, hexachlorobenzene, xylene, ethyl benzene, and dimethyl sulfoxide were listed as having induced histopathological changes during chronic toxicity studies on animals. Three compounds (chloroform, benzene, and carbon tetrachloride) were listed as carcinogens. The report (U.S. Environmental Protection Agency, 1972) recommended "that municipal water treatment plants install treatment facilities designed to obtain removal of organic contaminants and heavy metals ..." They suggested that "continuous use of activated carbon would probably be required to remove the trace organics in the water supplies." The treatment plants in the area studied did not introduce activated carbon treatment as recommended.

The case of New Orleans can be considered as a prime example of *indirect* wastewater reuse with few of the precautions that are now generally recognized as essential to remove deleterious chemicals that appear in the contaminated raw river water.

Although the results of the New Orleans study cannot as yet be considered as conclusive evidence that cancer is in fact caused by consuming such contaminated water, the implications of these very suggestive findings must be fully taken into consideration in any planned project for the renovation of wastewater for domestic consumption. Specific treatment processes having a proved capability of removing toxic organics including the carcinogenic chemicals will certainly be required elements of any such treatment.

F. The Removability of Refractory Organics

While some of the complex, nonbiodegradable synthetic compounds under discussion may be harmful to health at concentrations at the ppb level, if ingested for long periods, advanced wastewater renovation technology still cannot normally reduce total organic carbon (TOC) to an absolute zero concentration.

Such advance wastewater treatment processes as those used in the EPA-Blue Plains Plant in Washington D.C. (see Chapter 15) or at Lake Tahoe can successfully reduce the TOC to 1–2 ppm in the effluent by a series of biological and chemicophysical processes including passage through activated carbon columns and heavy ozonization. Current thinking in advance treatment technology is in the direction of even more extensive removal of organics to levels as low as 0.1 ppm by the use of additional treatment processes such as chemical oxidation with ozone or hydrogen peroxide, resins, membranes, and volatile stripping (U.S. Environmental Protection Agency, 1975).

The results of an experimental study in which CCE extracts of Mississippi water suspected of causing cancer in humans were injected into mice produced no evidence of carcinogenic properties in the water (Dunham *et al.*, 1967). The authors suggest that one possible explanation for the inconclusive results is that some of the potential carcinogens that may have been in the polluted river water were not adsorbed on activated carbon. This possibility must raise the question as to whether potentially hazardous organics are removed adequately even by activated carbon treatment which is certainly one of the most effective treatment processes available.

Borneff and Fisher (1962) have shown that polynuclear aromatic hydrocarbons, potentially dangerous carcinogens found in wastewater, are poorly removed by conventional biological treatment and sand filtration, but 99% are removed by activated carbon filtration if flow rates are low enough.

However, insufficient information is available today as to the ability of the various advanced wastewater treatment processes to remove the hundreds of different specific organics found in wastewater which may have a deleterious effect on persons exposed to low concentration of them in water over long periods of time (Ongerth *et al.*, 1973).

The International Meeting on the Health Effects Relating to the Reuse of Wastewater for Human Consumption (World Health Organization, 1975b) has recommended that this question be given the highest research priority in studies essential to evaluate the health effects of consuming renovated wastewater. Such information will be essential for planning water renovation and reuse for domestic purposes.

The same report emphasized that there is continued value in applying the use of a general test for total organics in water such as the TOC. They pointed out that good quality drinking water should usually contain no more than a few milligrams per liter of total organic carbon.

They concluded that they "strongly felt that at this stage with the tremendous gap of knowledge concerning the toxic effects associated with the organic components in renovated water or water produced from polluted sources, the most prudent policy would be to provide for optimal removal of total organic carbon to the lowest feasible level. As a tentative goal, T.O.C. levels under 5 mg/l should be strived for in the case of either direct or indirect water reuse." Several members of the group strongly recommended that in cases of direct planned wastewater reuse, where the organics present are known to be primarily derived from municipal and industrial wastes, the TOC level after treatment should be no higher than 1 ppm.

Whether or not TOC removals to 0.1 or 1 ppm can be fully achieved in practice is yet to be determined. The ability of available treatment processes to fully remove many known specific toxic chemicals is also a moot question.

III. HEALTH ASPECTS OF VARIOUS TYPES OF REUSE

Each form of wastewater reuse presents its own specific health problems. This section will review the problems associated with each type of reuse.

WHO (World Health Organization, 1973) has presented a comprehensive and authoritative report on the health aspects of various forms of reuse. The

Table I
Suggested Treatment Processes to Meet the Given Health Criteria for Wastewater Reuse[a]

	Irrigation			Recreation		Municipal reuse		
Health criteria[b]	Crops not for direct human consumption (A + F)	Crops eaten cooked; fish culture (B + F or D + F)	Crops eaten raw (D + F)	No contact (B)	Contact (D + G)	Industrial reuse (C or D)	Non-potable (C)	Potable (E)
Primary treatment	●●●	●●●	●●●	●●●	●●●	●●●	●●●	●●●
Secondary treatment		●●●	●●●	●●●	●●●	●●●	●●●	●●●
Sand filtration or equivalent polishing methods		●	●		●●●	●	●●●	●●
Nitrification						●		●●●
Denitrification								●●
Chemical clarification						●		●●
Carbon adsorption								●●
Ion exchange or other means of removing ions						●		●●
Disinfection		●	●●●	●	●●●	●	●●●	●●●[c]

[a]Reproduced with the permission of the World Health Organization from *Reuse of Effluents: Methods of Wastewater Treatment and Health Safeguards.* WHO Technical Report Series No. 517. Geneva, 1973.

[b]Health criteria: A, Freedom from gross solids; significant removal of parasite eggs. B, as A, plus significant removal of bacteria. C, as A, plus more effective removal of bacteria, plus some removal of viruses. D, Not more than 100 coliform organisms per 100 ml in 80% of samples. E, No faecal coliform organisms in 100 ml, plus no virus particles in 1000 ml, plus no toxic effects on man, and other drinking-water criteria. F, No chemicals that lead to undesirable residues in crops or fish. G, No chemicals that lead to irritation of mucous membranes and skin. In order to meet the given health criteria, processes marked ●●● will be essential. In addition, one or more processes marked ●● will also be essential, and further processes marked ● may sometimes be required.

[c]Free chlorine after 1 hr.

WHO report has suggested treatment processes to meet the given health criteria for wastewater reuse which is reproduced here as Table I. The rationale behind the health criteria and the treatment processes required to meet them is presented in the following sections.

A. Agricultural Reuse

The application of human feces as an agricultural manure has been widely practiced in the Far East for many centuries, while the reuse of municipal wastewater for agricultural irrigation is one of the oldest forms of water reclamation. At the end of the last century, major land-irrigation projects were developed in Germany and in England. It should be pointed out, however, that the primary motivation of these early projects was essentially treatment and disposal of municipal wastewater, rather than water conservation and recycling.

The first Royal Commission on sewage disposal in England concluded in its report in 1865 that "the right way to dispose of town sewage is to apply it continuously to the land, and it is by such application that the pollution of rivers can be avoided."

It is important here to review the possible health risks associated with various forms of water reuse in agriculture. The degree of risk involved may vary greatly. Such reuse may be directed solely to the irrigation of industrial or other crops not for direct human consumption, or, on the other hand, it may involve highly health sensitive crops such as fruits or vegetable that are generally consumed uncooked. In either case, the health risks to agricultural workers must be evaluated as well as the possible dispersion of aerosolized pathogens by spray irrigation in the vicinity of residential areas.

1. CONTAMINATION OF CROPS WITH PATHOGENS

Although public health authorities have long ago pointed out the risks of using human feces as a manure on vegetable crops, systematic scientific studies on the survival of enteric pathogens in soil and on crops began to appear in the literature only in the 1920s. One of the earliest studies on survival of enteric pathogens in soil was made in 1921 by Kligler who was the founder of the Department of Hygiene at the Hebrew University of Jerusalem. He showed that typhoid bacilli could survive for months in moist subsoil contaminated with feces, even though the movement of the bacteria through the soil was very restricted. McClesky and Christopher (1941) studied pathogen survival on strawberries, while Falk (1949) studied the survival of enteric bacteria sprayed on tomatoes. His work at Rutgers was followed up by the extensive classical study with his colleagues on the health risks associated with growing vegetables in sewage-contaminated soil (Rudolfs *et al.*, 1950, 1951).

These studies indicated that bacteria, protozoa, and helminths do not penetrate healthy undamaged surfaces of vegetables and die away rapidly on crop surfaces exposed to sunlight. However, pathogens can survive for extended periods inside leafy vegetables or in protected cracks or stem areas. We initiated our first studies with colleagues in Israel in the early 1950s on the health problems associated with sewage irrigation (Shuval, 1951). Our colleague Bergner-Rabinovitz (1956) detected numerous pathogens in the raw sewage of Jerusalem slated for irrigation including *Salmonella* sp., *Shigella dysenteriae*, parasitic eggs of *Ascaris, Trichuris, Trichostrongylus, Taenia hymonolepis* and cysts of *Giardia lamblia* and *Endomoeba*. In the effluent of a trickling filter plant used for one of our sewage irrigation studies (Rigbi *et al.*, 1956), the same pathogens were detected, but less frequently and in smaller numbers.

In these studies, the survival of *Salmonella tennessee* organisms inoculated into the effluent was studied in sewage irrigated soil. In the winter, the bacteria could not be detected by day 46 on the surface of the soil, but only disappeared from the moist subsoil on day 70. While in the summer they disappeared on day 23 on the surface and day 37 at a depth of 8 in. In all cases, there was about a 99% reduction within the first week.

Dunlop *et al.* (1951) found that although *salmonellae* could be recovered from a large number of samples of sewage-contaminated irrigation water, they were unable to recover these pathogens from samples of vegetables irrigated with this water.

In our own recent studies in Israel, we have investigated the survival of poliovirus in sewage-irrigated soil and on crops. We have been able to show that poliovirus inoculated into the sewage could be detected in the irrigated soil for 7 days and it was possible to recover the virus from the cucumbers 7 days after initial irrigation (see Fig. 1) (B. Fattal and E. Katzenelson, Environmental Health Laboratory, Hebrew University, Jerusalem unpublished data).

It is quite clear from the many studies to date that pathogens are present in sewage in great quantity and variety and can survive for periods in the soil or on crops (Benarde, 1973). However, the viability of such microorganisms varies greatly depending on the type of organisms and various environmental factors such as climatic conditions, soil moisture, soil pH, and the amount of protection provided by the crops.

In his review of this problem, Dunlop (1952) went so far as to say that despite the known presence and viability of pathogens in wastewater and soil, he knew of no disease outbreaks or epidemics that had been related to or were known to be caused by irrigation with properly treated sewage. He concluded that if effluents were properly treated, it was safe from the microbial point of view to harvest crops for human consumption in 4 hours of irrigation. Dunlop did not define the degree of treatment or disinfection he would require.

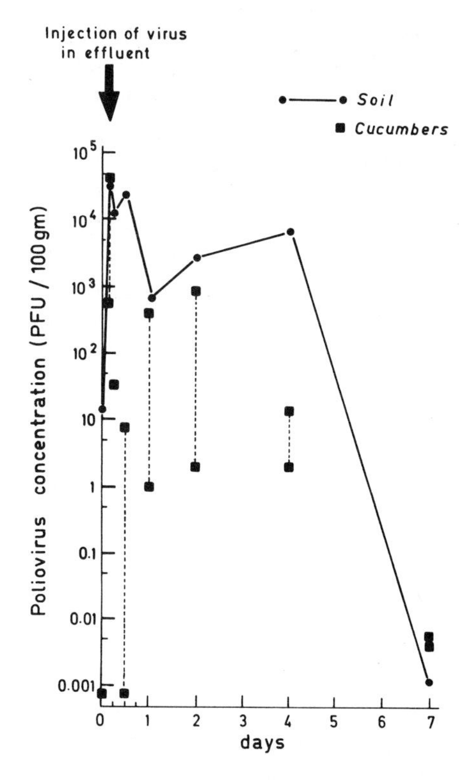

Fig. 1. Virus concentration in soil and on cucumbers irrigated with effluent inoculated with poliovirus.

A dramatic demonstration that pathogens in untreated sewage used for irrigation could, in fact, remain viable on the vegetables long enough to cause a cholera outbreak occurred in Jerusalem in 1970 (Cohen *et al.*, 1971). When cholera cases first appeared in the city, it was quickly ascertained that the drinking water supply derived from deep protected wells and chlorinated before distribution was not the vector. However, during the outbreak, we demonstrated the presence of *Cholera vibrio* of the *El tor* type in the main sewage lines in various parts of the city and in the soil of some agricultural plots illegally irrigated with raw sewage used for growing vegetables supplied to the Jerusalem market. The pathogens were later recovered from the sewage-irrigated vegetables. The illegal sewage irrigation was stopped and the epidemic quickly came to an end. The freshly sewage-irrigated leafy vegetables widely sold in Jerusalem undoubtedly provided the main secondary pathway for the spread of the disease after a few carriers or clinical cases entered the city from neighboring countries where cholera outbreaks were in progress.

From the foregoing, it is apparent that irrigation of health-sensitive crops, including fruits and vegetables eaten uncooked, with raw or partially treated wastewater can present real health risks. Even effluent from conventional biological wastewater treatment plants cannot generally be considered safe. For these reasons, health authorities in many countries have established regulations restricting the types of crops that can be grown with effluent that has not undergone a high degree of disinfection. Irrigation with nondisinfected effluent of industrial crops, seed crops, tree nurseries, and other crops not destined for direct human consumption is typically allowed.

The use of raw sewage is generally not allowed for irrigation af any kind, both for esthetic reasons and to avoid the presence of aggregates of fecal matter on the fields which may serve as a source of direct contamination of workers or mechanical transmission by flies and other vectors.

Even sewage irrigation of crops usually consumed cooked, such as potatoes or beets, should not be considered risk-free since the contaminated surfaces of the vegetables may introduce pathogens into kitchens where working surfaces and utensils may become contaminated and thus infect other foods.

2. UNRESTRICTED AGRICULTURAL UTILIZATION

Certain agricultural and economic conditions may warrant the treatment of wastewater to such an extent that it can be used for unrestricted irrigation of all agricultural crops. Farmers find some difficulty in carrying out normally required crop rotation regimes if restrictions on the types of crops that may be grown are too onerous. This also may become a serious public health problem, since it is not administratively possible in many cases to control the types of crops that are in fact grown by farmers who are supposedly required to limit themselves to certain "safe" crops. The economic temptation to grow high value salad crops even if the effluent quality does not warrant this is difficult to overcome. These considerations have led to the development of treatment procedures and standards to overcome such problems.

If wastewater is to be used for the irrigation of agricultural crops in an unrestricted manner, including fruits and vegetables usually consumed uncooked, a high degree of disinfection is necessary to inactivate the pathogens. Additional processes may be required to remove certain resistant protozoans or helminths. The State of California (1973) (see Chapter 8) has established standards which require that reclaimed water to irrigate food crops at all times must be adequately disinfected, filtered wastewater with a median coliform count of no more than 2.2/100 ml. The WHO (World Health Organization, 1973) meeting of experts on this subject has recommended that crops eaten raw should be irrigated only with biologically treated effluent that has been disinfected to achieve a coliform level of not more than 100/100 ml in 80% of the samples. It has been demonstrated that it is technically feasible to effectively disinfect a good quality wastewater effluent to achieve such low coliform counts (Shuval, 1975a).

WHO also states that, in certain situations, sand filtration or equivalent polishing methods may be required. This relates in particular to the need to remove helminths in those areas of the world where such parasitic diseases are endemic. A defined and tested technology for the removal of protozoans or helminths generally resistant to chemical disinfection is yet to be established, but microstrainers have been proposed to meet this requirement. Slow sand filtration should be effective but, in many situations, it may not be economically feasible.

A number of problems arise in meeting the objectives of effluent disinfection for unrestricted irrigation. The problem of the greater resistance of enteroviruses to chlorination has not been fully overcome. Our studies (Shuval, 1975a) have shown that it is quite feasible to achieve a coliform count of around 100/100 ml by applying as little as 5 mg/liter of chlorine to the effluent of a high-rate biological filter plant (see Fig. 2). However, when we inoculated poliovirus type 1 into the same disinfection system, only about a 90% reduction was achieved in one hour (Fig. 3). Figure 4 shows a comparison of concentration-time relationship required to achieve 99.9% inactivation of poliovirus, echovirus, and coliforms in the effluent by chlorine. From this it can be seen that with a 1 hour

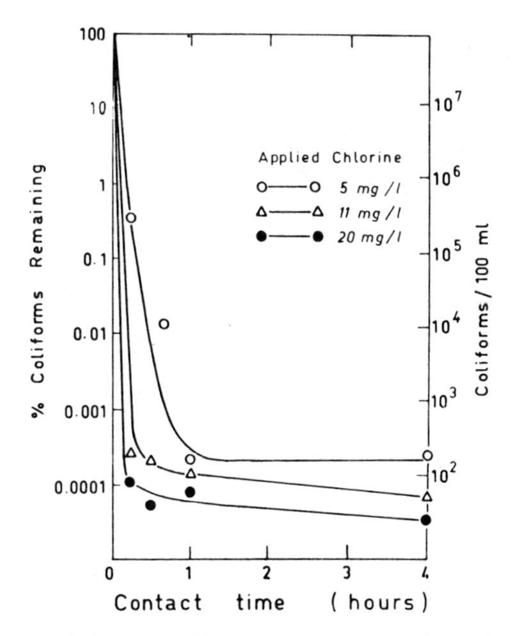

Fig. 2. Inactivation of coliforms in effluent at 20°C at varying doses of applied chlorine with residuals of combined chlorine.

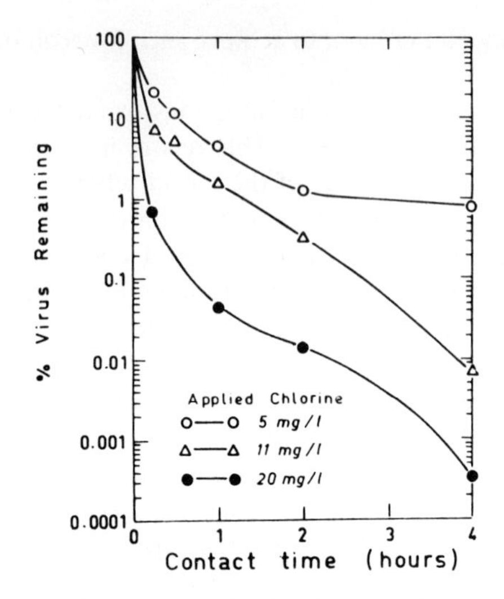

Fig. 3. Inactivation of poliovirus suspension inoculated in effluent at 20°C at varying doses of applied chlorine with residuals of combined chlorine.

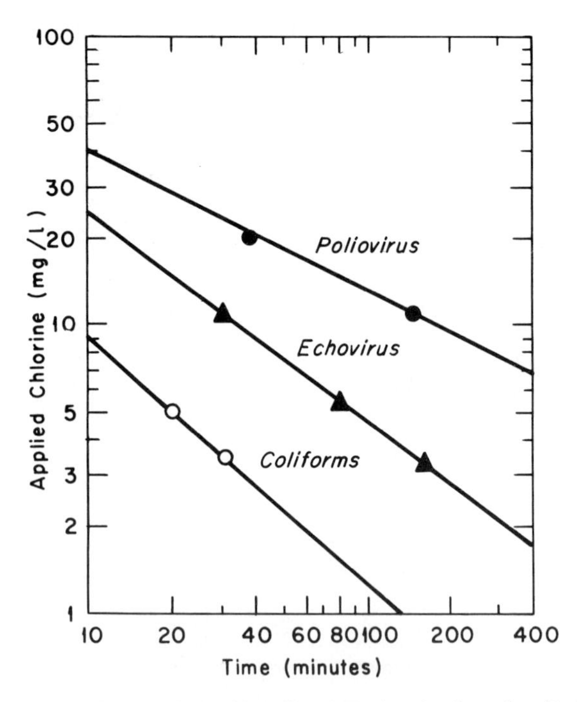

Fig. 4. Concentration-time relationships for 99.9% inactivation of poliovirus, echovirus, and coliforms in effluent at 20°C with varying doses of applied chlorine and combined residuals.

contact time, about 10 times as much chlorine is required to achieve the same degree of disinfection for poliovirus as is required for coliforms. Besides the cost, the formation of potentially toxic organohalide compounds by such high doses of chlorine might rule out such treatment (Bellar *et al.*, 1974).

Our recent studies on ozone (Katzenelson *et al.*, 1974) indicate that this powerful oxidant inactivates viruses many times more rapidly than chlorine and may hold promise in the effective disinfection required for reuse purposes.

Another problem is the regrowth of coliforms after chlorination. This phenomenon was first reported upon by Rudolfs and Gehm (1936). In our studies of disinfection of the effluent of a high-rate biofiltration plant in Jerusalem we were able to show that coliform counts lower than 100/100 ml could be achieved on a regular basis within 15 min after chlorination at doses of some 10 mg/liter. However, after storage for 3 days in an operational reservoir prior to agricultural irrigation, coliform counts increased 10-fold on average. In laboratory studies, we were able to demonstrate massive regrowth of coliforms and fecal coli 3–4 days after chlorination (see Fig. 5). The hygienic significance of the high coliform counts in such cases is difficult to determine since it is not clear whether pathogens can regrow after initial partial disinfection in a like manner. The State Health Department of California (1973) has recognized this problem and specifically states that the bacteriological standard is considered to be fulfilled if "at some point in the treatment process" the required coliform count is achieved. However, since information on the regrowth ability of enteric pathogenic bacteria in dilute substrates such as polluted water or wastewater effluent is lacking, the California formulation should be considered as a tentative one.

3. HEALTH OF WORKERS AND PUBLIC

Little attention has been paid in the past to the potential health risks to workers in wastewater irrigation projects or to the public who may live in adjacent residential areas or who may pass through on public highways.

a. Direct Contamination. One early study in the United States on the health of workers at sewage treatment plants did not reveal any excessive risks of communicable disease including infectious hepatitis in this group. However, it has been reported from India that hookworm infections are much more common among workers on sewage farms than among the farming population in general. The low levels of personal hygiene and the local custom of walking barefoot are undoubtedly major contributory factors to the disease situation among sewage farm workers in India.

The potential health risk to workers may be derived from direct contact with wastewater that may contaminate hands which later contaminate food.

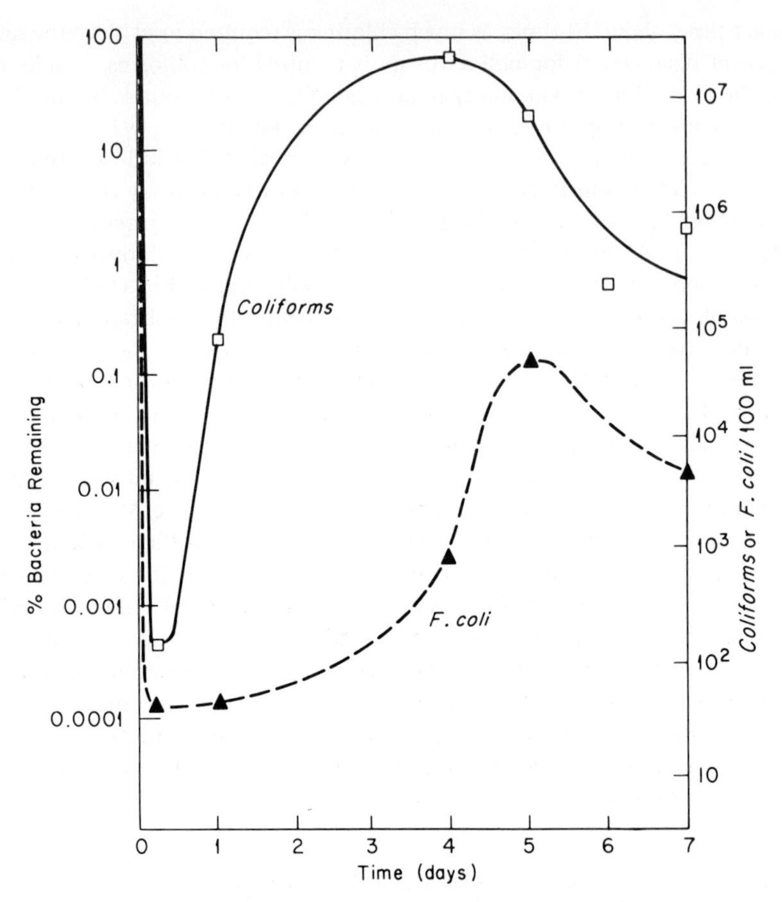

Fig. 5. Regrowth of coliforms and fecal coli in effluent at 20°C after inactivation with 5 mg/liter of applied chlorine and no dechlorination.

Appropriate sanitary facilities for washing and eating and good personal hygiene can go a long way to reduce this risk.

b. Dispersion of Aerosolized Microorganisms. Another problem is the possible inhalation of aerosolized sewage containing pathogens from spray irrigation. Our laboratory in Jerusalem has initiated studies to evaluate this problem (Katzenelson and Teltch, 1975) and in preliminary studies have been able to recover enteric bacteria including *Salmonella* sp. 100–350 m downwind of a field, spray irrigated with nondisinfected effluent. The size of the viable aerosol particles were determined with an Anderson Cascade Sampler and it could be shown that a significant percentage of the recovered bacteria were associated with particles in the 1–4 μm range which can be inhaled into the lungs and can be considered potentially infectious. Our estimates indicate that somewhere between 0.1 to 1% of the sewage sprayed into the air forms aerosols which are capable of being carried considerable distances by the wind. The rate of die-away and reduction in concentration of pathogens incorporated in the aerosols is a function of wind speed, temperature, relative humidity, UV radiation, and local topographic features. Sorber *et al.* (1974) made some theoretical calculations as to the potential dispersion of bacteria or viruses aerosolized by sewage spray and suggested that a buffer zone of up to one mile would be advisable to prevent infections in adjacent residential area. Although there is as yet no sound scientific basis for establishing such buffer zones, there is already sufficient data to indicate that an area of some 500 m from spray irrigation with sewage can carry infectious bacteria in the air. The limits of the buffer zone including some safety factor should surely be beyond this range.

The possible health risks from aerosols associated with the land disposal of the sewage of Muskegon, Michigan, by spray irrigation over a 10,500 acre area became a critical issue in the evaluation of that project by governmental authorities and by the public (Chaiken *et al.*, 1973).

From the preliminary findings of our recent epidemiological study in Israel it appears that members of Kibbutzim (agricultural cooperative villages) practicing sewage irrigation suffer two to four times higher rates of salmonellosis, shigellosis, typhoid fever, and infectious hepatitis than those living in communities not practicing sewage irrigation. The regulations of the State Health Department of California (1973) previously cited have already related to this problem. They require the same rigorous disinfection standard for effluent used for spray irrigation as that required for the irrigation of edible food crops. There appears to be strong logic in their case and it would appear advisable to consider heavy disinfection of all effluents used for any form of spray irrigation regardless of the crops irrigated or the proximity of residential areas.

There is undoubtedly need for further investigation of this question and such studies are in progress in a number of places. Until a more firm scientific basis is available, caution should prevail in the protection of agricultural workers and the public from any risks that may be associated with spray irrigation.

4. IRRIGATION OF PASTURE LAND

The risks to the health of animals and potentially to man associated with cattle grazing on sewage-irrigated pasture have been studied by a number of authors. Greenberg and Kupa (1957) in their review of the transmission of tuberculosis by wastewater point out that the wastes from institutions treating tuberculosis patients or industries such as dairies or slaughter houses handling tuberculous animals will almost always contain large numbers of tubercle bacilli. The bacilli will not be removed by conventional biological treatment, but only after very heavy chlorination.

Animals grazing on sewage-irrigated pasture or drinking such sewage can become infected. In another review, Greenberg and Dean (1958) point out that a number of authors reported that cattle grazing on pastures irrigated with sewage often show significant increases in beef tapeworm infections of *Cysticercus bovis* which can infect persons who consume the beef with the adult stage of the tapeworm called *Taenia saginata.*

The disease is widely distributed throughout the world in both animals and man and is still considered a serious health problem in many areas. Reports indicate that conventional sewage treatment is inadequate to completely eliminate tapeworm eggs from sewage or sludge. Sand filtration or microstraining are suggested processes for effective removal of the eggs (Silverman and Griffiths, 1955). Microstraining has been shown to remove about 90% of the *T. saginata* eggs.

Some regulations for sewage irrigation of pasture lands have recommended allowing the grazing of cattle after the fields are completely dry. The efficacy of this procedure is open to question in light of the findings of Jepsen and Roth (1952) who showed that the eggs of *T. saginata* may remain viable under natural conditions for months, "long enough to permit protracted contamination of fields and crops."

It appears that in areas where this disease is endemic, sewage irrigation of pasture lands should be avoided unless special treatment facilities for removal of the pathogens are provided for.

5. FISH PONDS

Wastewater has been used to add nutrients to fish ponds used for growing fish for human consumption in some areas. In an area such as Israel, where this is practised, the potential danger of shistosomiasis transmission to pond workers exists. The parasitic eggs are excreted by persons infected and the larvae which hatch in the water infect specific species of snails often found in fish ponds. These snails serve as the intermediate hosts and from them go forth the second stage larvae which can infect the workers by directly penetrating through the unbroken skin. The treatment of sewage in oxidation ponds provides good removal as does conventional biological treatment (Rowan, 1964). In addition, the control of the snails in the fish ponds can prevent the transmission of the disease.

Although fish cannot become biologically infected with human pathogens, they can become mechanically contaminated and in that way introduce pathogens into kitchens and cause human infections. Although there is little epidemiological evidence on actual disease transmission, the risk exists. Holding fish in clean-water ponds for some period before marketing may reduce the extent of contamination. The question of accumulation of toxic chemicals and off-tastes in fish grown in wastewater has not been studied, but reports on taste problems have been made.

B. Industrial Reuse

In many areas, the use of municipal waste water in industry has been successfully practiced, but a number of problems exist from a public health point of view. Possibly the main problem concerns the danger of cross-connecting pipelines carrying treated sewage and those carrying safe water for use in food processing or for human consumption. The careful color coding of pipes would be helpful in reducing such risks. Generally speaking, however, it would be sound policy to bring treated sewage into industrial plants only after it has been treated and disinfected to the highest possible degree and has achieved a bacteriological quality approaching that of drinking water. Such a high level of treatment would reduce the risk of a major outbreak of disease occurring as the result of an accidental cross-connection. If wastewater is used for cooling purposes only, very few additional health problems exist, although there are a number of engineering and hydraulic problems that must be overcome. For example, unless the sewage is adequately treated, the problem of slime control might become critical.

If the reclaimed wastewater is to be used as process water in industry, special consideration must be given to possible public health implications, and particular care will have to be taken if treated wastewater is to be considered for use in industrial food-processing plants. In such cases only water meeting the strictest standards for drinking waters should be used. One of the most effective and economical ways of using wastewater in industry is in the intraplant reuse of treated and recycled industrial effluents. Generally speaking, public health problems involved in recycling industrial effluents are less severe than those resulting from the use of municipal sewage.

Great care must be taken to prevent cross-connections with the general community water system which supplies industrial plants reusing wastewater. The arrangement of a total physical disconnection between the community water supply by an appropriate air gap is the safest. There are a number of tested "fail-safe" double check-valve arrangements which may be used if continuous maintenance and inspection can be assured.

Little consideration has been given to date on the possible health risks to the community through the use of wastewater for cooling towers. Studies have shown that spray and mist from large cooling towers can be carried for many miles. Little data are available on the possible dispersion of aerosolized bacteria and virus which can potentially be carried even greater distances. Cooling towers can, in effect, serve as huge aerosol generators which could spread pathogens still present in wastewater effluent used for cooling to nearby communities causing potential health hazards. Until this question has been fully evaluated, the only prudent policy would be to require that effluent used for cooling towers be disinfected to reduce pathogenic bacteria and viruses to the lowest feasible levels. It should be pointed out here that the use of heavily polluted river water in cooling towers may present no less a health risk than the case of overt, direct wastewater reuse. There is every justification to require the same bacteriological standard for such cases of indirect wastewater reuse with polluted river water.

C. Reuse for Recreational Impoundments

Although many rivers and lakes contaminated with varying degrees of raw or treated wastewater have been used for recreational purposes including body contact sports, planned, direct reuse for such purposes is relatively recent. The Santee Recreation Project and a similar one at Lancaster, both in California, successfully developed treatment processes that produced renovated wastewater for recreational impoundments meeting the most rigorous microbiological criteria (Askew *et al.*, 1965). After initial periods of careful monitoring for bacteria and viruses, body-contact sports were tentatively approved under carefully supervised conditions with no deleterious health effects being detected. Public reaction at these desert locations suffering from a shortage of water sport recreational facilities has been very favorable.

The evaluation of the health risks associated with bathing in polluted water has been a controversial subject for years, particularly since clear-cut epidemiological evidence associating contaminated bathing water with overt enteric disease transmission has been sparse. Moore and his committee (Medical Research Council, 1959), who studied this question at English beaches concluded, ". . . the risk to health of bathing in sewage contaminated seawater can for all practical purposes be ignored." Many public health authorities have not accepted these conclusions, however.

The potential health hazards that may be associated with bathing in contaminated recreational water includes waterborne enteric infections, as well as upper respiratory ear, eye, and nose infections. Ingestion of toxic chemicals or skin or eye irritations due to chemical industrial wastes must also be considered in cases of wastewater reuse.

It must be assumed that persons bathing in a recreational impoundment filled with renovated water may ingest from 10–50 ml of water. It also must be assumed that for certain enteroviruses the ingestion of one virus infectious dose is sufficient to cause infection in some percentage of the persons so exposed (Shuval, 1975b).

The risk of infection will also increase at times of epidemics of enteric disease in the community when a higher percentage of pathogens are being shed by the population. The relative risk will also be greater in those communities with high endemic rates of enteric disease.

Taking all these factors into consideration, a WHO working group on guides and criteria for the Recreational Quality of Beaches and Coastal Waters (World Health Organization, 1975a) concluded that there is indeed a potential health risk associated with bathing in sewage-contaminated water which justifies establishing broad microbial standards of quality. They recommended that recreational water of good quality should show an *Escherichia coli* count of under 100/100 ml, while *E. coli* counts above 1000/100 ml indicate an unacceptable level of pollution. The State Health Department of California (1973) is apparently one of the few authorities to establish specific standards for recreational impoundments using reclaimed wastewater. They require that the reclaimed water used as a source of supply in a nonrestricted recreational impoundment shall be at all times an adequately disinfected and filtered wastewater. Their regulations imply that an effective system of coagulation and filtration following secondary biological treatment precede the disinfection which should produce an effluent with a medium coliform MPN which does not exceed 2.2/100 ml. For restricted recreational uses, not involving body contact sports, the same bacterial standard is required, but the requirement for additional filtration after biological treatment is dropped.

These requirements are indeed more stringent than those that may be required of naturally polluted recreational areas, but are justified for both hygienic reasons and in light of the fact that the legal and moral responsibility in cases of direct wastewater reuse is a heavy one indeed—and one that falls directly on the shoulders of those operating or supervising such a project. Maximum feasible precautions should therefore be required.

D. Restricted Municipal Reuse

In this category there are two possibilities: One is the limited use of treated wastewater for certain restricted municipal purposes, such as fire-fighting, irrigation of parks, gardens and golf courses, and for street cleaning.

There is also the possibility of using treated wastewater in public buildings, or even in homes, for the purpose of flushing toilets. The cost of dual water systems might make this use uneconomical in existing built-up areas, but similar restricted utilization of wastewater might be worthwhile in new areas suffering from a severe water shortage. It should be assumed that even for limited municipal use, wastewater would have to be treated and disinfected to such an extent that it would be safe from a microbiological point of view, although it might not meet all the chemical standards usually desirable for drinking water. The specifications for treatment and disinfection would be rather strict because the danger of cross-connections or the possibility of accidental use of treated water for drinking purpose is quite considerable.

Okun (1973) has argued in favor of a concept of a hierarchy of water quality uses based on the policy for planned water reuse of the U.N. Economic and Social Council (1958): "No higher quality water, unless there is a surplus of it, should be used for a purpose that can tolerate a lower grade." He suggests that, rather than treat wastewater to meet the standards of drinking water, dual supply systems for wastewater reuse deserve careful study, since a relatively small portion of the total community water supply is required for drinking, cooking, or other uses that demand high purity. In most industrialized cities, this portion is only about 10% the remainder being used for industrial purposes, toilet and street flushing, public fountains, irrigation, and other purposes not requiring high-quality water.

Key (1967) has also supported this concept of preserving high-quality water for domestic consumption while using lower quality water, including reuse of wastewater, for other needs. Such dual water supply systems are in use in the Bahama Islands and in Hong Kong Island, where seawater is used for flushing toilets. Okun feels that the technology for safe management of dual water-supply systems now exists and that in new communities or new sections of existing cities the cost of such a dual distribution system would be only about 20% higher than a conventional one. He argues that, in the long run, the benefits may outweigh the disadvantages.

Considering the many problems associated with direct reuse of reclaimed wastewater for unrestricted urban use, which will be discussed later, the concept of dual, municipal water systems may indeed offer many important advantages and should be given careful study in new communities or even in existing communities suffering from very severe water shortages.

E. Unrestricted Municipal Reuse

The development of advanced wastewater treatment technology has in recent years brought the possibility of direct wastewater reuse for domestic consumption under active consideration in a number of water-short areas. In the process of developing wastewater treatment trains capable of removing much of the incremental inorganic and organic contaminants which might have deleterious effects on water courses, it has become apparent that in certain cases the effluent produced could be of a quality comparable to recognized drinking water standards.

In Windhoek, South West Africa, wastewater processed by such advanced treatment trains has actually served as a source of municipal water supply (Van Vuuren *et al.*, 1971) and the adequacy of treatment has been justified by comparing the effluent quality with drinking water standards (see chapter 13). A number of other cities, including Denver (Ogilvie, 1975) Dallas (Graeser,

1974), are in the preparatory stages for such planned reuse for domestic purposes.

At this point, it is appropriate to ask whether our knowledge of the toxicological and epidemiological implications of wastewater renovation for domestic consumption is sufficient. Is the public health adequately protected if processed municipal and industrial wastewater can be brought in line with today's conventional drinking water standards? Were these standards conceived with such a possible application in mind? If not, what remains to be done prior to giving the final green light to total direct wastewater reuse for all purposes including human consumption? We shall attempt to discuss these questions as well as propose possible approaches to answering some of the yet unanswered questions.

I. ADEQUACY OF DRINKING WATER STANDARDS

The U.S. Public Health Service (1962) Drinking Water Standards list 20 chemical parameters, only 9 of which serve as absolute grounds for rejecting a supply as unsafe for human consumption. The World Health Organization (1971) Drinking Water Standards contain a few more chemical parameters which may serve as grounds for rejecting a supply. EPA proposed drinking water standards will expand the list a bit further. None of these widely known and accepted standards list more than a few synthetic organic compounds, despite the fact that hundreds of such chemicals may find their way into municipal and industrial wastewater and many of them are known for their potential deleterious effects on human health. For that matter, neither do these standards exhaust the list of potential inorganic toxicants that may be found in industrial wastes.

It must be recognized that conventional drinking water standards were originally based on the assumption that water for human consumption would generally be drawn from groundwater sources or from the "best available" protected uncontaminated surface-water sources and that the limited number of chemical parameters included were adequate for most situations. This assumption is rarely true for most surface supplies of today. Can standards be developed to cover the wide range of contaminants that are actually found in wastewater destined for domestic consumption?

Several hundred threshold-limit values for such industrial chemicals in air have been established in the field of industrial hygiene, while some 100 maximum allowable concentrations for pesticides and other toxic chemicals have been established under food regulations. Stokinger and Woodward (1958) have shown how certain of these accepted threshold limit values of substances in air may be used to arrive at approximate limiting concentration in water. Although there is still little direct evidence that such chemicals in water have produced or are now producing widespread deleterious effects on health, it is not difficult to envision such a possibility as the percentage of wastewater in many surface supplies increases or if renovated wastewater is used for drinking without proper removal and control of such toxic materials. The experience in New Orleans cited earlier points in that direction.

The drinking water standards of the U.S.S.R. Ministry of Health are the first to have recognized the scope of the problem and now include some 400 chemical parameters with more under study. Certainly not everyone of the 400 or so chemical standards for drinking water established by U.S.S.R. health authorities is of toxicological importance, since many were established on organoleptic grounds alone. Nevertheless, the differences between the large number of standards in the U.S.S.R. list and the very limited number of standards presently considered adequate in the United States and many other countries indicate the extent of the problem that will have to be faced when the wastewater of a city including industrial wastes has to be processed to the point of becoming fit for human consumption.

The mere comparison of the quality of the final effluent against those standards currently listed in the USPHS, WHO, or EPA drinking water standards leave too many questions unanswered to be accepted as adequate evidence that such an effluent is completely safe from the public health point of view.

The WHO International Working Group (World Health Organisation, 1975b) concluded "... that conventional drinking water standards alone cannot provide a sufficient basis for the health evaluation of reused water for domestic consumption."

Not enough is known about the true identification of the residual microchemical pollutants both inorganic and organic, which of course vary widely from one situation to the next, depending on the nature of the industrial wastes that enter the sewerage system.

2. MICROBIOLOGICAL PROBLEMS

While most bacterial pathogens can be effectively inactivated by chlorination and other conventional disinfection procedures, there is ample evidence that some of the enterovirus are manyfold more resistant than bacteria to the same disinfection processes (Shuval, 1970). Infectious hepatitis has been transmitted by sewage-contaminated water on many occasions and, in some cases, the virus appears to have passed through conventional water-treatment plants which have included disinfection by chlorine.

Enteric virus levels in raw sewage have been shown to be about 10–100/ml, while their concentration in heavily contaminated river water, which may lead to epidemics of infectious hepatitis may be 1/liter, or even less. Laboratory techniques for detecting such low levels of viruses in water are still in the developmental stage and present one of the major obstacles in studying the efficiency of wastewater renovation processes for virus removal. Methods for concentration of and detecting viruses in water have been reviewed by Shuval and Katzenelson (1972). Another problem arises from the fact that it may take 5 days or more before the results of a virus assay of water can be completed. Effective methods of assuring virus inactivation in wastewater treatment still leave something to be desired, although there is growing evidence that advance waste-treatment methods followed by disinfection with adequate concentrations of free available chlorine (HOCl) with sufficient contact time can be highly effective in reducing enteric viruses levels. Ozone is particularly promising as a virucidal agent for water and wastewater treatment (Katzenelson *et al.*, 1974).

Nevertheless, the effectiveness of any proposed wastewater renovation treatment train must be fully evaluated as to its virus removal efficiency and ways must be found to monitor routinely such plants to assure their continued effectiveness in removing viruses.

3. TOXICOLOGICAL EVALUATION OF RENOVATED WASTEWATER

There are two possible approaches to the toxicological evaluation of renovated wastewater to be used as drinking water (Shuval and Gruener, 1973). The first would require the establishment of maximum allowable concentrations or limits for each of the potentially hazardous chemicals that may be found in renovated wastewater. The approach that has been developed by toxicologists in setting tolerance limits for food additives and chemical contaminants in food has been to establish acceptable daily intake (ADI) levels for man. These ADI are based on all relevant toxicological data available at the time of evaluation including data from cases of human exposure, which is usually very limited. After determining the "no-effect" level and introducing a certain safety factor, the ADI can be established. Such figures can then provide the toxicological basis for establishing tolerance levels or maximum allowable concentrations in food, water, and air which are based on known consumption patterns and realistic levels of contamination that are unavoidable. In establishing such figures for water, due consideration must be given to the total body burden from all environmental sources. Although much of the basic data may be available to assist in setting such standards for water, important information is still missing (World Health Organization, 1973).

The toxicological evaluation of chemicals found in the environment cannot be simplified to take into account acute or subacute effects alone. Today, such evaluation must include effects from long-term exposure and studies for carcinogenicity, mutogenicity, teratogenicity, and various biochemical and physiological effects. Even if the specific toxicity of defined industrial and agricultural chemicals is established, the possible toxic effects of their breakdown products may be more difficult to determine. Natural biodegradation or specific treatment processes such as chlorination may lead to the development of new compounds having toxic properties quite different from those of the parent compound. Work to identify these breakdown products and to study their toxic effects is required.

Another factor complicating the toxicological evaluation of heavily polluted water or renovated wastewater is the combined and possible synergistic effect resulting from the exposure to a mixture of toxic and nontoxic chemicals. Increased toxic impact of such combinations is known to occur under certain circumstances and the case of renovated wastewater must take into account such possibilities.

Although much is to be gained by establishing proper tolerance levels for many of the known toxicants that might appear in renovated water, this approach will take a long time to develop and even then will not cover all possible toxic effects as pointed out above.

For these reasons, it is felt that a second approach is required. A full toxicological evaluation should be carried out on the actual finished renovated water intended for human consumption with its real mixture of known and unknown residual chemicals remaining after treatment.

Long-term feeding experiments with more than one species of experimental animal should be required, as well as other more rapid toxicological screening tests using bacteria or cell cultures. Such studies would include testing concentrates of the residual chemicals in the final processed water, as well as the normal unconcentrated effluent. Concentration techniques used must avoid being selective as is the case with activated carbon and must not lead to the breakdown of the chemicals involved by overly harsh treatment, such as high temperature distillation. Reverse osmosis and lyophilization might offer possible approaches to this problem.

Most governments require the full toxicological evaluation of any new drug or food additive before allowing its commercial use. The requirements for evaluating renovated wastewater with its many and often complex unknowns should be, at least, as rigorous. WHO reports on this matter (World Health Organization, 1973, 1975b) have emphasized the need for complete toxicological evaluation of the actual water planned for reuse as an essential step in evaluating the safety of such reuse.

4. BUILDUP OF DISSOLVED SOLIDS IN RECYCLING

In multiple recycling of wastewater, there will be a buildup of those dissolved solids not removed or only partially removed by the wastewater treatment plants unless specific demineralization processes are included. However, the buildup in concentration will not be infinite, since there will be fresh makeup water added in each cycle to compensate for water losses that do not appear as sewage flow. In most cities, these losses normally range from 10–20% (Metzler *et al.*, 1958).

Thus, if we assume that in each recycle 90% of the water input into the community appears as wastewater which will be processed for recycling, the concentration say, of, sodium chloride will increase with recycling until it reaches an equilibrium 10 times greater than its original concentration in the wastewater (Long and Bell, 1972). The same might be true for certain refractory organics of potential public health danger. Under such conditions, partial demineralization would be necessary to keep dissolved inorganics at acceptable levels. The problem of toxic trace organics and inorganics may be more difficult to deal with since these substances may not all be removed with equal effectiveness by some demineralization processes which may be selective. The buildup of such toxic materials on multiple recycling would be very undesirable. Until such time as complete information is available on the removability of the various toxic organic and inorganic trace elements that may appear in the wastewater stream, it might be prudent to reduce the possibilities of buildup by providing additional dilution from freshwater sources. For example, with a reuse factor of only 30%, the maximum concentration at equilibrium of a chemical not removed at all will be only 40% greater than the original concentration in the wastewater stream, while with a reuse factor of 50%, equilibrium will be reached with the concentration of the refractory compound of twice its original concentration. Such dilution with freshwater would normally avoid the need to include an expensive demineralization step in the treatment and would provide an additional safety factor against the buildup of compounds whose removability by demineralization or other processes may not as yet be known.

5. EPIDEMIOLOGICAL EVALUATION

No matter how thorough a toxicological evaluation is made, there always remains the problem of extrapolating the findings with laboratory animals to fit the human situation. With drugs of potentially great medical importance, human trials are held after completion of the toxicological evaluation. In the case of new food additives that are usually less essential to human welfare than drugs, negative findings in the toxicological evaluation do not automatically mean the new chemical will be allowed for use in human food. It must be demonstrated that the chemical will make a significant contribution to improving the quality or preservation of the foods in which it will be used. In water-short areas, water renovation can often be justified as being of potentially great importance to human welfare as well.

A considerable body of information has been built up concerning the effect on human health of many of the environmental contaminants discussed here as a result of direct exposure of humans under various industrial situations. Further information has been gained by the accidental exposure of humans to certain toxic materials.

If wastewater renovation for domestic consumption is ever to become widely accepted, there will be a need at some stage to carry out a full-scale epidemiological evaluation of the impact of such reuse on the health of the population exposed. It may be difficult to choose an appropriate population group for such a study, but to the extent that certain communities in water-short areas have already gone ahead with wastewater reuse for drinking water, every effort should be made to carry out a thorough epidemiological evaluation. Such a study should include baseline health evaluation of a sample population before the introduction of renovated water and then a follow-up of the same group, as a panel study, over a 5- or 10-year period. Such opportunities will be few and far between, and every effort should be made to gain as much data from each case as is possible.

A promising alternative to such a study with a population exposed to planned wastewater reuse would be a series of studies of populations exposed to indirect or unintentional wastewater reuse. Such population groups are easier to identify than might be imagined, since many millions of people throughout the world are, in fact, consuming renovated wastewater everyday and have been doing so for years. Some 15–20% of the flow of the Ohio River is fully or partially treated municipal or industrial wastewater (Cleary *et al.*, 1963). Millions of people in the Rhine River Basin consume water from the river which, at times of low flow, may contain as much as 40–50% of industrial and municipal wastewater.

It must be recognized that unintentional and, in many respects, uncontrolled reuse of wastewater is now very widely practiced and provides a basis for evaluating the health impact of such use as well as the expected impact of fully engineered and carefully controlled direct wastewater reuse of the type under discussion.

Although such prospective epidemiological studies are expensive and take many years to complete, it is essential that they be made even if planned direct wastewater reuse were not under consideration. Such studies are essential to evaluate present environmental exposure from consuming water from polluted sources, which will become even more polluted in years to come. Epidemiological studies may shed light on the need to make major improvements in present-day water-treatment technology, which has been demonstrated as being relatively ineffective in removing many of the refractory organic toxicants that appear in increasing quantities in polluted water. The New Orleans study with all of its limitations illustrates the importance of this approach. The findings will also be of vital importance in planning future wastewater renovation programs, where it can surely be expected that treatment trains of demonstrated efficiency will be utilized to remove potentially hazardous chemicals or pathogenic micro-organism to the lowest possible levels.

6. MONITORING THE REUSE OF WASTEWATER FOR DOMESTIC CONSUMPTION

The nature of a monitoring or quality-control program for products produced for human consumption should vary according to the degree of risk to health involved. Conventional water-supply monitoring programs have in the past assumed that the product is basically a safe one and that it can be supplied to the consumer directly after processing without waiting for the results of quality-control tests. Bacteriological test results are usually available 24–48 hours after sampling, while routine tests for toxic chemicals, when they are made, may be available only after many days. The water tested has usually been consumed by the population by that time.

Drugs and food additives and many processed foods are tested routinely and are released for use in batches only after the test results indicate no positive findings.

In the case of a plant for renovating wastewater for human consumption, it would appear necessary to require a more rigorous monitoring and quality control regime than that currently practised by the water supply industry. Technical breakdowns and human failures at such a plant might lead to major health hazards. It would not be illogical to require that renovated water be fully tested and certified as safe before its release to the general water-supply system. With improved bacteriological techniques, results can be obtained in under 24 hours as can the results of most of the important chemical tests, many of which can be automated. Ways of carrying out rapid toxicological evaluation of the finished water with bioassay techniques should be developed. For the moment, virus assays require at least 5 days for completion, but, here too, more rapid assay techniques are under study and may become available. Katzenelson (1975) has developed a rapid virus-detection method using fluorescent antibodies that can provide a qualitative answer in 10 hours and a quantitative one in 24 hours.

Renovated water could be produced and held in batches until completion of the quality-control tests, before being released. This will add additional costs to wastewater renovation plants, but the additional safety obtained would justify the expenditure. Certainly, such precautions should be practiced in all early plants until it can be demonstrated that less stringent quality-control measures are adequate.

Many might agree that the proposed monitoring regime should be applied to any case where heavily polluted surface water is the source of drinking water supplies. Such supplies may be an even greater risk than planned direct reuse programs.

7. POLICY CONSIDERATIONS

The approach presented here concerning water reuse for domestic consumption may appear to be overly cautious and place too heavy a burden on future wastewater renovation programs. In answer, it must be stated that criticism of current drinking water standards applies as much to any case where polluted surface water serves as a source of drinking water as to the special case of direct wastewater renovation. In fact, indirect, unplanned wastewater reuse may well be a greater risk than planned direct reuse, which would include treatment processes more capable of coping with the organic pollutants found in wastewater. Unplanned or covert wastewater reuse is far too widely practiced today, with too few controls, to allow one to feel complacent.

However, planned direct wastewater reuse for domestic consumption carries a heavy responsibility with it, since it involves full engineering and health responsibility from the beginning to the end, without the intervening hand of "nature." The fact that nature provides little protection in heavily polluted rivers whose self-purification capacity is overtaxed gives little justification for a similar lax approach in a planned direct reuse project. In such a project, the designers, operators, and health authorities who must give their approval must carry the full responsibility of any adverse health effects which may result, even if it can be shown that communities consuming polluted surface water are exposed to equal or greater risks.

Many such communities are indeed exposed to undesirable health risks and, thus, a subsequent equal tightening up of standards, treatment procedures, and quality control for all cases of wastewater reuse whether direct or indirect is fully justified.

It still remains to be demonstrated that water treatment technology can overcome the many problems involved in processing wastewater with its many complex components and rapid fluctuations in quality to achieve a uniform end product meeting the health requirements for wholesome and safe drinking water. However, further work in this direction will be very important, whether it be applied to direct water reuse projects or to the more urgent and widespread cases of indirect, covert wastewater reuse that exists so widely today in communities drawing water from the polluted lower reaches of the great rivers of the world.

Another consideration that cannot be overlooked is that of public attitudes toward water reuse for domestic purposes. The strong public opposition which, in many cases, thwarted efforts to introduce fluoridation despite strong technical evidence and the support of the scientific community is an illustration of an aroused public. Water reuse for domestic consumption may not be easily accepted by the public, even if all the precautions outlined above are taken. The study of Bruvold and Ward (1972) in 10 towns in California indicated that out of 25 forms of possible wastewater reuse only 11 would be likely to receive no public opposition. These include such items as golf course irrigation, commercial air conditioning, and hay, alfalfa, or orchard irrigation. Over 50% of those interviewed opposed reuse for domestic purposes. Any planned programs for reuse must give careful consideration to this question from the very beginning, or they may find years of scientific and technical effort vetoed by the public.

In the final analysis, direct planned wastewater reuse for human consumption may well become feasible through the development of advanced wastewater treatment systems with a demonstrated fail-safe capability of removing the hundreds of potentially toxic inorganic and organic chemicals that appear in today's wastewater streams. A major combined effort of developing appropriate advanced technology and health effects evaluation will certainly be required to achieve this goal.

F. Groundwater Recharge

The use of treated wastewater for groundwater recharge is practiced in a number of areas. In some cases, the sole objective has been to build up a barrier to prevent salt-water intrusion into coastal areas where groundwater withdrawals have been excessive. If no direct withdrawal of the groundwater and, with it, the recharged wastewater is practiced, it will flow to the sea and create few if any public health problems. However, if the recharge occurs in areas where groundwater pumping takes place, the effects of the effluent on the quality of the groundwater withdrawn may be considerable. The main factors that must be considered are the nature of the aquifer, the mean residence time betweeen recharge and withdrawal, withdrawal rates, and finally the degree of dilution obtained with the surrounding groundwater.

In uniform sandy aquifers, a high degree of microbial removal can be anticipated. Studies have shown that within a distance of a few hundred meters from the point of recharge, effective removal of viruses and bacteria can generally be achieved. Long residence times of several hundred days in the aquifer may also prove effective in the removal of viruses and bacteria through dieaway. However, in the case of nonuniform aquifer formations of gravel or karst limestone, there may be little or no microbial removal over extensive distances.

Inorganic and organic chemical removal will be a function of the adsorption and ion exchange characteristics of the aquifer which may, under certain circumstances, provide a considerable degree of removal, while, in other cases, such chemicals may travel over great distances with little or no reduction in concentration. Even when studies indicate a degree of chemical removal by filtration through the aquifer, there is the possibility that once the adsorptive or ion exchange capacity is exhausted there will be a breakthrough of chemical contaminants which may appear suddenly and possibly in high concentrations at the withdrawal wells. This can present a serious threat to the quality of the reclaimed water.

In areas where groundwater recharge with treated wastewater is planned, a major factor in determining the degree of pretreatment required is the ultimate use of the water withdrawn. If only agricultural or industrial utilization is planned, it will usually be possible to meet the health requirements for such use without too much difficulty, or at most, by additional disinfection of the pumped well water. However, if the water is destined, all or in part, for municipal use including domestic consumption, all of the limitations mentioned previously for unrestricted municipal reuse must be applied, unless very high rates of dilution with pure groundwater can be assured.

Effective removal of toxic organics and heavy metals must be assured prior to the recharge operation, although some dilution effect and actual removal may be obtainable by aquifer filtration.

The water renovation project in Orange County, California (Cline, 1975), is based on full tertiary treatment including multiple stages of biological and physicochemical treatment and disinfection prior to recharge to meet the strict State of California Health Department requirements for recharge with wastewater. Even with all of that, the California Health Authorities are yet to approve of the reclaimed groundwater for domestic consumption. They will require that the safety of the reclaimed water be fully demonstrated by careful toxicological testing.

In Israel, the Dan Region Water Reclamation Project will provide full biological treatment including nitrification and denitrification, as well as excess lime treatment prior to recharge of the effluent into the sand dune area south of Tel Aviv (Shuval, 1975c). The designers' original plans called for producing a reclaimed effluent withdrawn after being recharged, which could be used for all purposes including domestic consumption. The minimum retention time in the aquifer will be 400 days. The recharge area will be completely surrounded by a ring of recovery wells controlled and operated by a single authority. In the final stage, the recovery wells will be pumping almost 100% recharged wastewater.

However, our preliminary studies have indicated that the sand dune filtration provides only partial removal of the dissolved organics. The effluent before recharge contains about 40 ppm of TOC, while water withdrawn from the aquifer shows TOC levels of about 15 ppm, with high concentrations of ABS detergents that reached the 13 ppm level. The Health Authorities in Israel have not approved this water for domestic consumption as yet, and will most likely require considerable additional treatment to remove dissolved organics and any remaining traces of toxic chemicals prior to domestic use. For the moment, the plan is to restrict the use of the reclaimed water to agricultural and industrial purposes, until a procedure is developed that will provide potable water of a demonstrated safe quality.

Groundwater recharge prior to reuse for domestic consumption certainly provides many advantages and a considerable safety factor as a result of the buffering effect of long retention and groundwater dilution, as well as a degree of removal of microbial and chemical pollutants. It also provides for an excellent opportunity to enable complete monitoring of water quality prior to withdrawal, since observation wells between recharge areas and withdrawal wells can be used to test water quality months before it is withdrawn from the aquifer. Just such a monitoring program has been included in the Dan Region Water Reclamation Project.

WHO (World Health Organization, 1973) has pointed out in its report that "groundwater recharge involving extended periods of underground storage can provide a considerable safety factor in wastewater renovation." However, careful planning and control of such recharge programs is essential to ensure that the full benefits of such a strategy are obtained.

REFERENCES

Askew, J. B., Bott, R. F., Leach, R., and England, B. (1965). Microbiology of reclaimed water from sewage for recreational use. *Amer. J. Pub. Health* **55**, 453–462.

Bellar, T. A., Lichenberg, J. J., and Kroner, R. C. (1974). The occurrence of organohalides in chlorinated drinking waters. *J. Amer. Water Works Ass.* **66**, 703–706.

Benarde, M. A. (1973). Land disposal of sewage effluents: appraisal of health effects of pathogenic organism. *J. Amer. Water Works Ass.* **65**, 432–440.

Bergner-Rabinowitz, S. (1956). The survival of coliforms *S. faecalis* and *S. tennessee* in the soil and climate of Israel. *Appl. Microbiol.* **4**, 101.

Borneff, J., and Fisher, R. (1962). Part IX: Investigations of filter activated-carbon after utilization in waterplant. *Arch. Hyg. Bakteriol.* **146**, 1–16 (in German).

Bruvold, W. H., and Ward, P. C. (1972). Using reclaimed wastewater—public opinion. *J. Water Pollut. Contr. Fed.* **44**, 1690–1696.

Bunch, R. L., Barth, E. F., and Ettinger, M. D. (1960). "Conference on Biological Treatment." Manhattan College, New York.

Chaiken, E. I., Poloncsik, S., and Wilson, C. D. (1973). Muskegon sprays sewage effluents on land. *Civil. Eng. Amer. Soc. Civil Eng.* **45**, 49–53.

Cleary, E. J., Horton, R. K., and Boes, R. J. (1963). Re-use of ohio river water. *J. Amer. Water Works Ass.* **55**, 683–686.

Cline, N. M. (1975). "Wastewater Reuse in Orange County, California," Proc. Workshop on Research Needs for Municipal Wastewater Reuse. U.S. Environmental Protection Agency, Washington, D.C.

Cohen, J., Schwartz, T., Kalazmer, R., Pridan, D., Ghalayini, H., and Davies, A. M. (1971). Epidemiological aspects of cholera El Tor outbreak in a non-endemic area. *Lancet*, **ii**, 86–89.

Dunham, L. J., O'Gara, R. W., and Taylor, F. B. (1967). Studies on pollutants from processed water: collection from three stations and biologic testing for toxicity, and carcinogenesis. *Amer. J. Pub. Health* **57**, 2178–2185.

Dunlop, S. G. (1952). The irrigation of truck crops with sewage-contaminated water. *Sanitarian, Los Angeles* **15**, 107.

Dunlop, S. G., Twedt, R. M., and Wang, W. L. L. (1951). *Salmonella* in irrigation water. *Sewage Ind. Wastes* **23**, 1118.

Epstein, S. S. (1974). Environmental determinants of human cancer. *Cancer Res.* **34**, 2425–2435.

Falk, L. (1949). Bacterial contamination of tomatoes grown in polluted soil. *Amer. J. Pub. Health* **35**, 1338–1342.

Graeser, H. J. (1974). Water reuse: resource of the future. *J. Amer. Water Works Ass.* **66**, 575–578.

Greenberg, A. E., and Dean, B. H. (1958). The beef tapeworm, measly beef, and sewage: a review. *Sewage Ind. Wastes* **30**, 262–269.

Greenberg, A. E., and Kupka, E. (1957). Tuberculosis transmission by wastewaters—a review. *Sewage Ind. Wastes* **29**, 524–537.

Harris, R. H. (1974). The Implications of Cancer Causing Substances in Mississippi River Water." Environmental Defence Fund, Washington, D.C.

Hueper, W. C., and Conway, W. D. (1964). "Chemical Carcinogenesis and Cancers." Thomas, Springfield, Illinois.

Jepson, A., and Roth, H. (1952). Epizootiology of *Cysticercus bovis*—resistence of the eggs of *Taenia saginata. Rep. Int. Vet. Congr., 14th, 1949* Vol. 2, 49.

Jolly, R. L. (1973). Chlorination Effects on Organic Constituents in Effluents from Domestic Sanitary Sewage Treatment Plants, Publ. No. 55. Environ. Sci. Div., Oak Ridge Nat. Lab, Oak Ridge, Tennessee.

Katzenelson, E. (1976). A rapid fluorescent antibody method for qualitative isolation of viruses from water. *Arch. Virology* **50**, 197–206.

Katzenelson, E. and Teltch, B. (1976). Dispersion of enteric bacteria in the air as a result of sewage spray irrigation and treatment processes. *Water Pollut. Contr. Fed.* **48**, 710–716.

Katzenelson, E., Kletter, B. and Shuval, H. I. (1974). Inactivation kinetics of viruses and bacteria in water by use of ozone. *J. Amer. Water Works Ass.* **66**, 725–729.

Kehr, R. W., and Butterfield, C. T. (1943). Notes on the relationship between coliform and enteric pathogens. *Pub. Health Rep.* **58**, 589–607.

Key, A. (1967). "Raw Water Quality in Relation to Water Use," Proc. Symp. on the Conservation and Reclamation of Water, pp. 37–53. Inst. Water Pollut. Contr., London.

Kligler, I. J. (1921). Investigations of soil pollution and the relation of various types of privies to the spread of intestinal infections. International Health Board Monograph 15 p. 1. Rockerfeller Institute of Medical Research, New York.

Linstedt, K. D., Bennett, E. R., and Work, S. W. (1971). Quality considerations in successive water use. *J. Water Pollut. Contr. Fed.* **43**, 1681–1694.

Long, W. N., and Bell, F. A. (1972). Health factors and reused water. *J. Amer. Water Works Ass.* **64**, 220–225.

McClesky, C. A., and Christopher, W. (1941). The longevity of certain pathogens on strawberries. *J. Bacteriol.* **41**, 98.

Medical Research Council. (1959). Sewage contamination of bathing beaches in England and Wales. *Med. Res. Counc. (Gt. Brit.), Memo. 37.*

Metzler, D. F. Gulp, R. L., Stoltenberg, H., Woodward, R., Walton, G., Chang, S., Clarke, N., Palmer, C., and Middleton, F. (1958). Emergency use of reclaimed water for potable supply at Chanute, Kansas. *J. Amer. Water Works Ass.* **50**, 1021.

Middleton, F. M. (1960). *Proc. Conf. Physiol. Aspects Water Qual.*, U.S.P.H.S Wash. D.C.

Ogilvie, J. L. (1975). "Wastewater Reuse as a Water Resource—Denver Experience," Proc. Workship on Research Needs for Municipal Wastewater Reuse. U.S. Environmental Protection Agency, Washington, D.C.

Okun, D. A. (1973). Planned water reuse. *J. Amer. Water Works Ass.* **65**, 617–622.

Ongerth, H. J., Spath, D. P., Crook, J., and Greenberg, A. E. (1973). Public health aspects of organics in water. *J. Amer. Water Works Ass.* **65**, 495–497.

Plotkin, S. A., and Katz, M. (1967). Minimal infective doses of virus for man by the oral route. *In* "Transmission of Viruses by the Water Route" (G. Berg, ed.), Wiley (Interscience), New York.

Rigbi, M., Amramy, A., and Shuval, H. (1956). Efficiency of a small high-rate trickling filter plant at Jerusalem, Israel. *Sewage Ind. Wastes* **28**, 852.

Rook, N. J. (1974). Formation of haloforms during chlorination of natural waters. *J. Soc. Water Treat. Exam.* **23**, Part 2, 234–243.

Rowan, W. B. (1964). Sewage treatment and schistosome eggs. *Amer. J. Trop. Med. Hyg.* **13**, 572–576.

Rudolfs, W., and Gehm, H. W. (1936). Sewage chlorination studies. *N. J. Agr. Exp. Sta. Bult.* **601**.

Rudolfs, W., Falk, L. L., and Ragotzkie, R. A. (1950). Literature review of the occurrence and survival of enteric, pathogenic and related organisms in soil, water, sewage and sludges, and on vegetation. I. Bacterial and virus diseases; II, Animal parasites. *Sewage Ind. Wastes* **22**, 1261 and 1417.

Rudolfs, W., Falk, L. L., and Ragotzkie, R. A. (1951). Contamination of vegetables grown in polluted soil. I. Bacterial contamination. *Sewage Ind. Wastes* **23**, 253.

Shuval, H. I. (1951). Public health aspects of sewage irrigation. *J. Isr. Ass. Architect. Eng.* p. 36.

Shuval, H. I. (1962). The public health significance of trace chemicals in waste water utilization. *Bull. W.H.O.* **27**, 791–799.

Shuval, H. I. (1970). Detection and control of enteroviruses in the water environment. *In* "Developments in Water Quality Research" (H. I. Shuval, ed.), p. 47. Ann Arbor (Humphrey), Sci. Publ., Ann Arbor, Michigan.

Shuval, H. I. (1975a). Disinfection of wastewater for agricultural utilization. *Progress in Water Tech.* **1**, 857–867.

Shuval, H. I. (1975b). The case for microbial standards for bathing beaches, Proc. Int. Symp. on Discharge of Sewage from Sea Outfalls. Pergamon, Oxf ord.

Shuval, H. I. (1975c). Evaluation of the health aspects of wastewater utilization for domestic purposes in Israel, Proc. Workshop on Research Needs for Municipal Wastewater Reuse. U.S. Environmental Protection Agency, Washington, D.C.

Shuval, H. I., and Gruener, N. (1973). Health considerations in renovating wastewater for domestic use. *Environ. Sci. Technol.* **7**, 600–604.

Shuval, H. I., and Katzenelson, E. (1972). Detection of enteric viruses in the water environment. *In* "Water Pollution Microbiology", (R. Mitchel, ed.), 47–71". Wiley, New York.

Silverman, P. H., and Griffiths, R. B. (1955). Review of methods of sewage disposal in Great Britain with special reference to the epizootiology of *Cysticercus bovis*. *Ann. Trop. Med. Parasitol.* **49**, 436.

Sorber, C. A., and Guter, K. J. (1975). Health and hygiene aspects of spray irrigation. *Amer. J. Pub. Health* **65**, 47–52.

Sorber, C. A., Schaub, S. A., and Bausum, H. T. (1974). Virus survival following wastewater spray irrigation of sandy soils. *In* Virus Survival in Water and Wastewater Systems, Univ. of Texas, Austin.

State Health Department of California. (1973). "Statewide Standards for the Direct Use of Reclaimed Water for Irrigation and Recreational Impoundments" Calif. Admin. Code, Title 17, Public Health.

Stokinger, H. E., and Woodward, R. L. (1958). Toxicologic methods for establishing drinking water standards. *J. Amer. Water Works Ass.* **50**, 515–529.

U.N. Economic and Social Council. (1958). "Water for Industrial Use," U.N. Rep. No. E/3058 ST/ECA/50. United Nations, New York.

U.S. Environmental Protection Agency. (1972). "Industrial Pollution of the Lower Mississippi River in Louisiana," Rep. EPA Reg. VI, Dallas, Texas. USEPA, Washington, D. C.

U.S. Environmental Protection Agency. (1975). "Proceedings of the Workshop on Research Needs for Municipal Wastewater Reuse." USEPA, Washington, D.C.

U.S. Public Health Service. (1962). "Drinking Water Standards." U.S. Dept. of Health, Education and Welfare, Washington D.C.

U.S. Public Health Service. (1970). "Evaluation of Environmental Carcinogens," Report to the Surgeon General U.S.P.H.S., ad hoc Committee on the Evaluation of Low Level of Environmental Carcinogens. U.S. Dept. of Health, Education and Welfare, Washington, D.C.

U.S.S.R. Ministry of Health. (1961). "Standards for Drinking Water." Moscow.

Van Vuuren, L. J. R., Henzen, M. R., Stander, G. J., and Clayton, A. J. (1971). The full-scale reclamation of purified sewage effluent for the augmentation of the domestic supplies of the city of Windhoek. *In* "Advances in Water Pollution Research" Vol. I, (S. H. Jenkins, ed.) Pergamon, London.

World Health Organization. (1964). World Health Organ., Expert Committee, *Tech. Rep. Ser.* **276**.

World Health Organization. (1971). "International Standards for Drinking Water." World Health Organ., Geneva.

World Health Organization. (1973). Reuse of effluents: Methods of wastewater treatment and health safeguards, *Tech. Rep. Ser.* **517**. World Health Organ., Geneva.

World Health Organization. (1975a). Working Group on Guides and Criteria for the Recreational Quality of Beaches and Coastal Water. World Health Organ., Copenhagen.

World Health Organization. (1975b). "Report of the International Working Meeting on Health Effects relating to Direct and Indirect Reuse of Wastewater for Human Consumption." International Reference Center for Community Water Supply, Amsterdam.

Yoshpe-Purer, Y., and Shuval, H. I. (1970). *Salmonellae* and bacterial indicator organisms in polluted coastal water and their hygienic significance, *In* "Marine Pollution and Sea Life" (M. Ruivo, ed.) p. 574–580. Fishing news Books, London.

3

The Use of Wastewater for Agricultural Irrigation

Josef Noy with Akiva Feinmesser

I. INTRODUCTION

A. The Motivation for the Use of Treated Wastewater for Irrigation

The advantages in the use of treated wastewater for irrigation are (a) low-cost source of water, (b) an economical way to dispose of wastewater to prevent pollution and sanitary problems, (c) an effective use of plant nutrients contained in wastewater, and (d) providing additional treatment before being recharged to the groundwater reservoir.

The principles of wastewater utilization discussed in this chapter are based on experience that has been accumulated in Israel and some investigations reported in the literature. The factors affecting wastewater utilization, its composition, and possible applications undoubtedly vary from place to place. Untreated wastewater is considered a health and aesthetic nuisance. The responsibility for wastewater treatment lies with the wastewater "producer" which may be a municipality, an industrial plant, or an agricultural settlement.

Wastewater, even after some degree of treatment, can cause the pollution of waterways, rivers, and groundwater reservoir. The problem of wastewater disposal becomes more severe with increasing population and industrialization. The quantity of municipal wastewater in Israel is between 70 and 220 liters per capita per day, according to the standard of living, climate, and season. In the agricultural settlements, wastewater quantities are between 150 and 300 liters per capita per day (Hershkovitz and Feinmesser, 1967).

Wastewater is usually the cheapest water in arid areas. In some cases, it is the only water available for irrigation. In calculating the cost of water one has to consider that the treatment of wastewater is required in any case, mainly because of health regulations. Moreover, the required health standard for irrigation quality is usually less strict than health standards for other purposes. The cost of water should therefore be based only on the cost to transport it to the irrigated plots.

Another advantage is the addition of nutrient elements required by growing crops. The increments of nutrients in municipal wastewater in Israel are about 50 gm N, 15 gm P_2O_5, 30 gm K_2O, and 150 gm organic matter per cubic meter (Hershkovitz and Feinmessev, 1967). In addition, there are small amounts of micronutrients present.

The application of wastewater in irrigation brings about the renovation of the percolating water through the soil profile, especially in the presence of growing plants. The root zone is considered a "living filter" and the "effective renovation" has been defined as (Kardos, 1967):

$$\frac{\text{kg of nutrients used by plants}}{\text{kg of nutrients added in irrigation water}} \times 100$$

It has also been found that phosphate was reduced in the percolating water by reaction within the soil, and nitrogen was reduced due to uptake by plants. In this manner groundwater pollution and eutrification of rivers and lakes by wastewater are drastically decreased.

B Possible Disadvantages in the Use of Treated Wastewater for Irrigation

When wastewater is used for irrigation, a number of possible disadvantages have to be considered:

a. The supply of wastewater is continuous throughout the year, while irrigation is seasonal and dependent on crop demands.

b. Treated wastewater may plug nozzles in irrigation systems and clog capillary pores of heavy soils.

c. Some of the soluble constituents in wastewater may be present in concentrations toxic to plants.

d. Health regulations restrict the application of wastewater to edible crops.

e. When wastewater is not properly treated, it may be a nuisance to the environment.

1. THE COORDINATION OF WASTEWATER SUPPLY WITH UTILIZATION

Wastewater supply is continuous. However, the rate of flow is not uniform. The quantities of domestic effluents are expected to be greater in the dry season than in the wet one. Also, rainwater will contribute to waste flows. There are seasonal variations in the water use of some industrial plants, such as fruit and vegetable canning. Therefore, there is a need for effluent storage, on both an operational and seasonal basis.

Treated wastewater may be stored for operational purposes together with water from other sources, such as floodwater or supply water; in this way, the total quantity of water to be applied in irrigation is increased and its quality improved. In small treatment plants, near villages and agricultural settlements, the oxidation ponds may also serve as storage ponds.

Seasonal storage for treated wastewater requires a large volume. Water from other sources, such as flood water, may also be ponded together with effluents, as previously mentioned. Another solution is recharging groundwater, which may be reused in irrigation by pumping. Groundwater replenishment causes a significant improvement in water quality because percolation through the soil causes a reduction of some undesirable suspended and soluble constituents by adsorption or precipitation. There is also a reduction in pathogens.

2. THE CLOGGING OF SOILS AND IRRIGATION SYSTEMS

Most of the suspended solids in raw wastewater are removed by proper treatment. However, some solids may be found in the effluent. Effective filtration and the use of large nozzles are then required to avoid plugging in sprinkler or trickle irrigation systems. The contribution of treated domestic effluent is about 150 gm/m^3 of organic matter (Hershkovitz and Feinmesser, 1967). In the irrigation season, when 10,000 m^3/hectare of treated effluent are applied, the addition of organic matter may reach 1500 kg/ha. Such a quantity, even when partly decomposed during the season. may bring about some changes in the physical properties of the soil, which may be favorable in sandy soils.

However, in heavy soils, organic matter may clog capillary porces, mainly in the upper soil layer, and thus bring about a decrease in the rate of infiltration. Suspended solids may also clog capillaries deeper in the soil profile where, under anaerobic conditions, decomposition of organic matter proceeds at a very low rate and the reduction in soil permeability to water persists.

Breaking up the surface crust and ploughing the deeper layers are sometimes necessary to improve infiltration rates.

3. TOXIC CONSTITUENTS

Treated domestic effluent, and certainly industrial wastes, may contain soluble constituents at concentrations toxic to plants. Domestic effluent pickup between 50 and 100 mg/liter of chloride and of sodium ions each. These may concentrate in the root zone and harm sensitive crops. The concentration of sodium may increase in particular where water softeners are in use. The negative effect of sodium on the soil is the deflocculation of clay particles which causes an unfavorable soil structure. This then decreases water and air permeability. Boron concentration increases in effluents mainly due to the use of washing powders that contain perborates. The pickup of 1 ppm boron in sewage effluent is common. This concentration may harm sensitive crops.

Some industries may add heavy metals at concentrations toxic to plants or animals feeding on plant material. The common heavy metals are: Zn, Mn, Cr, Cd, Ni, Pb, Hg. Due to their chemical properties the larger part of these elements are found in sewage sludge and their concentration in the effluent is small. However, their build-up in the soil or in ground water may reach hazardous concentrations. In light soils their hazard may be greater because in heavy soil, heavy metal fixation to insoluble forms renders them unavailable to plants.

Other wastes may contain organic compounds, such as organic acids and phenols, that may restrict biological activity in the root zone.

Mixing of such wastewater with other wastes dilutes the toxic constituents to a tolerable concentration. In some cases, such wastes have to be separated to avoid their utilization in irrigation.

4. HEALTH RESTRICTIONS

Wastewater may contain pathogenic bacteria, parasite eggs, cysts, and viruses which are carried by human excreta. The treatment of wastewater brings about a decrease in their number. A decrease of 99.9% in the number of coliform bacteria may be achieved by treatment. However, the remaining concentration of coliform bacteria may still exceed a 10^5/100 ml. (The standard allowed for coliform bacteria in wastewater used for irrigating certain edible crops is sometimes set at 100/100 ml or less.)

Health authorities maintain strict standards in the use of wastewater. Thus, irrigation with effluent is usually restricted to fiber and other industrial crops, seed crops, ornamentals, and sometimes fodder crops and vegetables that are consumed only after cooking.

5. AESTHETIC NUISANCE

Treatment, transmission and use of wastewater may cause unpleasant odors. The dissipation of such odors is felt more strongly when the sprinkler method is used. There is also a greater hazard of mosquito and housefly breeding from small sewage puddles. The higher the grade of treatment, the less the nuisance. Also, the breeding of mosquitoes and flies has to be controlled in irrigated areas and sewage channels.

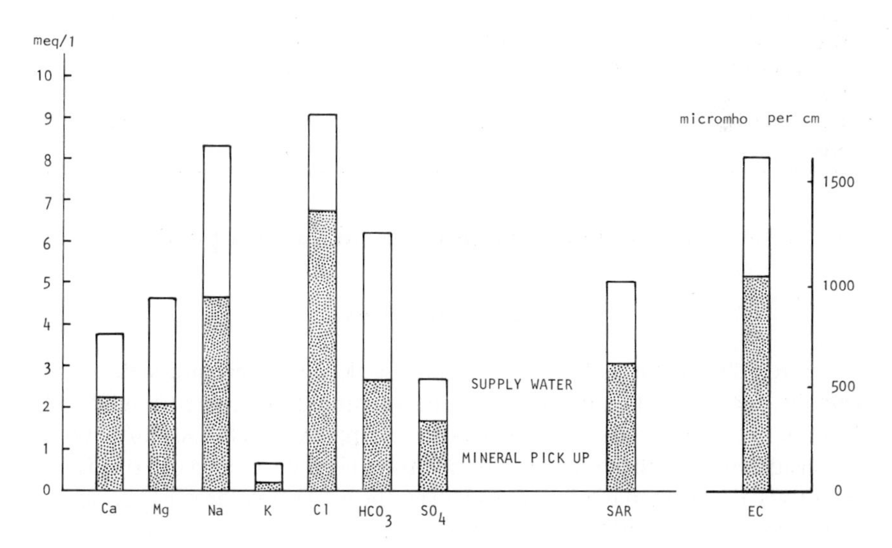

Fig. 1. The resultant composition of wastewater from Kinneret supply water and average mineral pickup. Open bars, supply water; striated bars, mineral pickup.

II. TREATED WASTEWATER PROPERTIES IN RELATION TO IRRIGATION

Wastewater is classified according to the source: domestic, industrial, and agricultural. Effluents from industrial plants are not uniform in their composition and may contain toxic compounds, as mentioned previously. Such effluents may have to be disposed of. Domestic effluents are more uniform in their composition and the mineral pickup found in such effluents can be estimated. A sample composition is presented in Table I and Fig. 1. The supply water source is from Lake Kinneret, which is a major water reservoir in Israel, and the mineral pickup composition is an average of domestic effluents in Israel (Water Commission, 1963). Accordingly, the resultant wastewater composition is always made up by contributions of both supply water concentrations and mineral pickup during use.

Table I
The Resultant Composition of Wastewater from Kinneret Supply Water and Average Mineral Pickup

Composition	Supply water (Kinneret)	Mineral pickup (Average)	Resultant Waste-Water Composition
Ca (mEq/liter)	2.43	1.41	3.84
Mg (mEq/liter	2.19	2.54	4.73
Na (mEq/liter)	4.72	3.63	8.35
K (mEq/liter)	0.13	0.48	0.61
Cl (mEq/liter)	6.77	2.32	9.09
HCO_3 (mEq/liter)	2.70	3.55	6.25
SO_4	1.61	1.06	2.67
EC (micromhos/cm)	1050	560	1610
SAR	3.1	2.0	5.1

A. The Effect of Wastewater on Soil Properties

1. SOLIDS IN WASTEWATER

Treated effluents usually do not contain settleable solids, although they do contain a certain amount of suspended organic and inorganic solids. The amount of organic matter found in effluents is approximately 150 gm/m^3 which is considered a significant contribution to the soil when applied in irrigation. In light soils, these organic solids may increase the water-holding capacity, silt and clay content, cation exchange capacity (CEC), and organic matter content. See Table II, in which average values from a number of plots are presented (Hershkovitz *et al.*, 1969).

Table II
Soil Properties in Plots Irrigated with Effluents and Nonirrigated Plots

	Irrigated with effluent			Nonirrigated		
Depth (cm):	0–30	30–60	60–90	0–30	30–60	60–90
Sand (%)	88.7	92.9	91.8	97.0	95.5	97.0
Silt (%)	10.2	5.0	5.0	0	0.5	0.5
Clay (%)	11.1	2.1	3.2	3.0	4.0	2.5
CEC (mEq/100 gm)	11.0	9.5	6.6	4.8	2.3	2.4
Organic matter (%)	2.1	0.3	0.14	0.9	0.3	0.07

The effect was measured after 25 years of wastewater irrigation. The effect on the physical and physicochemical properties of the soil is to be related not only to the organic matter contributed directly by the effluent applied in irrigation, but also to the organic matter synthesized by plant and microbiological growth. The decomposition of the organic solids of the effluents proceeds at a fast rate judged by the rate of decrease in BOD levels of the effluent. However, plants irrigated with effluents leave behind organic residues, thus the measured increase is also indirect because of the stimulation of vegetative growth.

In heavy soils, there is a danger of soil-pore clogging. This is not only a mechanical process, but also a biological phenomenon due to the fast rate of algae development on the organic particles that precipitate in the small soil pores. Thus, a crust is formed in the upper soil layer.

The rate of infiltration of a sandy loam was measured in a citrus orchard that has been irrigated with treated effluent for a number of years. The measurements included an irrigated undisturbed surface and an irrigated disturbed surface, which had been tilled prior to the infiltration test. A third measurement was made in a neighboring orchard with the same soil type, except that it received a regular supply of water. The results are presented in Fig. 2 (Noy and Brum, 1973). Therefore, it can be seen that the restricting surface can be rendered more permeable when mechanically disturbed.

Another example is groundwater recharge by spreading basins in Israel, in which the infiltration rate decreases with time due to the clogging of the surface soil layer. Water is then transferred to another basin, so that the soil surface is dried out. It is then tilled and, in turn, receives effluent again.

Some accumulation of organic products, mainly due to anaerobic decomposition, may cause the clogging of soil pores in deeper layers. Decomposition of organic matter, enhanced when aerated, will clear these pores of such clogging.

The favorable effect on the physical properties of soil can be found in light soils, while in heavy soils a temporary decrease in permeability of water and air may be anticipated.

2. THE EFFECT OF SODIUM ON SOIL AGGREGATION

Clay particles are generally aggregated in soils and thus coarse pores can be found in between them. The composition of the exchangeable cations, adsorbed to the clay particles, affects aggregation. When the exchangeable sodium percentage ESP (exchange Na/CEC) $\times$ 100 in. mEq/100 gm soils exceeds 15, clay particle, flocculation is avoided, packing of particles becomes denser, coarse porosity decreases, and permeability to water and air is drastically reduced. This hazard is limited to medium and heavy soils. In light soils, which contain little clay, the sodium hazard is much less pronounced. The desirable ESP value should be less than 10. The quantitative relation between exchangeable cations is dependent on the quantitative relation between the cations in the soil solution, which will depend on the composition of applied irrigation water. The sodium adsorption ratio (SAR) which is $Na/(Ca + Mg/2)^{1/2}$ in mEq/liter of solution is related to ESP when soluble and exchangeable cations are at equilibrium, see Fig. 3 (Richards, 1954). Thus, the critical SAR value of the soil solution is about 13 and the desirable value should be less than 9.

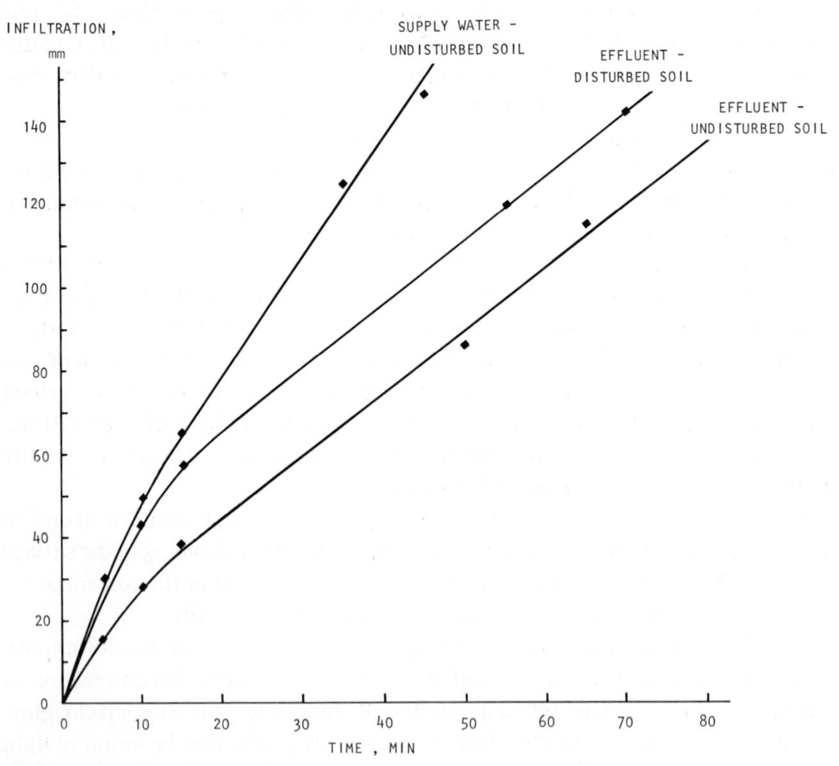

Fig. 2. The infiltration into a sandy loam through soil surfaces irrigated with treated wastewater and supply water.

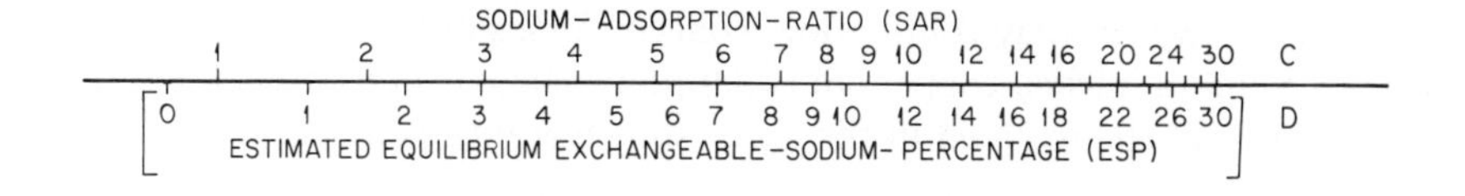

Fig. 3. ESP and SAR values of soil at equilibrium.

Treated wastewater has a high concentration of bicarbonate ions. As the water content of soil decreases, due to plant use and evaporation from the soil surface, the ion concentration increases. The solubility product of calcium and carbonate ions is low and calcium carbonate precipitates out. As a result of the reduction in soluble calcium, SAR values increase and the effected ESP of the soil increases beyond the expected value from the original composition of irrigation water.

The SAR pickup of domestic sewage is about 2. In industrial wastes, it may be higher where water softeners are in use. When an additional pickup of 3.5 mEq/liter or more of bicarbonate ions is considered, the possible hazard of sodium is increased.

There are several claims that have not been substantiated: first, that critical SAR values in wastewater can be higher than in supply water because sodium is less effective in avoiding clay particle flocculation in the presence of other flocculants carried by wastewater (Hershkovitz *et al.*, 1969).

When the expected sodium hazard is great, for any of the following parameters—SAR and bicarbonate content of the wastewater, the soil texture, and the amount of water applied in irrigation—a soluble calcium salt, such as gypsum $CaSO_4$, may be applied to reduce the effective SAR value. The amendment is applied directly to the soil or with the irrigation water.

3. EXCHANGEABLE CATIONS

The relationship between the exchangeable and the soluble cations in the soil has been discussed in relation to sodium. Potassium ion concentration increases in the soil solution after treated effluent is applied to the soil (see Table I) and so does the exchangeable potassium percentage EPP. This contributes to the fertilizer value of wastewater.

The magnesium increment during use is usually greater than that of calcium, increasing the magnesium-calcium ratio in wastewater. In heavy soils, magnesium is weaker than calcium in maintaining clay particles flocculated; thus, the deflocculating effect of the increased sodium concentration may be more effective.

Industrial wastes may contain low concentrations of heavy metals such as zinc, chromium, nickel, copper, and cadmium. These cations are strongly adsorbed and will be retained in the upper soil layer when effluent is applied in irrigation. The percolating effluents lose most of these cations and will contain smaller concentrations of the above cations upon reaching groundwater. Some pollution of the root zone must therefore be anticipated, although it is diminished by the precipitation of insoluble salts and the uptake of these ions by plants.

B. The Effect of Effluents on Plants

Wastewater is useful primarily as a source of water for irrigation and other purposes. This is especially felt in arid areas where water is scarce and its cost high.

The concentration of salts in supply water of arid areas approaches the limit tolerated by plants in many cases, so that treated wastewater does not effectively deteriorate the existing supply water. Wastewater contains, as mentioned previously, soluble constituents, of which some serve as plant nutrients and some may have a specific detrimental effect on plants.

1. PLANT NUTRIENTS IN EFFLUENTS

The amount of nutrients contained in treated domestic wastewater in Israel are about 50 gm N, 15 gm P_2O_5, and 30 gm K_2O/m^3 (Hershkovitz and Feinmesser, 1967). These must be assumed to save fertilizers when applied in irrigation. A number of trials have shown an increase in yield due to the fertilizer effect of the effluents used in irrigation. Wachs *et al.* (1971). Weighed *Avena* plants that were grown in pots and were irrigated with supply water and treated effluent, with and without fertilizer application. See Table III.

Table III

Yield, Grams per Pot, of Avena Irrigated with Supply Water, Effluent, And Different Amounts of Fertilizer [N as $(NH_4)_2$ SO_4 and P + K Fertilizers]

$(NH_4)_2SO_4$ equivalent to kg/hectare	Supply water		Effluent	
	Without P + K	With P + K	Without P + K	With P + K
0	1.54	1.44	3.90	4.10
150	1.70	1.99	3.86	4.14
300	1.77	2.61	3.94	4.06
600	2.21	3.26	4.14	4.19
1200	2.46	4.36	4.19	4.11

Note that the yield of plants irrigated with effluent was about 4 gm per pot, with and without the addition of phosphorus and potassium. When the high level of the nitrogen fertilizer was applied to the plants irrigated with supply water, with the addition of phosphorus and potassium, the yield of 4 gm per pot was also obtained. The addition of fertilizer to the treated, effluent-irrigated pots did not increase their yield. It may be concluded that the effluent contained sufficient nutrients to satisfy the full requirement of the plants.

Day and Tucker (1960) compared the effect on three crops (barley, rye, and wheat) in the field, which received different treatments:

a. Well water with no fertilizer (control).

b. Well water with recommended fertilizer (100 lb N, 75 lb P_2O_5, 0 lb K_2O/ acre).

c. Well water with synthetic sewage (200 lb N, 150 lb P_2O_5, 100 lb K_2O/ acre).

d. Sewage effluent with no additional fertilizer (see Table IV).

The effect of sewage water on yield increase was not the same for the three grains. The yields of cotton in a heavy soil, irrigated for 5 yr with municipal sewage water (1963–1968) did not show an increase in yield compared to neighboring plots irrigated with supply water. All plots received the same amounts of water and fertilizers (Noy and Kalmer, 1970). Since 1968, there has been a lower yield of cotton in the effluent irrigated plots compared to the neighboring plots receiving supply water. The lag in yield may be a result of a more vigorous vegetative growth, observed in the plots irrigated with effluents, which repressed the yield of fiber.

It is evident that the response of plants to nutrients contained in treated effluents applied in irrigation depends on the kind of plant, its nutrient requirements, and the fertility level of the soil. The economical value of the nutrients in wastewater must be related to crop response and not to the nominal contents in the effluent.

In some cases, yields from plots irrigated by effluents were higher than could be expected due to the nutrients contained in them. These increments in yields

Table IV
Hay Yields (Air-dry in Tons/Acre) for Three Small Grains Grown under Different Irrigation and Fertilizer Treatments

Irrigation and fertilizer treatment	Barley	Oats	Wheat
a. Well water with no fertilizer	2.42 a[a]	2.13 a	2.76 a
b. Well water with recommended fertilizer	5.64 b	2.73 b	5.47 b
c. Well water with synthetic sewage	7.15 c	4.27 c	5.93 bc
d. Sewage water with no additonal fertilizer	5.88 b	6.05 d	6.43 c

[a]Yield values followed by the same letter for each crop are "equal."

Table V
Protein Percentage and Digestible Laboratory Nutrient (DLN) Percentage in Barley Forage Grown with Different Irrigation and Fertilizer Treatments

Irrigation and fertilizer treatment[a]	Protein	DLN
a. Well water with no fertilizer	11.1	72.7
b. Well water with recommended fertilizer	15.0	72.9
c. Well water with synthetic sewage	19.9	71.7
d. Sewage water with no additional fertilizer	20.0	71.1

[a]See Table IV.

may be obtained by improved conditions of plant absorption due to chelation of cations, or to specific effects of organic compounds. An example cited from G. Tewes *et al.* (1953) (cited in: Sklute, 1956) is the polyuronic acids formed from carbohydrates in dairy wastewater. Some workers point at changes in the chemical composition of yields from effluent applications. An example is presented in Table V (Day *et al.*, 1961).

Citrus plots were irrigated with effluents and supply water for a period of a number of years. Fertilizers were applied equally to all plots (Noy and Amichai, 1968). See Table VI. There was no significant difference between the plots receiving effluent compared to plots receiving supply water in yield, size of fruit, juice content, sugar to acid ratio, and mineral composition of the leaves; except for a slight increase in sodium and chloride leaf content and a decreased potassium content in the plots irrigated with effluent compared to the plots receiving supply water. Differences may become apparent after more years of the same treatments.

It may be concluded that the nutrients in treated wastewater are valuable. Some fertilizer amounts may be saved by the use of effluents for irrigation. The

Table VI
Yield, Juice Content, Sugar to Acid Ratio, and Mineral Content of Leaves in Citrus Plots on Two Soils Irrigated with Effluents and with Supply Water

Soil	Irrig. water	Yield kg per tree	Juice content %	Sugar to acid ratio	Mineral content of leaves, %					
					N	P	K	Mg	Na	Cl
Sandy clay loam	Supply	90	43.6	6.9	2.3	0.13	0.82	0.40	0.19	0.33
	Effluent	101	43.5	7.0	2.4	0.12	0.71	0.44	0.21	0.36
Sand	Supply	90	44.3	7.1	2.3	0.13	1.01	0.36	0.37	0.56
	Effluent	91	44.6	6.9	2.3	0.11	0.78	0.27	0.32	0.59

increments in yields and in certain plant constituents may be significant in soils of low fertility. However, the benefit from the nutrients in wastewater is not uniform and cannot be expressed in fixed monetary values.

2. ADVERSE EFFECTS ON CROP DEVELOPMENTS

The solutes contained in wastewater may restrict crop development either by the increase in osmotic pressure of the soil solution in the root zone, due to the accumulation of soluble salts, or by the specific effect of soluble constituents toxic to plants.

The critical concentrations detrimental to plant development depend on a number of parameters, such as soil permeability, precipitation, quantities of water applied, methods of irrigation, and crop tolerance. There are some indications that the critical levels of the detrimental constituents in treated wastewater are higher than those for the same constituents in supply water because of the moderating effect of organic matter and nutrient elements present (Hershkovitz *et al.*, 1969).

The permissible ranges in wastewater suggested in Israel are presented in Table VII. These are working limits and may be used more liberally when the soil is highly permeable and the climate wet.

The sensitivity of crops to salinity varies. The list of crop tolerance to salinity may be found in publications of the United States Salinity Laboratory at Riverside, California (Richards, 1954). Accordingly, sensitive fruit crops are avocado, citrus, and deciduous trees, while date palms are tolerant to salinity. Sensitive field crops are beans and Ladino clover, while cotton, sugar beet, barley, and Rhodes grass are tolerant.

Boron has a special place among the elements effecting plant growth. It is one of the essential elements; however, at concentrations exceeding 1 ppm boron may be toxic to sensitive crops. Although boron compounds may be highly soluble, it will be partly held by soil particles. Boron reaches waste-

Table VII
Suggested Critical Limits for Domestic Effluents Used for Irrigation

Composition	Sensitive crops	Tolerant crops
Electrical conductivity (micromhos/cm)	2000	3000
Chloride (mg/liter)	200	450
Sulfate (mg/liter)	300	500
Boron (mg/liter)	0.7	2.5
SAR value	8	15

water mainly from soap powders. In Israel, the incorporation of boron into soap powders and other detergent formulations has been restricted to diminish its concentration in wastewater.

The sensitivity of plants to boron varies. A list of tolerance values for plants to boron has been published by the United States Salinity Laboratory at Riverside, California (Richards, 1954). Accordingly, avocado, citrus, and deciduous fruit crops are the most sensitive. Sugar beet, alfalfa, gladiola, onions, and potatoes are among the most tolerant crops.

Industrial wastewater may contain heavy metals in concentrations exceeding the permissible limits for use in irrigation. These can be found in a report of the Committee on Water Quality Criteria of the U.S. Water Pollution Control Administration (1968). Accordingly, limits of chromium, lead, lithium, and zinc are 5 ppm; manganese, 2 ppm; aluminum and arsenic; 1 ppm; nickel, 0.5 ppm; cobalt and copper; 0.2 ppm; and cadmium, 0.005 ppm. When used occasionally, the limiting concentration is higher.

The heavy metals may be diluted by effluents from other sources, so that their concentration does not reach the prohibited values. When industrial wastes contain excessive quantities of heavy metals toxic to plants, they should be separated, usually into special evaporation basins or into the sea, and not used for irrigation.

III. TECHNICAL ASPECTS REGARDING IRRIGATION WITH TREATED WASTEWATER

A. Conveyance of Treated Wastewater from the Treatment Plant to the Irrigation Area

When designing the conveyance system of effluent from the treatment plant to the irrigated area, the topographical conditions have to be considered and, where conditions permit, every effort should be made so that the effluent will be conveyed by gravity flow. Only when topography is unfavorable should a pumping station be erected to pump the effluent to the irrigated fields.

The pipes commonly in use in Israel are asbestos cement pipes or steel pipes, usually with cement lining. Since the raw sewage flow varies during the day, it is necessary to provide a storage reservoir to equalize the flow and to ensure a constant supply of effluent for irrigation. The reservoir is also necessary to store the effluent during night hours and in situations where irrigation is being carried out at intervals of a few days.

The usual arrangement in Israel for conveying treated effluent is to build a pumping station adjacent to an oxidation pond or another treatment plant, which pumps the effluent (under pressure) to the irrigated area (in case the topography is not suitable for gravity flow) (see Fig. 4).

The pump, in most cases, is a centrifugal vertical pump to ensure a convenient and simple operation as well as one that can be started easily without priming. Whenever it is feasible, the pumps are driven by means of an electric motor, which is easy to operate and maintain. Only in cases when electric power is not available, or the connection to the electric network is economically unjustified, are the pumps driven by diesel or other internal combustion engines.

B. Irrigation System

In many countries, sprinkler irrigation is commonly used. When irrigating with treated effluent, plugging of sprinkler nozzles should be considered. The raw sewage, after receiving secondary treatment, contains practically no settleable solids and little suspended solids. Therefore, the possibility of plugging of sprinkler nozzles is slight. As a precautionary measure, a gravel filter or a strainer can be installed at the outlet of the oxidation ponds or other treatment plant (see Fig. 5) to prevent the clogging of water meters, irrigation devices, and sprinkler nozzles. In addition, the diameter of the nozzles, preferably, should not be less than 5 mm. These strainers should be cleaned and the lateral pipes flushed out periodically.

Furrow irrigation with treated effluent does not essentially differ from irrigation with water from wells or streams. However, land leveling should be carried out carefully to avoid puddles of a stagnant treated effluent.

Trickle irrigation is now being used more extensively in a number of countries. This irrigation method has proved to be efficient in arid areas, for it saves water consumption and increases yield of crops. Trickle irrigation is based on applying the water to the land continuously by means of tricklers or small diameter pipes having small ports at regular intervals. Tricklers have very small flow of some 2–15 liters hr. The diameter of the passage of the trickle devices is very small, sometimes less than 1 mm. Attempts have been made in Israel to irrigate with treated effluent by this method, and the results have been quite encouraging (Sadovsky *et al.*, 1973). The advantage of trickle irrigation with effluent is that the contact between the treated effluent and the crop is less than with other methods, which reduces the possible contamination by pathogens.

C. Corrosion Problems

Domestic sewage that has received secondary treatment usually does not cause corrosion of the irrigation facilities, such as pumps, pipes (including

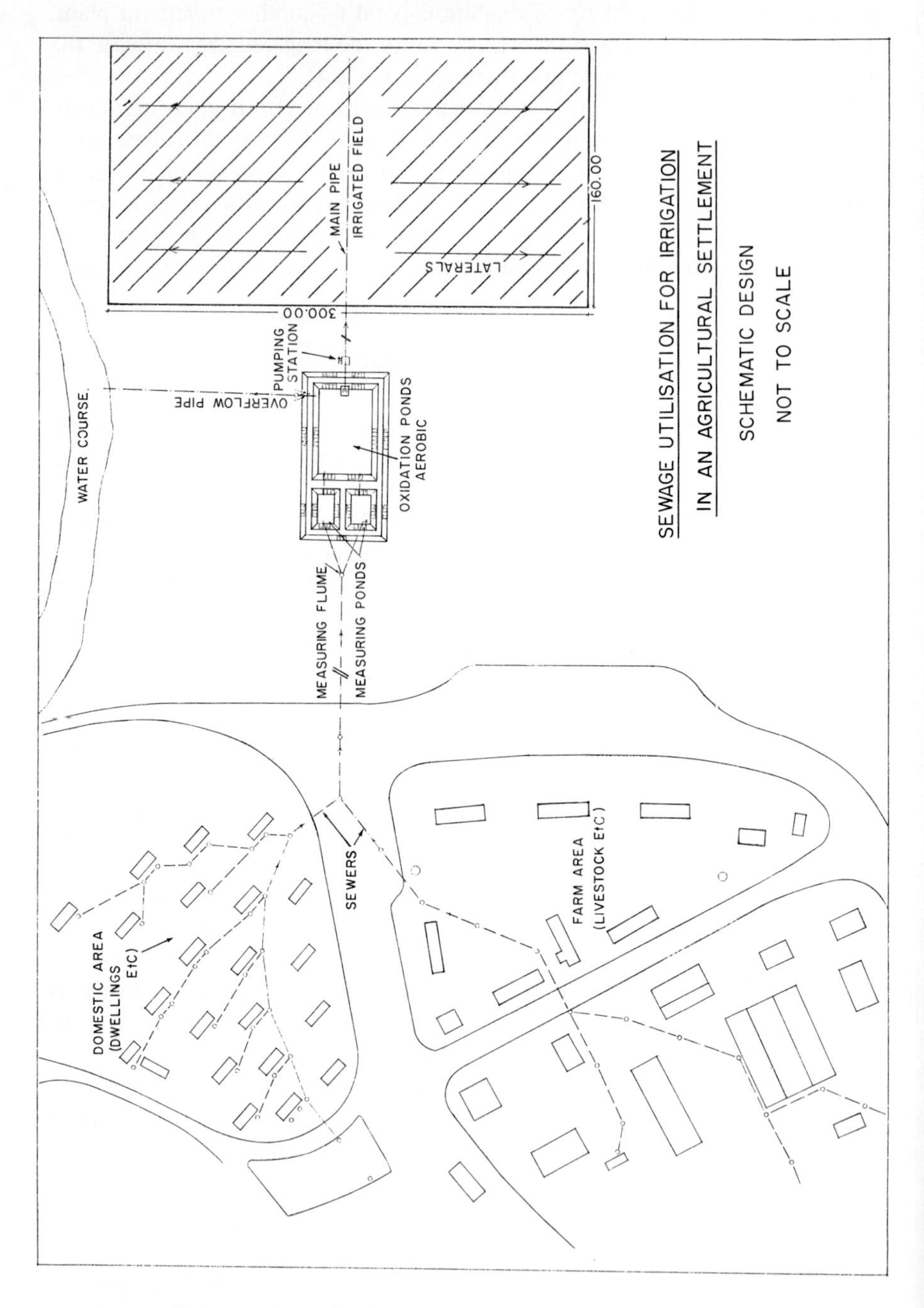

Fig. 4. Sewage utilization for irrigation in an agricultural settlement.

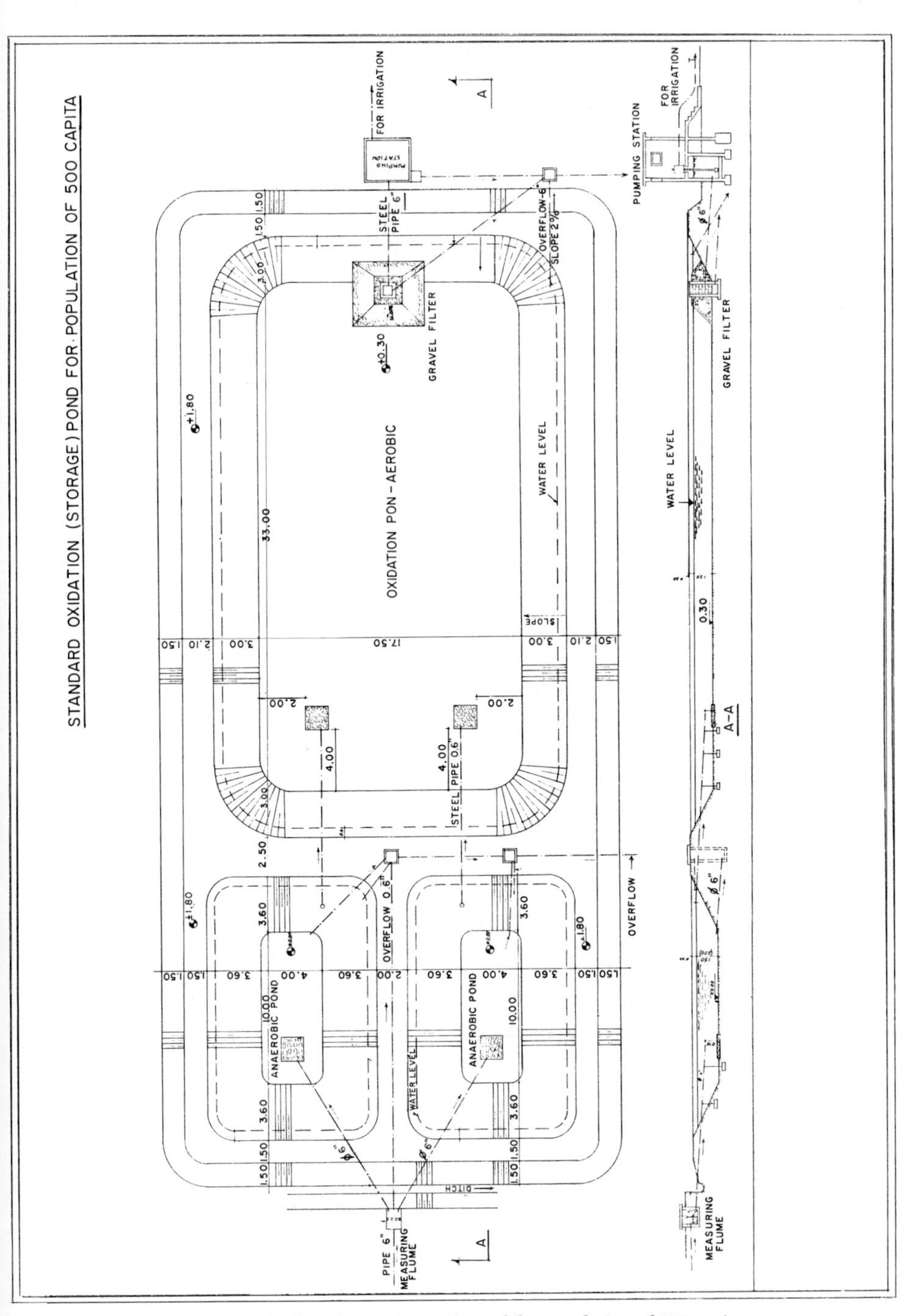

Fig. 5. Standard oxidation (storage) pond for population of 500 capita.

aluminum pipes), etc. This was reported in literature "Corrosion Problems Connected with Sewage Effluent May be Minimum" (Cantrell *et al.*, 1968). In Israel, corrosion problems are minimal. However, based on experience, the life expectancy of aluminum pipes used for irrigation with treated effluent is 10–20% less than when such pipes are used for irrigation with supply water.

On the other hand, industrial wastes are liable to be very corrosive. In certain cases, as was mentioned previously, these kinds of wastes should be segregated to prevent their detrimental effect on the sewer pipe, on the irrigation system, and on the irrigated crop.

D. Storage Reservoirs

As was mentioned previously, to make efficient use of treated effluent for irrigation, storage reservoirs are needed. Actually, such ponds act as polishing ponds and further reduce the BOD and pathogenic organisms (Cantrell *et al.*, 1968). Oxidation ponds can also serve as a dual purpose facility. In semiarid and arid climate, a seasonal storage of effluent is necessary to store effluent during the rainy season so it will be available for use during the dry season. Storing the effluent during the wet season will also prevent it from flowing into water courses, which would create sanitary and pollution problems (Finemesser, 1971).

The relatively inexpensive earthen reservoirs are 3 to 5 m deep and have sufficient capacity to store unutilized treated effluent during the wet period. There is a possibility that during such a long time of storage, anaerobic conditions might develop that will cause sanitary nuisances, such as odor dissipation. To prevent such undesirable anaerobic conditions from developing in the reservoir, artificial aeration is required.

E. Oxidation Ponds

Oxidation ponds to treat sewage are in wide use. This means of treatment has justified itself in Israel, especially in agricultural settlements and small- or medium-sized towns. The long hot summers enhance the development of algae on which this treatment method is based. The building, operation, and maintenance of these ponds are relatively inexpensive when compared with conventional treatment plants. The arrangement generally adopted is to convey raw sewage through two parallel anaerobic ponds where alternate cleaning occurs and then through an aerobic pond for final treatment (see Fig. 2). The design criteria, which are recommended today are as follows: anaerobic ponds: depth, about 2.5 m; detention time, about 1 day; load, 150–200 gm BOD/m^2/day. These values have recently been adopted to assure anaerobic conditions throughout the whole depth of the pond.

Aerobic ponds: depth 1.2–1.5 m for good light penetration; detention time, 7–10 days; load, about 15 gm BOD/m^2/day. This value has been adopted to assure aerobic conditions even during the winter, when the solar radiation is comparatively low. Oxidation ponds have, however, a disadvantage that cannot be disregarded in comparison with conventional treatment plants because they require considerable areas of land. When land value is low, this disadvantage is not particularly significant, but as towns develop and the price of land around them increases, the land value then becomes a factor that has to be taken into consideration. In cases where land values are a factor, aerated ponds are being introduced in Israel, which allow higher rates of organic loading and, consequently, require less land.

ACKNOWLEDGMENT

Thanks are given to Seymour (Shimon) Cohen for his critical reading of the manuscript and for his welcome comments.

REFERENCES

Cantrell R. P., Wilson C. W., Beckett F. E., and Calvo, F. A. (1968). *In* "Municipal Sewage Water for Irrigation" (C. W. Wilson and F. E. Beckett, eds.), pp. 135–156. Louisiana Alumni Found., Ruston.

Day, A. D., and Tucker, T. C. (1960). Hay production of small grain utilizing city sewage effluent. *Agron. J.* **52**, 238–239.

Day, A.D., Vavich, M. G., and Tucker, T. C. (1961). Protein and digestible laboratory nutrients (D.I.N.) in forage using sewage water as a source of irrigation water. *Barley Newslett.* 5, 3–5.

Feinmesser, A. (1971). Survey of sewage utilization for agricultural purposes in Israel. *Proc. Int. Conf. Water Pollut. Res., 1970* p. 33/6–7.

Hershkovitz, S. Z., and Feinmesser, A. (1967). Utilization of sewage for agricultural purposes. *Water Sewage Works* **114**, 181–184.

Hershkovitz, S. Z., Mor, A., Noy, J., Feinmesser A., Fleisher, M., and Kishoni, S. (1969). "Utilization of Sewage for Crop Irrigation" (in Hebrew). Water Commission, Ministry of Agriculture, Jerusalem, Israel.

Kardos, L. T. (1967). Waste water renovation by the land. *In* "Agriculture and the Quality of our Environment," Publ. No. 85, p. 000. Amer. Ass. Advan. Sci., Washington, D.C.

Noy, J., and Amichai, M. (1968). "The Use of Ra'anana Treated Waste Water for the Irrigation of Citrus" (in Hebrew). Soil and Irrigation Field Service, Ministry of Agriculture, Jerusalem, Israel.

Noy, J., and Brum, M. (1973). "Crusting of the Soil Surface by Treated Waste Water applied by Irrigation" (in Hebrew). Soil and Irrigation Field Service, Ministry of Agriculture, Jerusalem, Israel.

Noy, J., and Kalmar D. (1970). "Irrigation of Cotton with Treated Waste Water" (in Hebrew). Soil and Irrigation Field Service, Ministry of Agriculture, Jerusalem, Israel.

Richards, L. A. (1954). Diagnosis and improvement of saline and alkali soils. *U.S. Dep. Agr., Agr. Handb.* **60**.

Sadovsky, A., Goldberg, D., Halperin, R., Ozrad, M., and Gornat, B. (1973). "Irrigation of Truck Crops with Treated Waste Water by the Trickle Method" (in Hebrew). Hebrew University, Faculty of Agriculture, Rehovoth.

Sklute, B. P. (1956). Irrigation with sewage effluents. *Sewage Ind. Wastes* **28**, 36–43.

U.S. Water Pollution Control Administration. (1968). "Report of the Committee on Water Quality Criteria," USWPCA, Washington D.C.

Wachs, A. M., Avnimelech, Y., and Sandbank, E. (1970). "Experimental Determination of Stabilization Pond Effluent's Fertilizer Value." Technion-Israel Institute of Technology, Haifa, Israel.

Water Commission. (1963). "Survey of Waste Water Utilization" (in Hebrew. Ministry of Agriculture, Jerusalem, Israel.

4

Water Reuse in Industry

Lawrence K. Cecil

I. INDUSTRIAL WATER USE

Water reuse does not change the use by industry, only the source. The many industrial uses of water have their specific quality requirements. Water from any source can be treated for these qualities. The selection of the source water is based primarily on economics. If it costs less to reuse wastewater than to buy "fresh" water from an outside source, an industry's wastewater will become one of its sources. Such a source has one extraordinary advantage. It is dependable.

All natural waters are polluted. Rain absorbs atmospheric pollutants such as carbon dioxide, sulfur dioxide, ammonia, various nitrogen oxides, organic vapors, and particulate matter. On the surface, the slightly acidic rain dissolves contaminants from vegetation and from the soil; it also carries silt particles into the receiving body of water. Percolation through the soil carries these same dissolved surface pollutants, plus those dissolved from rocks and soil, into the subsurface water strata. Domestic and industrial wastes from point and non-point sources enter these same water bodies. Rarely can industry find a natural water supply that meets the quality requirements for its simplest uses. For more sophisticated uses, treatment is always necessary.

A. What is Industrial Water Reuse?

For our purpose, we will limit direct water reuse to waters that have been polluted by human activities and have not been blended with natural waters. An industry taking water from a natural source, either surface or subsurface, that has been contaminated by human activity is not directly reusing wastewater. It is taking water from a polluted source which is a form of indirect reuse. Use of wastewater from another industry or a municipality is direct water reuse, if it has not been blended with natural water. If the original owner retains control of its wastewater and delivers it undiluted by natural waters to the new user, that's direct water reuse.

Public control authorities have made industry responsible for the pollution picked up on its own plant area by surface runoff. If this water is contained and used, this is industrial wastewater reuse.

Recycling is a special case of water reuse. It is upgrading wastewater from a use and returning it to the same use. Recycle is not synonymous with reuse.

B. Complete Industrial Water Reuse

Water reuse deals only with the recovered water fraction of the wastewater. It does not involve reuse or disposal of the removed pollutant fraction. Modern environmental controls by public bodies limit, and sometimes prohibit, the disposal into the environment of these pollutant fractions. Public Law 92–500 of the United States 92nd Congress, S. 2700, enacted October 18, 1972, sets a national goal that the discharge of pollutants into the navigable waters be eliminated by 1985. The 1985 no-pollutant-discharge is not a law and may never become one.

Consideration of water reuse should evaluate the advantages of complete water reuse. If water reuse and pollutant reuse or disposal are considered as different systems, the cost may be much greater. When all factors are considered, complete water reuse, or zero discharge, may be an economic asset instead of an economic liability for industry. Reuse of the pollutant fraction should be an integral part of all water reuse planning.

C. Process Design for Minimizing Production of Pollution

Some products can be manufactured by different processes from the same feedstock. The capability of producing a minimum of polluted water and having this wastewater amenable to treatment for reuse should be one of the important variables used in making the selection of the manufacturing process. Often the product can be manufactured from different feedstocks, and again, the potential for wastewater reuse should be one of the variables considered. This is well exemplified in the case history where a plant was being designed during that time period when wastewater pollution standards were being imposed. The design engineers proceeded on the old basis, with no regard to the cost of disposal. The plant produced over 500,000 gal/day of 3000 mg/l biochemical oxygen demand (BOD) wastewater. The cost for treatment to disposal or to reuse standards was so high that the entire manufacturing operation would have been uneconomical. A disposal well could have been drilled in this area (it would actually have taken three such wells), but, again, the cost was too high. A comparatively inexpensive redesign of the distillation section of the plant reduced the amount of wastewater to about 50,000 gal/day of 250 mg/l BOD. The water was amenable for treatment for reuse at a nominal cost. The recovery of additional product in the manufacturing end of the system made the redesign profitable.

In this dynamic world, industry is faced with a tremendous increase in the cost of energy. Energy has now become one of the factors that must be carefully evaluated.

II. PROPER WASTEWATER PLANNING

A. The Market Survey

A company may be planning to abandon certain products or to add others. The design engineers should confer with management to verify the identity of the product line when they make long-range plans for water reuse. So much time elapses between the start of planning and the actual construction and start-up of the water reuse plant that it could be obsolescent because of manufacturing changes.

It is essential to determine the volume of wastewater and the content of pollutants for each manufactured product. Some may be contributing an unreasonable share of the pollution load. Determine the cost, both capital and operating, for manufacturing process changes to reduce the pollution load from these products. Management can decide if the product profitability justifies the manufacturing changes to reduce the production of so much pollution, or to accept the cost of handling the pollution load as is, or of discontinuing the product.

B. The Use Survey

1. VOLUME

Does the plant actually need all the water now used? While I was walking through a petroleum refinery with the superintendent, he stopped at a wide-open rising stem valve in a 6-in. water line and commented, "I'll change this to a 3-in. line and put the 6-in. pipe back in stock. The smaller line will furnish all the water this unit needs. That valve is supposed to be throttled. My men will use all the water they can get, regardless of need." Use of unneeded water is a double cost, first of the input water, and then in the wastewater system.

2. QUALITY

There is a tendency to treat all the incoming water to quality standards suitable for all plant services. Many processes do not need such high quality water. It may not be economical to install several treatment plants, but it may be practical to install a cascade system. The pollution added in one process often leaves the quality satisfactory for use in another process. Sometimes the cascade system can handle three or four processes before quality upgrading is needed.

Even then a simple treatment may make it useable for more cascade steps before it goes to the general wastewater treatment plant.

Sequential treating processes will be needed to produce treated wastewater of high enough quality to be returned to the use system. Each succeeding step increases the quality and cost of the effluent. If the designer has determined the actual quality requirements of the various uses, he may find water can be taken from an intermediate step for use in some of these processes. The succeeding wastewater treating processes can be smaller, at a considerable saving in overall cost, both capital and operating.

3. CONSUMPTION

Use does not imply consumption. An industry that takes in 10 mgd of freshwater and discharges 10 mgd of wastewater has not "consumed" any water. The two major ways in which water is consumed by industry are incorporation into the product and evaporation in heat-exchange systems. The latter is a fairly transient consumption. The water is changed in phase only, from liquid to vapor, which will soon return to earth as rain. Unfortunately for the industry that produced the water vapor, the rain falls on a plain far away. Some systems are under development that will enable in-plant recovery of most of the vapor as liquid with concurrent utilization of the low-level heat. They are not well enough developed to incorporate into our water management scheme.

Incorporated into the product, the water may be unchanged as a carrier, such as in packaged drinks, or lose its identity, as when reacted with coal to produce methane. Sometimes the water hydrates the product. Other times the water moistens the product to prevent dust in transportation and use. Whatever the means by which the water is incorporated into the product, when it leaves the plant with the product, it has been consumed.

Manufacturing plants with small water consumption have a more severe problem in water reuse system design than do plants with a large consumption. If the use and waste volumes are near balance, the introduction of surface runoff may upset the system. Perhaps some of the water must be treated to disposal quality. There is also the possibility of adding to the factory a process that will consume water.

4. STEADY-STATE OPERATION

A use and reuse system in balance will be a steady-state system only if the reuse treatment system removes all the pollutants added during use. This is improbable. Some of the compounds not removed in the treating system may in time build up to concentrations that will prevent use of the treated water. All treating systems have some liquid blowdown in removal of the concentrates.

Thus, the maximum concentration of objectionable compounds not removed in treatment will be related to the amount of concentrate blowdown. The volume of blowdown may be increased, but this will make more expensive the ultimate disposal of the concentrate. Selection of the treating system should be designed so that any substance that could cause this difficulty will be removed. It's poor engineering if something has to be added later.

Most, but by no means all, industrial plants have cooling requirements. Water will be evaporated, but the solubles will be left behind. These accumulated solids added to pollutants from chemical process operations make more concentrates for disposal than the nonconsumptive plants. One special type of cooling, quenching of coal ash or coke and slag granulation, evaporates the water, but leaves the solubles in the solids being cooled. If the increased ash content of these cooled solids is not objectionable, concentrates from wastewater-treating operations can be used as cooling water. If excess water is not used for cooling, this is a good method of concentrate disposal.

The acid rain in the Scandinavian countries, presumably from the sulfur dioxide from combustion of coal by industries in Scotland, is a striking example of water pollution caused by too high a concentration of industry. It is indirect, air pollutants transferred to water by rain. This does not alter the fact it ends up as water pollution. Many processes for removal of the sulfur dioxide by scrubbing the stack gases with aqueous slurries of various alkalies are under development. All of them consume water and leave a residue of the salts originally in the water plus the absorbed sulfur dioxide.

These combustion-gas scrubbing systems are in competition with processes for desulfurizing the fuel before combustion. Regardless of the method of sulfur control that is finally adopted, much water will be used and consumed. There will be a massive residue of mineral salts.

Incineration of various slurries and solids such as sewage sludge, municipal garbage, and various organic industrial wastes produces air pollutants such as hydrochloric acid and heavy metals. These must be removed by water scrubbing of the exit gases. Here again water is consumed, leaving a residue of mineral salts.

C. Good Housekeeping Benefits

Careless use of water in the manufacturing plant can waste a lot of water. Leaving a hose running, forgetting to turn off a valve, letting water sample lines run continuously, and lavishly using washup water are samples of poor housekeeping that greatly increase water use. Not all careless water wastage is in small increments. A phosphate fertilizer manufacturing plant was getting very rapid filling of the expensive gypsum pond. A water use survey found one place an inexpensive process change could save 100,000 gal/day. That's a lot of water, but it did not come close to explaining the pond fillup rate. Finally an 80 psi 4-in. freshwater line, with wide-open valve, was found discharging needlessly into the pond.

Small product leaks may seem so insignificant to the careless operator that he does not bother to stop them. Products are calculated in percent. Wastewater contaminants are determined in milligrams per liter. Remembering that 1% is 10,000 mg/liter, it is easy to see how a few "insignificant" process leaks can deal havoc to a wastewater treatment and reuse plant.

A careful, methodical survey of housekeeping methods in a plant will point up places that can reduce water use and decrease both the volume and the pollutant concentration of the wastewater.

1. SURGE CAPACITY

The need for use water and the production of treated wastewater are not in complete balance. There are product spills, batch operations that take large volumes of use water in a short time without concurrent production of wastewater, intermittent large volumes of surface runoff, and a variety of unexpected conditions that dictate the need for surge capacity of both untreated and treated wastewater.

The water management survey of both use and wastewater volumes should indicate both the size and plant location of the surge basins. It is not economical to install a waste treatment plant of sufficient size to handle the peak flow of wastewater. Such a plant would be operating at a small percentage of capacity most of the time. The balancing of the capacities of surge basins and treating plant is an intricate exercise in engineering and economics.

There is a growing tendency to allocate to an industrial plant the total amount of pollutants that can be discharged each day. This is a departure from the older standards of a specified percentage reduction of pollutants or a maximum concentration of pollutants without defining the maximum volume of treated wastewater. The control authorities are not interested in just what portion of the plant produces the pollutants. It's the total amount of pollutant entering the aqueous environment that's monitored. On this basis, suface runoff pollution is just as much the responsibility of the industry as is the process generated wastewater.

The industry must dike its whole plant area to contain rainwater falling directly on it and to exclude runoff from adjacent land. In making the economic evaluation of water reuse versus disposal, the engineer must decide what maximum rainfall, both in duration and intensity, to use as a basis for calculations. Will it be a 10-year, a 50 year, or a 100-year rain? Remember that a 100-year rain does not necessarily come 100 years from the last one. It may come next week. Rate of precipitation of a storm is not uniform throughout the duration of the storm. Four in. of a 2-day, 6-in. rain may come in the first hour. A 12-in. deluge in 1 hour on a small area (but larger than an industrial plant) is not unknown.

Pollutants on a paved area may be so thoroughly removed in the first hour of a rain that the balance of the stormwater can be allowed to escape without objectionable damage to the aqueous environment. This is not true with unpaved areas. The engineers in a petroleum refinery on the west coast of the United States were evaluating this surge capacity problem when there came a 2-day, 8-in. rain. To their consternation, the pollutant concentration in the last hour was as high as the first hour. So they introduced into their calculations the cost of paving the whole plant area to reduce the amount of pollutants that could enter surface runoff. Surface runoff during a storm is the wild card. Process wastewater is being produced continuously during the rain. A high rate treatment plant to produce use water from the rainwater must have a place to put this water, so it is not a substitute for untreated surface runoff surge capacity.

Where land values have become very high, it may be necessary to install underground surge tanks. Light-weight surface structures such as parking lots, roadways, or one-story buildings can divide the cost of the surge area. Extra-deep cooling tower basins or firewater storage basins with liberal ullage in normal use may be cheap surge capacity.

Untreated wastewater surge basins must be equipped with some means of preventing anaerobic decomposition of the pollutants. Some of the organic degradation products are foul smelling. Sulfate-splitting bacteria produce hydrogen sulfide from the ubiquitous sulfates. Hydrogen sulfide escaping from the liquid surface will be absorbed in the water droplets condensing on the roof of the surge basin. Another type of bacterium converts hydrogen sulfide to corrosive dilute sulfuric acid. Nearly all anaerobic decomposition products are acidic, so this must be considered in selecting the basin structure.

A surge basin must be considered as a pollutant-equalizing basin also. Too great a variation in the concentration and type of pollutants in the influent of the treating plant imposes control problems on the treating system. Mechanical mixing systems, supplemented by aeration devices, can give the desired combination of equalization of pollutant concentration and maintenance of aerobic conditions.

Aeration may be objectionable if the wastewater contains volatile organic compounds that would produce pollution in the escaping air. In this case, the basin should be covered and the effluent air passed through activated carbon filters, or perhaps water-scrubbed to return the organics to the basin. In special cases, it may be best to use a small excess of chlorine instead of aeration.

It is almost an axiom that it is best to overbalance the surge system in favor of storage of treated water.

2. INSTALLING THE SURGE SYSTEM

Planning the surge system is so uniquely related to each plant that only general suggestions can be given here. Careful attention in the survey should be given to fluctuations in both use and production of wastewater. If these volume and/or pollutant concentration fluctuations are unreasonably wide, it may be possible to suggest changes in some of the manufacturing operations to smooth out the fluctuations.

There must be a large margin of safety in storage of use water. Careless operation may produce treated wastewater unsuitable for use. It must be returned to some wastewater storage for retreatment, and the deficiency made up from use water storage. Normal maintenance or breakdown of some part of the treating equipment will upset the approximate steady state system you have devised. Shutdown of a water-consuming process will overbalance the production of wastewater compared to use water. A careful analysis of the history of such plant operating fluctuations and its projection into the future is one of the best bases for sizing surge basins.

When the surge system gets too far out of balance, it may be necessary to buy use water or, much worse, discharge improperly treated water to the environment. One of the important intangible benefits of a closed water system is freedom from governmental control. More than one failure to keep the system closed may bring continuous governmental regulation, which is a cost you must evaluate. This portion of your evaluation will aid in getting management acceptance of what otherwise may seem to be an unreasonably high surge system cost.

3. END-OF-PIPE SYNDROME

The simplest way to design a wastewater reuse system is to measure the variable volumes and pollutant concentrations of wastewater at the end of the pipe, determine the treatment that will upgrade the wastewater for the major use needs, establish the surge capacities, and design the system. Of course, there must be an economic evaluation of reuse versus disposal.

The end-of-pipe syndrome might be described as a desire to maintain, or at least take advantage of, the status quo. Case histories of water management surveys, some of which will be outlined later, have shown how the volume, and sometimes the pollutant concentration at the end of the pipe, can be minimized. If you proceed with an end-of-pipe design, a later survey may show that you have wasted most of the cost in unnecessary structures and operating expense.

4. CURING THE END-OF-THE-PIPE SYNDROME

A typical case is the fish canning complex in the United States that was prohibited from continuing to discharge its wastewater into the bay where it was located. The nearby municipality was preparing to expand its wastewater treatment plant to accept the 25 mgd of very highly polluted wastewater from the fish canning complex, at a ruinous cost to the complex. A consulting engineering organization familiar with factory design and operation as well as treatment of industrial wastewater reorganized operating procedures. Internal water reuse, largely cascading systems, reduced the volume to 3 mgd of moderately polluted wastewater. Recovery of oil and protein formerly discharged with the wastewater added a profitable cattle feed product for sale. Rearranging the cycle of manufacturing operations freed so much floor space that construction of a planned storage building was abandoned. The residual 3 mgd of wastewater could have been treated for reuse, but under present conditions it was profitable to pay the city to accept it. If disposal conditions ever become adverse, the industry management knows it can go to complete wastewater reuse.

Some of the factors that made reuse so profitable for this canning complex were the inefficient operations so typical of an old industrial plant. Most end-of-the-pipe projects an engineer is called upon to evaluate are in old factories. In many cases, the cost of end-of-pipe treatment will force closing of the factory. An engineering analysis of the processes that produce the polluted wastewater often will enable the mangement to comply with all environmental controls at a profit.

5. REUSE VERSUS DISPOSAL

Management has the choice of treating wastewater for reuse or disposal. The decision should not be made on balancing costs and benefits that can be quantified. A value must be placed on intangibles, which often will control the final decision. These basic elements apply to all cost/benefit studies:

Costs

- Buying input freshwater for use
- Preparing the water for use
- Preparing the wastewater for reuse
- Preparing the wastewater for disposal
- Disposing of treatment concentrates
- Containing surface runoff
- Installing surge system
- Modifying plant processes
- Making inplant structural changes

Preparing various reports for all governmental control agencies
Training plant personnel in use of new system

Benefits

Reducing purchase of freshwater
Eliminating reports for governmental control agencies
Saving in cost of treating for reuse instead of disposal
Improving housekeeping
Improving public relations
Improving plant personnel morale
Establishing independence

III. COMPETITION FOR WATER

There are already signs of fierce competition for water to be used in the production of energy as against its use in the production of food. Energy, of course, is also an important item in food production. In the western part of the United States, the water needed for strip mining of the low sulfur coal in Wyoming (including that water needed for reclamation of the strip-mined area) is already being fairly well used for farming and cattle raising. The same situation exists in the Dakotas for the semibituminous coal deposits there. It would be more economical to convert the semibituminous coal of the Dakotas to pipeline gas or to electrical energy adjacent to the mine. This, however, requires very large additional volumes of water in an area that is already water short and increases the intensity of the competition for water.

The building of new power plants with the general elimination of single pass cooling would involve the installation of very expensive cooling towers or the usually less expensive cooling ponds. The large area of land, usually productive farmland, for the building of the cooling pond generates strong objections from the farming community that has to give up the land. If the plant site happens to be in an area where water is used at some time of the year for crop irrigation, the competition for water is unusually strong.

In spite of our decreasing birthrate, certain industrial areas are growing in population. In the San Francisco Bay area, there is a rather restricted amount of very high-grade water, not enough for the increasing population and increasing industry. As a result of public pressure, the industries are reusing their own wastewater to minimize the purchase of fresh water. Better yet, there are regional municipal wastewater plants upgrading their effluent for sale to the industries to further reduce the industrial use of fresh water. Many of these industries are water consuming, so the reuse of their own wastewater does not begin to supply their water needs, and the treated sewage is the best answer to satisfying public demand.

Industry does get a public relations "good neighbor" benefit from the industrial use of treated sewage because it decreases or eliminates the discharge of municipal sewage into public water bodies, which makes these public waters available again for recreation.

There is a similar competitive situation in which an industrial city wants to increase its water supply in order to increase its industrial base. Here again, the creation of additional lakes takes farmland out of production. Reuse of the existing industries' wastewater plus the availability of treated municipal sewage as source water for new industries will solve the problem of permitting the city to increase its industrial base without taking farmland out of production to create a new water supply lake.

The public is determined to have clean streams, and the two things that will produce these clean streams are industry reusing its own wastewater and accepting municipal sewage as source water.

A. Source Water

Does the industry get water direct from natural sources? Does it actually have a legal right to it, or are there water laws that have not been invoked, but under which its use can be challenged? Are there laws that will prevent another industry from locating upstream and taking so much water that there will not be enough left for existing industry? Local and state water laws must be checked carefully. In time of severe drought, the courts are sure to protect municipal over industrial needs as a public policy.

If the source is indirect, as from a city, a private water company, a water district, or even a private landowner, industry may not be protected from price fluctuations (always up), or a restriction on the volume below its needs.

One petroleum refinery used water reuse as a means of getting a very favorable water rate from its supplier. The refinery, which used 10 mgd of water, installed a 2 mgd wastewater recovery system. The supplier, to prevent an expansion of the reuse system, offered such a reduction in price that the refinery shut down the reuse plant and agreed not to reuse its wastewater for the duration of the new contract.

An industry must verify whether it has the right to reuse its own wastewater. The laws of the state of Colorado prohibit water reuse. No doubt these laws will be revoked now that Colorado has joined the ranks of water-scarce areas. The City of San Bernardino, California, which gets its water from a nearby well field, planned to treat its wastewater and spread it on the well field area to increase the field capacity to meet the growing needs of the city. Down-valley users sued to prevent this, with the claim that they had built up communities and industries based largely on the availability of San Bernardino wastewater.

They won. It could happen to a large industry. The city of Phoenix, Arizona, which buys its water from a Water District, started selling its wastewater for irrigation and planned to sell to a new power plant under design. The Water District claimed title to the wastewater, and asserted that the city only had the right to the use, but not the reuse, of the purchased water. The city won. Here was an industry, planning to use cheap wastewater, that might have had to go to a different and more expensive source.

The study should evaluate the cost and availability of other freshwaster sources. At some future time an industry's present source may be so restricted that it must go to another source. Balanced against these new freshwater sources can be wastewater from others, both municipalities and other industries. Case histories will be given later of industries that have switched from freshwater to wastewater from both municipalities and other industries. In some areas with limited supplies of high-quality freshwater, environmentalists are pressuring industry to switch to municipal wastewater.

B. Preparing Wastewater for Use or Disposal

There are no special water-treating methods for water reuse. The quality requirements for use in any industry are established without regard to the source of the water. There are well-developed methods for upgrading water of any pollutant concentration to meet these specifications. Water for disposal must meet quality standards established by governmental control agencies. Again, there are well-developed methods for treating polluted water to meet these standards.

Wastewater for reuse is only one source of use water. We establish the cost of preparing both fresh and wastewater for the desired uses. These costs are added to the cost of bringing either water to the point of use, and the difference is a plus or minus for selecting the wastewater as a source of use water. The cost of treating the wastewater to disposal standards is only one of the items in our overall cost evaluation for disposal.

A long discussion of methods for treating both source and wastewater does not belong here. Nevertheless, certain suggestions are in order. We must dispose of the water, either reuse it or throw it away, and we must also dispose of the treating concentrates. Nearly all chemical treating methods increase the mass of the concentrates. Ion exchange systems more than double the soluble chemicals in the concentrate. Precipitation reactions increase both the dissolved and suspended solids. Physical methods such as centrifugation, membrane filtration for suspended solids, and filtration through semipermeable membranes (reverse osmosis) or electrochemical processes such as electrodialysis, plating out of metals on an electrode, and adsorption of solubles on various regenerable solids remove the pollutants without increasing the mass of the pollutants. Oxidation of organic matter by ozone is a special case of chemical treatment that does not add to the pollutant mass. All these methods deserve careful consideration in your selection of treating methods. (For details on treatment methods see Chapters 1 and 6.)

One serious uncertainty clouds our cost estimates for preparing wastewater for disposal. Although governmental executive agencies may establish disposal quality standards in accordance with legislative mandates, these standards are subject to challenge. Public hearings prior to establishing standards do not always give results that the polluter thinks are reasonable. He can invoke the judicial process, and with a series of appeals through our tortuous legal proceedings delay the final decision for years. During this time, dissemination of information about the economic and social impact of the standards may change the legislative action, and we start all over again. The engineers and economists who are conducting the water management study must guess at the disposal quality standards to which the treated wastewater must conform. Management must concur with this guess before the decision of reuse or disposal is made.

IV. CHANGING REGULATIONS TO STIMULATE WATER REUSE

The case history of an industrial city in the east-central part of Illinois typifies the changing conditions that point inexorably to industrial wâter reuse. The water system, including the supply lake, is privately owned. The sewer lines are city owned. The interceptor lines and the sewage treatment plant are owned and operated by the sanitary district. The city has a sewage service charge of 38% of the water bill. All industries must discharge their wastewaters into the sanitary sewage lines; discharge into the storm sewers or into surface runoff is not permitted. In addition to the 38% of the water-bill sewer-service charge, industries are charged for their proportionate share of the cost of operation of the sewage treatment plant. The charges are based on the volume, the amount of suspended solids, and the concentration of biochemical oxygen demand. These charges are calculated each year on the basis of the proportionate share of the load on the sewage treatment plant for the previous year.

In order to minimize the industrial load on the sewage treatment plant, industries have recently been notified that they are required to pretreat their industrial waste to domestic raw sewage standards, namely 300 mg/l BOD and 350 mg/l suspended solids. There is consideration of reducing these figures further to 200 mg/l each. Faced with the necessity for installing pretreatment equipment, the various industries considered manufacturing process changes and internal water reuse. One industry, a large grain-milling plant, switched from a wet-milling to a dry-milling process. This made a very substantial reduction in both the amount of water purchased (thus, cutting the 38% sewer service charge) and the amount and quality of the wastewater discharged to the sanitary sewer, which considerably reduced their share of the cost of operating the sewage treatment plant. Certain relatively minor internal reuse systems further reduced the total amount of purchased water and of wastewater discharged.

Another large industry produces about 40 to 50 tons of SO_4 per day as a waste product. This had been going into the sewage system without any special charge for the sulfate content because it did not come within the biochemical oxygen demand and suspended solids categories. A few years ago, the large concrete interceptor sewer line accepting this wastewater collapsed because of the action of anaerobic bacteria producing hydrogen sulfide, which was converted to sulfuric acid at the top of the sewer line. This stimulated the industry to arrange for recovery instead of disposal of the sulfate. The amount discharged has been reduced to about 12 to 14 tons/day, with the balance being sold largely as ammonium sulfate. The sanitary district has notified the industry that it must reduce the sulfate effluent to 4 tons/day. This recovery of waste sulfate with concurrent internal water reuse has made a very large reduction in the amount of water purchased and the volume of wastewater discharged at this plant. Studies are continuing to make more effective reuse of wastewater.

The city is negotiating for the purchase of the water company, which will include the source lake. This lake is very badly silted up, and a considerable proportion of the storage capacity has been lost. Consideration is being given to building a lake on another branch of the stream, served by a different watershed, for both water supply and recreation. The very large recent reduction in industrial purchase of fresh water plus the probability of further reductions because of increased industrial water reuse have minimized the need for the additional lake as a source of water supply for the city. Environmentalists who do not want a lake on the second branch of the stream and farmers who do not want the loss of productive farmland are focusing attention on the desirability of dredging the existing lake. Strip mining for coal has been practiced in the immediate vicinity for more than 70 years, with no effort made to reclaim the land. This gives an excellent location for disposal of the dredged material from the lake.

The sanitary district has designed an expansion and upgrading of the sewage treatment plant to include tertiary treatment facilities that will produce an effluent very satisfactory for industrial source water. As the industries now installing pretreatment equipment begin to evaluate the small additional cost for improving these pretreatment facilities to reuse facilities, there is likely to be another large increase in the amount of industrial wastewater that is reused.

All of these factors point strongly toward better utilization of the water resources of the area, plenty of source water for a large expansion of the industrial base of the city, reclamation of some of the abandoned strip-mined areas by the spoils from dredging from the lake, maintenance of the existing flowing stream so ardently sought by environmentalists, and a distinct financial profit to the industries by reusing the wastewater instead of conforming to all of the costs of disposal.

This rather complex history of a small industrial community of about 50,000 population is being repeated with variation throughout all of the country. Industrial water reuse is profitable not only to the industry, but to the overall quality of life of the residents of the community.

V. RENTING WATER REUSE EQUIPMENT

Small industries may find it difficult to raise the capital for the purchase of equipment for treating wastewater either for reuse or disposal. The treatment systems are just as complex for small volumes as for large volumes. They are just smaller in size. Small industries often do not have the technical personnel competent to operate such equipment. There will, in time, be a very large service industry that will furnish equipment, and operate it if desired, for industrial water reuse and will take the concentrates to a central point for consolidating with concentrates from a variety of industries for ultimate disposal. At least one manufacturer of electrodialysis systems now furnishes such a service. Cleaning of electrodialysis stacks is not only a tedious business, but also one requiring considerable skill. This company will sell or lease the electrodialysis equipment and will furnish its own trained operators to service the stacks as needed to keep the equipment in satisfactory operation. Another company, specializing in activated carbon absorption equipment, will furnish a full service for all phases of industrial wastewater treatment, either for reuse or disposal. Thus, small industries will have the capability of treating their wastewater for reuse instead of disposal, if that appears to be the most profitable solution, without the necessity for a large capital investment.

The regional industry accepting concentrates of all kinds, both liquids and solids, from a variety of industries, will develop its own water reuse systems, which may be more complex than those of the average industry.

VI. DISPOSING OF TREATMENT CONCENTRATES

It is necessary to emphasize again and again that the objection is not to water discharged into the environment. It is to the pollutants contained in the water. Our treating processes transfer the pollutants into a smaller volume called concentrates. They must go someplace where they will not pollute the environment beyond legal limits. We talk glibly about "zero discharge of pollutants" and "ultimate disposal of contaminants" without really knowing what they mean.

On the basis of our present knowledge we try to convert the pollutants into three forms:

1. Compounds indigenous to the environment, with our addition being so small that it will not have any measurable impact
2. Compounds so inert and of small enough mass that they will be acceptable in the environment
3. Compounds that can reenter the marketplace

A. Compounds Indigenous or Beneficial to the Environment

Biological processes convert the carbon, nitrogen, hydrogen, and oxygen of organic compounds to carbon dioxide, nitrogen gas, and water. Combustion converts these elements of organic matter into carbon dioxide, nitrogen oxides, and water. Three of these four products are present in nature in such quantities that our addition from disposal of our concentrates is acceptable. Nitrogen oxides are not so easily incorporated into the biosphere.

Many organic compounds are not amenable to biological conversion. Humus (the dark organic material in soils, produced by the decomposition of vegetable or animal matter, and essential to the fertility of the earth) is organic matter not subject to further biological degradation, but acceptable in the environment. Others may be converted by chemical oxidation, or adsorption with subsequent combustion, or fractionated into salable fractions.

Alkali and alkaline earth salts of chlorides and sulfates can be introduced into saline waters, usually the ocean, if they are in reasonable balance with the ratios of similar salts in the receiving water. Unfortunately, most such salts are removed from concentrates at inland locations where the cost of transportation is prohibitive.

Calcium carbonate, with its varying percentage of magnesium carbonate and silica, is so widely distributed in the soils of the earth that sludges from lime treatment are usually acceptable (and sometimes beneficial in acidic soils) for land disposal. The sludge must be analyzed to be sure no objectionable heavy metals or organic compounds have been incorporated in the treating process. Calcium sulfate is beneficial for land disposal if soils have an adverse sodium/calcium ratio.

Organic sludges high in nitrogen, phosphorus, and potassium are desirable for land disposal if they do not contain too much toxic metal and organic constituents.

B. Inert Compounds

Sand, gravel, and grit are solids that contaminate both fresh and wastewaters. Properly cleaned, usually by water washing or burning, they can go back into the environment without adverse effect. They can go onto the land or the stream bed, if well enough distributed. Some of the inorganic residues from intermediate processes of concentrate disposal can enter the environment under the same limitations. Calcining of lime sludges to recover the lime values for reuse produces residues of magnesium oxide, aluminum oxide, silica, calcium phosphate, and iron oxide that are so inert that they can safely reenter the environment. Shale particles from water used in coal washing and some minerals belong in this class. Some industrial operations produce so much inert solid residue that the inert solids from our contaminants can be incorporated with no important impact on the total mass.

C. Compounds that Can Reenter the Marketplace

Water pollutants are sometimes called resources in the wrong place and in the wrong concentration. The physical treating processes previously listed concentrate the pollutants to the point that they may become usable in manufacturing operations. If treatment of wastewaters by freezing and eutectic freezing of concentrates from any purification process lives up to its promise, we will have another physical treating process that may convert our pollutants into process feed materials.

Organic compounds that normally would be converted to environmentally acceptable compounds by the conventional processes of biological and chemical oxidation or by adsorption on activated carbon may, by use of physical treating processes, reach such concentrations that they have recovery value. Once they have reached proper concentration, they become susceptible to the usual operations of fractional distillation. This may not be adequate in many cases, and more sophisticated methods will be required.

Inorganic compounds such as the mineral salts of the alkali and alkaline earth metals can be easily, if not cheaply, fractionated into marketable products by the usual method of evaporative crystallization. The developing process of fractionation by successive eutectic freezing may turn out to be competitive with evaporation for both recovery of the residual water and the purity of the solid salts.

The residues from treatment processes for water reuse are extraordinarily complex. It seems probable that with the restriction of deep-well injection or ordinary disposal sites a really new separation science will be developed. Many laboratories are currently making such evaluations. A typical example is the work of the Institute of Paper Chemistry at Appleton, Wisconsin, on the use of ultrafiltration and reverse osmosis to fractionate and concentrate marketable wood chemicals from spent sulfite liquor. Engineers interested in fractionating their treatment residues should start with Li's two-volume *Recent Developments in Separation Science* (1972).

D. In-Plant Structural Changes

To convert an existing plant from a buy-use-water-discard-used-water system to a water reuse system will involve many plant structural changes. Diking of the plant area to contain surface drainage, installing surge basins with the necessary pump stations and interconnecting piping, modifying piping for cascade systems, installing the necessary monitoring and control systems for the treating plants, and building the one or more treating plants involve problems in geography and geometry. In areas where there is a serious land shortage, it may be necessary to install multistory units instead of spreading out in the conventional manner.

Recycling is the ideal system for water reuse insofar as inplant structural changes are concerned. Treat the wastewater as produced and return it the short distance to the use end. Very small surge basins, really only pump sumps, are required.

VII. TRAINING PLANT PERSONNEL IN USE OF THE NEW SYSTEM

No matter how well one designs and builds a water system, its success depends on good operation. Systems are designed and operating instructions are written by engineers. Plants are operated by laborers. They speak different languages. The engineer knows things he assumes the operator knows (but does not), so he leaves them out of the instructions. Instructions should be written in operator language, so that he understands them.

In training courses one must explain the importance to the company of successful operation of the water system. Once operators understand that their work is just as important as that of the manufacturing process operators, their pride will assure good results.

Include a session with all manufacturing plant personnel to explain the capabilities and limitations of the water system. This will help avoid incidents like the operating failure of an activated sludge phenol removal unit in a petroleum refinery. Investigation disclosed the failure was caused by the cooling tower operator diverting the chromate-containing blowdown to the phenol unit because he was afraid the regular blowdown line would freeze. He ended his statement with a defiant, "That water treatment plant is supposed to handle anything we send down there."

It is not always wise to select engineers as operators to assure good operation. Engineers know too much to be good operators. They are always experimenting to try out their own ideas and the treating plant never settles down to regular operation. Intelligent technical operators tend to run the plant according to the instructions and will deliver the expected results.

VIII. PREPARING VARIOUS REPORTS FOR ALL GOVERNMENTAL AGENCIES

In the United States there is a many-tiered governmental bureaucracy involved in water pollution control operations. The oceans are usually considered the ultimate depository of all stable pollutants. Efforts are under way to establish various worldwide pollution control organizations with authority over ocean pollution. The most obvious ocean pollution is direct dumping of oil, various toxic chemicals, and sludges into the ocean, and these will come under control first. Later, land-based pollutants eventually reaching the ocean will come under joint ocean-continent pollution control groups. This may result in stricter standards for streams in control of pollutants that are not harmful in flowing streams, but may accumulate in objectionable concentrations in the ocean.

Federal, state, county, municipal, sanitary district, and interstate compacts like the very successful Ohio River Valley Water Sanitation Commission are the major (but not the only) governmental bodies to which a polluter must report. The cost of the multiplicity of reports necessary, frequently with varying standards, must be thoroughly evaluated, if an industry goes for disposal instead of reuse.

For initial construction, an Environmental Impact Statement is required. It is liable to cost 5% of the total project cost on small jobs, and then graduating down to 2% of large jobs. For operations, there will be at least one, and probably several, frequent continuing reports. One should not underestimate the costs of all these reports. Industries must learn what reports are not required if they opt for complete water reuse instead of water disposal.

IX. BENEFITS

Many of the benefits that can be quantified have been discussed under the costs section. The benefits that are likely to influence the decision to reuse or dispose are more difficult, but not impossible, to quantify. It probably is not within the province of the systems engineer to make this numerical evaluation.

In those areas where the engineer cannot reach a reasonably firm figure, he should describe the benefit and leave the quantification to management.

A. Reducing Purchase of Freshwater

An industry that uses, but does not consume, large quantities of water may so reduce the amount of outside water required that it can switch to a different and more favorable source. The increased need for water, brought about by the changing pattern of producing energy plus the increasing demand, will soon revolutionize the patterns of water management. Prices will increase so drastically that management, particularly in water-consuming industries, will be reevaluating the whole water program.

One new source of input water is municipal wastewater. Once an industry has designed equipment and processes to upgrade the quality of its own highly polluted wastewater, it will be easy to incorporate municipal wastewater into the same system. Such a source will be more dependable than present freshwater sources.

B. Eliminating Reports for Governmental Control Agencies

The large cost figure for preparation of reports to governmental pollution control agencies for disposal of wastewater to the environment is almost eliminated in a complete water reuse system. Once a simple environmental impact statement that there will be no discharge of water or its pollutants to the environment has been submitted, there should be no further reporting to the government, if the reuse system works. If in the reuse study you have put a true evaluation of the total cost of preparing these reports, there will be a large plus in favor of reuse.

C. Saving in Cost of Treating for Reuse Instead of for Disposal

With known water quality conditions both for use and for reuse or disposal, reasonably exact costs can be developed. There are two areas of uncertainty: One is the changing standards of quality for disposal. As new studies establish the adverse effect of certain compounds in water, standards are upgraded. Already power plants are in trouble because of the use of chlorine to control biogrowths in the heat-exchange system. Even the metal corrosion products from the same heat exchange system are under suspicion. As of the present, the water-planning engineer cannot make an assured evaluation of the future cost of treatment for disposal. The second area of uncertainty is in industry's own use requirements. In this area, an industry can have control and can make a reasonably sound estimate of future problems if it has had good cooperation from the manufacturing process engineers.

D. Improving Housekeeping

The benefits of improving housekeeping to minimize the volume and pollutant concentration of wastewater extend to overall manufacturing operations. Cleaner and drier floors reduce the probability of accidents. The better maintenance of process equipment extends its useful life and reduces the chance of shutdown. Although it is difficult to apply a numerical value to these intangible benefits, there is a real profit in better housekeeping.

E. Improving Public Relations

Pollution control laws are actually made by environmentalists whose decision to go after certain laws is actuated by a belief that this is the only way to get the quality of environment they want. Industry characteristically reacts after the fact, instead of anticipating the fact and getting credit instead of blame. If an industry is discharging its waste into a stream, and something goes wrong in the stream, automatically the finger of blame points at the industry. No matter how innocent the industry may be, the bad name cannot be erased.

An industry cannot get proper benefit from a reuse program without a well-planned program to tell and show the public what it is doing to improve the environment. The program should be tried on the plant personnel, all of whom are environmentalists to some degree, before an industry goes to the general public.

If the product is sold direct to the public, a "Good Neighbor" campaign based on complete water reuse may bring a substantial sales increase. A properly planned and executed campaign will lessen the public's suspicion of industry and decrease the probability of environmentalists going after more restrictive laws. It may be difficult for the engineer to put a numerical value on these benefits. Management may have to make the evaluation.

F. Improving Plant Personnel Morale

The community status of an industry's employees is related to the community's attitude toward the company. The improvement of community feeling about the company engendered by its reuse public relations campaign will give employees a feeling of pride that will result in overall better work in the plant. Although even management cannot quantify this benefit, it is important.

G. Establishing Independence

With the plant operating a complete water reuse system, management personnel are free from involvement with pollution control agencies and can apply all of their efforts to manufacturing a product for sale at a profit.

In some industries, where competitive know-how is an important factor in company profits, the elimination of pollution control inspector visits does away with one possibility of inadvertent disclosure of the know-how.

The ability to locate an industry close to its market may depend on the availability of the required amount of freshwater and the ability to dispose of wastewater. Complete water reuse frees the industry from these restraints.

About 1950, a pulp and paper manufacturer started on a program to build tissue mills close to their market. It would be much cheaper to ship pulp to a tissue mill located at the market than to manufacture the tissue products at the site of the pulp mill and ship the finished product to market. Tissue mills use, but do not consume, large volumes of water and produce a similar volume of highly polluted wastewater. Many good market areas do not have such large volumes of fresh water at reasonable cost. Usually these areas do not have wastewater facilities large enough to handle such a high additional pollution load. If complete or almost complete water reuse could be developed, a tissue mill could locate at a big market area anyplace in the world.

The first installation developed an 85% reuse factor. There are certain processes in tissue manufacture that require very high quality water. The wastewater equipment that was installed could not produce this quality treated water. Local facilities could furnish the 15% make-up water and handle the 15% wastewater.

Further studies demonstrated that the 15% of waste water could be treated to the special use needs, but at a cost higher than justified at this particular mill. The original thesis that a tissue mill could be almost independent of make-up water and disposal of wastewater was proven.

X. LITERATURE SURVEY

There is an enormous literature on all phases of wastewater reuse. There is one thing that must be considered in making literature studies: it is difficult to get people to write about their failures. Sometimes a report on a failure is of great value to the reader, who, at least, can see how it was not done and avoid making the same mistake. Unfortunately, people want to write about their successes, not their failures.

The Office of Water Resources Research, now the Office of Water Research and Technology of the U.S. Interior Department, has published four volumes of a continuing series entitled "Water Reuse. A Bibliography." Each volume contains several hundred abstracts, most of which deal with reuse of waste water by industry.

There is a real gold mine of information in three publications of the American Institute of Chemical Engineers, New York, New York. *Water Reuse*, Symposium Series 78, volume 63, deals primarily with reuse in nine of the major industrial countries of the world. Two later volumes, *Proceedings of the First National Conference on Complete WateReuse: Industry's Opportunity and Proceedings of the Second National Conference on Complete WateReuse: Water's Interface with Energy, Air, and Solids*, are largely oriented toward articles on industrial water reuse. The more than 250 water reuse articles contained in these two volumes contain detailed discussions of all phases of industrial water reuse.

The increased rate of publication of articles on industrial water reuse assures a continuing supply of information that will give the diligent searcher help on almost any water reuse problem his plant may develop.

REFERENCES

American Institute of Chemical Engineers. (1967). "Water Reuse," Symp. Ser. No. 78, Vol. 63. Amer. Inst. Chem. Eng., New York.

American Institute of Chemical Engineers. (1973). "Proceedings of The First National Conference on Complete WateReuse: Industry's Opportunity." Amer. Inst. Chem. Eng., New York.

American Institute of Chemical Engineers. (1975). "Proceedings of The Second National Conference on Complete WateReuse: Water's Interface with Energy, Air, and Solids." Amer. Inst. Chem. Eng., New York.

American Institute of Chemical Engineers (1976). "Proceedings of the Third National Conference on Complete WateReuse: Symbiosis as a Means of Abatement for Multi-Media Pollution," Amer. Inst. Chem. Eng., New York.

Li, N. N. (1972). "Recent Developments in Separation Science," 2 vols. Chem. Rubber Publ. Co., Cleveland, Ohio.

Office of Water Resources Research. (1973). "Water Reuse Bibliography," Vol. 1, No. PB-221-998. Off. Water Resour. Res., Washington, D.C.

Office of Water Resources Research. (1973). "Water Reuse. A Bibliography," Vol. 2, No. PB-221-999. Off. Water Resour. Res., Washington, D.C.

Office of Water Resources Research. (1975). "Water Reuse. A Bibliography," Vol. 3, No. PB-241-171. Off. Water Resour. Res., Washington, D.C.

Office of Water Resources Research. (1975). "Water Reuse. A Bibliography," Vol. 4, No, PB-241-172. Off. Water Resour. Res., Washington, D.C.

5

Reuse of Water for Municipal Purposes

G. J. Stander

I. INTRODUCTION

Wastewater reclamation is nothing new; in fact, the diluting and self-purification capacities of rivers and lakes have been, to this day, purifying sewage and industrial wastes for more than a century (Fish, 1967; Downing, 1968). In earlier times, such rivers and lakes were not regarded as sources of public water supply, and watersheds that served as sources of public water supplies were zealously protected against pollution.

Since then, population explosion and industrial development have exhausted, in many areas, the available resources of protected catchments, and recourse has had to be made, to an ever-increasing extent, to the purification of polluted waters to satisfy demand. Records indicate that, as from about 1948 onward, strict antipollution legislation to control pollution of rivers and lakes has been invoked practically throughout the world. In spite of marked improvements due to these measures, serious pollution of basic water supplies is prevalent and the gap between the supply and demand of clean water is narrowing. The result is that thousands of conventional water treatment plants in many countries today are producing public water from polluted sources; thus, indirect wastewater reuse for municipal purposes is widely practiced today.

Because of increasing public health requirements, the quality criteria for drinking water are currently under worldwide review.

The stark reality facing mankind today is that the freshwater resources of many countries are in dire danger. However, during the past decade, the efforts of scientists and engineers, motivated by the growing imbalance between demand and supply of clean water, by the economic and many other advantages of water reuse, and by pollution control, have led to remarkable achievements in wastewater technology in restoring the usefulness of freshwater supplies, in reclaiming wastewater for reuse, and in the treatment of wastewaters before discharge to rivers and lakes.

As anticipated, these achievements have led to new concepts on conventional water purification and wastewater reclamation and reuse, in addition to creating new horizons of thinking as well as challenges to scientists and engineers engaged in wastewater management and pollution control. From the extensive data so far published, it is obvious that the basic challenge facing scientists and engineers in many countries is the optimization of the freshwater resources at man's disposal by the most effective routes, not only to meet a growing demand, but also to solve water pollution problems aggravated by progressive industrialization and concentration of population masses. Therefore, in the treatment of the subject matter of this chapter, all polluted waters are regarded as wastewaters, whether from a river or lake, or as an effluent from an industry or from a municipal sewer.

II. CURRENT STATUS OF WATER RECLAMATION FOR REUSE

The current status of wastewater technology, particularly wastewater reuse, is extensively described in readily available publications, as well as in the relevant chapters of this book. Although a review of the examples of reuse covered in the literature would prove superfluous, the decision maker nevertheless needs in-depth facts about the potential of wastewater technology to produce water for municipal reuse to purify polluted river waters to meet public health requirements for drinking purposes and to control pollution.

An endeavor will therefore be made, here, to bring into focus basic principles, expertise, and concepts that have emerged under the collective pressures of demands for quality criteria for the many uses of water. These factors will be examined specifically in regard to the key role they will play in giving proper direction to the critical planning of water pollution control, wastewater reclamation, water reuse, and, in general, the optimum use of water resources.

A. Water Quality

The growing scarcity of usable water in countries once blessed with adequate basic freshwater resources and the concomitant reactions of various sectors of water users and of the public to the destruction of the usefulness of water resources—particularly where public health, aesthetics, and the many amenities of rivers and lakes are concerned—have resulted in a critical awareness of water quality. As a consequence, water quality criteria and standards have been developed; and the experience gained by their application to water pollution control, public water supply, and wastewater reuse has aroused a worldwide critical appraisal of their reliability and accuracy to detect and measure:

1. Changes across the whole spectrum of water quality requirements to safeguard public health, lake and stream life, and the specific needs of public and industrial water supply and of the many other sectors of water users
2. The efficiency of conventional water purification plants and of wastewater reclamation installations in the production of treated water meeting the quality requirements for the particular use to which the water is put
3. The pollution of rivers and lakes

The tremendous growth in the production of synthetic chemicals and new products for various uses has introduced into the environment sophisticated toxic, carcinogenic, and subtle pollutants. Many of these have proved intractable to degradation by the self-purification processes of rivers and lakes and biological treatment and, thus, can pass through conventional water purification plants to reach the consumer. Furthermore, the ever-increasing proportions of domestic wastewater—containing these pollutants—reaching rivers and lakes have effected marked changes in the quality of the receiving waters.

The cumulative adverse effects of these pollutants on water quality, public health, and aquatic life have exposed an urgent necessity to revise and extend criteria and parameters in order to judge accurately the quality and safety of water for various uses and to monitor the performance of the associated water treatment installations. The complexity posed by this situation brought in its wake the necessity to develop and perfect analytical techniques and automatic monitors, to quantify the various parameters of water quality, particularly viruses and pathogens; to identify specific pollutants such as carcinogens, intractable organic residues, biocumulative chemicals, and substances known to be inimical to human health and animal and plant life; and to effect "fail-safe" control of water treatment plants. The situation is becoming even more complex due to the increasing concern of conservation and recreational interests to preserve the aesthetic quality of the natural water environment.

The increasing presence of micropollutants responsible for tastes and odors in water has become a matter of grave concern to water supply undertakings. Water quality parameters to ensure accurate measurement of these qualities are of critical importance due to public sensitivity—in fact, this problem, rather than that of viruses and carcinogens, is currently a much stronger driving force behind the inclusion of tertiary processes in conventional water treatment plants and behind the application of tertiary treatment to wastewaters before their discharge to rivers and lakes.

Methods for the identification of viruses, particularly the hepatitis virus, are urgently needed—in order to stop speculation more than anything else. While it is true that a zero *E. coli* count in 100 ml of reclaimed water does not guarantee the absence of virus, it is equally true that conventionally purified water to a zero *E. coli* standard is not necessarily free of pathogens; and when one deals with larger quantities of water, e.g., the entire supply, one can be no more certain no one is going to become ill by drinking the conventionally purified water from polluted rivers than one can be certain that on one will become ill by drinking water reclaimed for drinking purposes by advanced techniques (Van Vuuren *et al.* 1971). No one can deny that the problem of viruses in reclaimed water used for drinking has to be resolved; but it is just as much a problem in the thousands of conventional water purification plants treating polluted river waters.

In view of these considerations, exacting water quality criteria and parameters are the key to dynamic progress in water pollution control, to wastewater reclamation, both for discharge to rivers and lakes and for domestic and industrial use, and to the upgrading of existing, as well as future, water purification facilities for public water supply. There is no question that the public's and industry's ever-increasing demand for better quality water, and the consequent increase in treatment costs to achieve this, will accelerate the pace of applying tertiary treatment to wastewater, first, for reuse or before discharge to rivers and lakes and, second, for upgrading conventional water treatment plants. Municipalities and industries will soon feel the burden of increased costs for public water supply and effluent disposal into rivers and lakes; needless to say, the pressures of the politician and the shareholder will necessitate a reappraisal of wastewater management, within the municipal and industrial household, on the basis of a cost-benefit analysis; this, in turn, will crystallize the critical plans, which will delineate (1) the exact economic benefits and procedures whereby these objectives could be achieved, (2) the facts and figures for a far-sighted research and development program to ensure that rivers and lakes will be of the required quality, that wastewater reuse will be an integral part of the water balance, and that optimum pollution-control will be established to obtain more "mileage of use per 1000 gallons" of water.

B. Unit Processes and Operation

The physical dilution and self-purification capacities of rivers and lakes have been exploited as a process in the purification of wastewater, ever since the industrial revolution; and today, in spite of a greater understanding of the limitations of these processes, man still exploits them for the disposal of sewage and industrial wastes. The modern wastewater biological purification plant constitutes effectively the harnessing of these unit processes under controlled engineering operation. In this man-made wastewater purification system, the physical and self-purification capacities of rivers and lakes are not only accelerated manifold in the unit processes of the plant, but the variables can also be controlled. Moreover, the significance of the dilution capacity of the receiving body is minimized.

With the increasing proportions of wastewater being discharged into rivers and lakes, it has become obvious that there is an overconfidence in the natural capability of rivers and lakes to maintain the quality of the water at a level suitable for all recognized uses. Furthermore, experience with biological wastewater purification plants, even allowing for their capability to accelerate the natural purification rate of rivers, has exposed notable limitations of natural biological purification processes. This experience has also provided confirmatory evidence that the self-purification processes in rivers and lakes are equally limited in their capability of degrading micropollutants, carcinogenic compounds, and toxic and complex organic substances; in the destruction of viruses, pathogens, and parasites; and in the removal of biocumulative substances, phosphates, and nitrogen. Scientists and engineers are very much aware of the limitations of secondary biological treatment processes as a pollution control measure and, during the past two decades, have diligently applied themselves to a concerted research and development effort to sharpen the tools of existing wastewater technology in three main directions:

1. Reviewing manufacturing processes. (a) Eliminate leaks into wastewater circuits of chemical substances listed as prejudicial to advanced wastewater treatment processes that degrade or remove pollutants before effluents are discharged into the water environment and, subsequently, used for public supply, or before the effluents are reclaimed for industrial or domestic reuse. (b) Replace wet processes used for cleaning and conveyance of waste materials and by-products by dry unit processes.

2. Upgrading wastewater biological treatment plants to remove phosphates, nitrogen, and nitrates; to improve their efficiency in destroying or

removing viruses, pathogens, and parasites; and to degrade organic substances listed as toxic, carcinogenic, and biocumulative.

3. Developing new and refining existing physical and chemical unit processes (a) to produce water of a suitable quality for internal recycling in industry or for augmenting supplies of water for industry; (b) as supplementary unit processes to biological wastewater treatment systems, as well as to conventional water purification plants having polluted water intakes, to produce water conforming to higher quality requirements for the various uses of water and for the control of pollution.

The objective of the foregoing detailed considerations is to dispel the many misconceptions regarding advanced wastewater reclamation as evidenced in the available literature. A proper understanding of the objectives of advanced wastewater treatment by all concerned is absolutely necessary for the immediate exploitation of the experiences and lessons of Chanute (Metzler *et al.*, 1958), Lake Tahoe (Sebastian, 1970), and Windhoek (Van Vuuren *et al.*, 1971). It will be observed from what is reported in this chapter and elsewhere that the Windhoek project is being handled in a most responsible manner, and results will progressively be made known in all parts of the world (see Chapter 13).

The selection of advanced wastewater treatment processes is largely dependent on the character of available raw waters and the quality requirements of the uses envisaged for the reclaimed water. From a technological point of view, the predominance of specific pollutants plays an important role in process selection and in combinations of processes. An indication of the available unit processes and their potential application is given in Fig. 1.

For industrial process water and for internal reuse of water, it is evident that by the judicious selection and combination of unit processes it is possibly to produce water of virtually any desired quality. In the case, however, of domestic recycling of reclaimed water to augment water supplies or for recycling in food processing, there are a number of requirements that dictate the selection and operation of unit processes:

1. Reliability of unit processes or combinations thereof to produce water or desired quality at all times
2. Automation of unit processes operations and operational flexibility to compensate for raw-water quality changes or changes in performance of process units
3. Standby process units to ensure continuity and stability in plant performance
4. Continuous monitoring of certain parameters requiring rapid analytical procedures
5. Rapid identification and quantification of pollutants listed as prejudicial to the particular reuse
6. Fail-safe removal of viruses, harmful substances, etc.

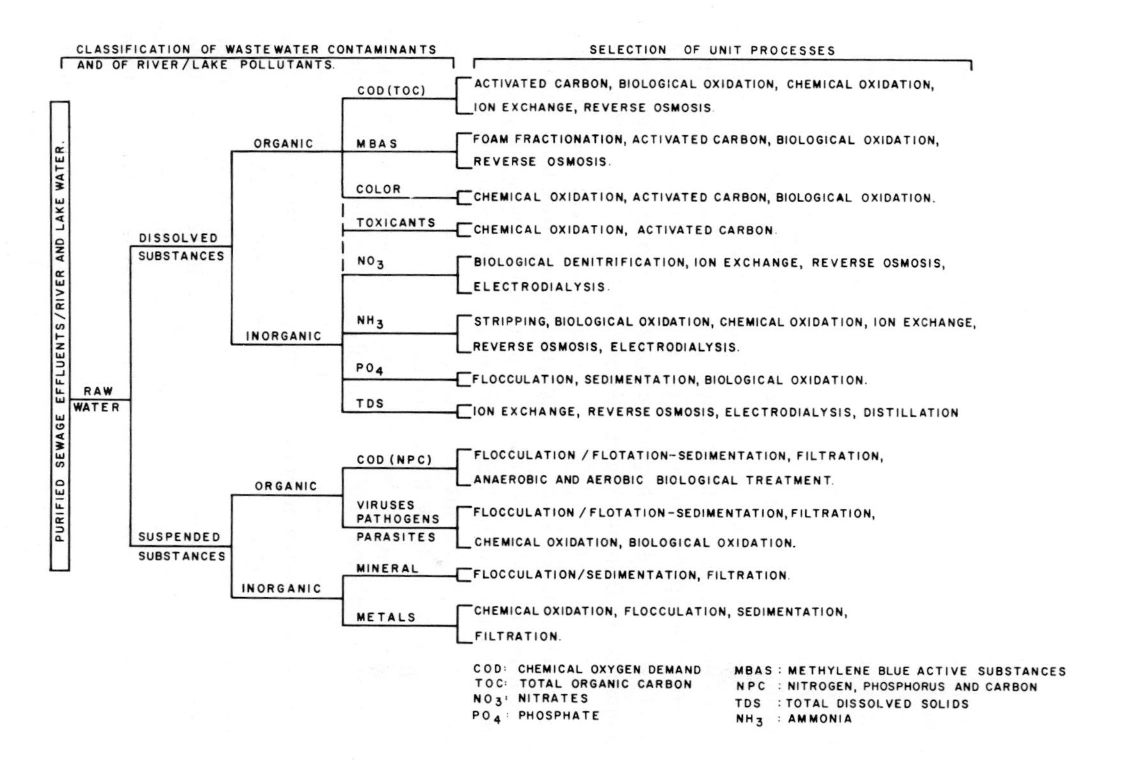

Fig. 1. Classification of wastewater contaminants and selection of process units.

It should emphatically be pointed out that these requirements also apply to indirect reuse: conventional water purification plants that have polluted water intakes.

It is worthwhile to comment here on some specific problems—not only in wastewater reclamation for domestic recycling, but also in the treatment of polluted river water and lake water for public supply—and to elaborate on unit processes which, with further research and development, could offer solutions:

1. The efficiency of virus removal by chlorination is largely dependent on the turbidity and organic content of the water. The chlorination of untreated polluted waters most certainly will not destroy viruses. Chlorination is a final unit process in any water or wastewater purification system; and its efficiency in achieving the desired bacterial and viral quality is determined by the proper selection and operation of the preceding unit processes. There seems to be complete unanimity at all responsible levels of professional thinking that, by proper selection and operation of unit processes to produce a water within the desired turbidity range of 0.1 to 1.0 Jackson units, chlorination to the extent of 1 mg/liter free chlorine residual, after 30 min contact, will ensure a safe water. It stands to reason that in order to ensure virus inactivation, breakpoint chlorination should be applied to give a free available residual chlorine at a pH value that will ensure a free hypochlorus acid concentration. The latest "European Standards for Drinking Water" issued by WHO Regional Office for Europe (World Health Organization, 1971) recommends a concentration of 0.5 mg/liter free residual chlorine for 1 hr for the inactivation of virus. It is mentioned that a redox potential of 650 mV will cause almost instantaneous inactivation of even high concentrations of virus. Further, enteroviruses are the most resistant to chlorination, and if they are absent from chlorinated water, it can be assumed that the water is safe to drink. It further states: "There must be some reservations about the virus of infectious hepatitis, since it has not so far been isolated, but in view of the morphology and resistance of enteroviruses, it is likely that if they have been inactivated, hepatitis virus will have been inactivated also."

2. Ozonization is regarded as a unit process of great potential in the purification of polluted waters, as well as in the reclamation of wastewaters for industrial and domestic reuse, as soon as the problems in connection with maintaining residual protection and with its rapid decomposition have been overcome. Ozone, like chlorine, is a powerful virucide and its concomitant potency in destroying organic residues, tastes, and odors is claimed to be most effective.

3. The removal of intractable organic residues that may be carcinogenic, biocumulative, and organoleptic (e.g., synthetic detergents) has been a most vexing problem for the water supply and pollution control authorities. The problem has also had a most restricting effect on wastewater reclamation for reuse. (See Chapter 2.)

It can now be confidently stated that experience (see Chapter 1) with activated carbon as a unit process has clearly shown its remarkable potentialities for the removal of refractory organics. In conjunction with certain polyelectrolytes and ion-exchange materials (Gaffney *et al.*, 1969), the possibilities are numerous. There is, however, a serious note of warning that emerged from the Windhoek project (Nupen and Stander, 1972): activated carbon and ion-exchange filters must not be used as unit processes for the removal of viruses and pathogens.

4. Desalination offers great promise as a unit process in wastewater reclamation for reuse. The buildup of dissolved mineral salts that can be tolerated varies with different uses. In a single cycle of domestic use, the total dissolved solids increment is of the order of 200–300 mg/liter, and only fractional desalination is necessary to balance this increment when reclaimed wastewater is recycled. In this type of application, desalination should be motivated on a proper cost/benefit analysis; the reduction of total dissolved solids (TDS) permits increased "mileage of use per 1000 gallons" of water; and, therefore, through the diffusion of desalinated water into a reticulation system to maintain a predetermined TDS, the costs can be reduced to a level that makes desalination economical by increasing the number of reuse cycles.

C. Direct and Indirect Reuse of Wastewater

Except in food-processing industries, the internal direct reuse of wastewaters is worldwide practice, and wastewater technology is sufficiently advanced to produce industrial process waters of any desired quality. There is an increasing awareness in industry of the water qualities that can be tolerated in unit processes, of the economic benefits and contributions to water-pollution control that may accrue from water reuse, and of the long-term value of efficient use of available water resources.

In regard to domestic use of reclaimed water, however, an objective appraisal of the question of direct and indirect reuse of wastewaters is very necessary, since this is a vexing and controversial issue which, in some countries, has precipitated a stalemate in pollution control, in wastewater reuse, and in the upgrading of conventional water purification facilities.

It is necessary to state emphatically that as far as domestic reuse is concerned, wastewaters must undergo effective oxidative biological stabilization by any of the recognized systems prior to reclamation. No responsible scientist could, in the light of current status of wastewater technology, consider the use of raw wastewater as intake to such a tertiary plant. As a note of warning, intakes from polluted rivers and lakes can quite easily permit raw wastewater that has escaped dilution and self-purification to pass directly into the conventional water purification plant; the results of many studies on mixing and dilution in rivers and lakes, as well as on the role of density currents, prove conclusively the validity of this observation. Therefore, it is obvious that the intake waters to water purification plants drawing from such sources must contain raw wastewater that has escaped the factors of time-lapse and dilution.

With a clear understanding of the direct and indirect routes that wastewater must follow before being reclaimed for domestic supply, we can now proceed to discuss what the reasonable alternatives are.

1. In the light of recent trends of pollution, munipical water treatment authorities will have to install additional unit processes to produce a finished water that is safe and aesthetically pure. In the course of time, conventional water treatment plants may be no different from advanced wastewater treatment plants. This would indeed be a sad situation. Surely then, it is imperative to restore the cleanliness of the natural water environment—the longer rivers and lakes are exploited as process units to absorb wastewaters, the longer the back door to increase pollution will be kept open.

2. The reservations that have been expressed regarding wastewater reclamation for domestic reuse, namely, the risk of (a) infections, (b) chronic toxicity, (c) carcinogenic effects, (d) sex hormone effects, and (e) radiological effects, can also be placed at the doorstep of many conventional water purification plants treating polluted water. A major proportion of current scientific effort assigned to studying the aforementioned issues of wastewater reclamation could profitably be diverted to a critical evaluation of many conventional water treatment plants that have polluted water intakes; this would enable the practical data so urgently needed by the decision maker to be gathered much sooner.

It will certainly go a long way to vindicate the current negative speculations that confuse the decision maker. After all, the responsible scientist and engineer concerned with research and development on advanced wastewater treatment do not arbitrarily simply want to force the use of reclaimed water down the public's throat—for, they are already aware that this is already being done by thousands of simple water purification systems with the concomitant risks to public health. Rather, they want to develop ways to restore the cleanliness of the natural water environment which, after all, is man's rightful heritage.

3. The Windhoek (Van Vuuren *et al.*, 1971) and Lake Tahoe (Sebastian, 1970) wastewater reclamation projects have unquestionably set the course that decision makers could follow, the objectives of the scientist and engineer concerned with wastewater reclamation are crystal clear. A full description of the Windhoek reclamation project is provided in Chapter 13.

III. DISCUSSION AND CONCLUSION

It is unfortunate that advanced wastewater treatment and the reuse of reclaimed water have been almost inseparably linked with direct recycling to the public. This impression is given front-page publicity in an overwhelming proportion of publications, which has helped crystallize microconceptions in the minds of the public and the decision maker.

The real objectives of wastewater reclamation are (a) the development of techniques to produce potable water conforming to the same standards set for public water supply derived from polluted sources, in order to solve the ever-narrowing gap between water demand and supply in developing areas, as in the case of Windhoek, and, in areas where economics of new water supplies are unfavorable or not available, to plan wastewater reuse as an integral part of the water economy; and (b) the exposure of the incapability of current conventional water purification facilities to meet the public health requirements for drinking water.

The public is lulled into a false sense of security by the belief that the self-purification processes in rivers and lakes provide a more effective safety barrier against viruses, pathogens, carcinogens, organic intractables, toxic and biocumulative chemicals than a properly designed and operated biological purification plant preceding an advanced wastewater reclamation plant. Every responsible scientist and engineer is aware that today this is no longer the case. Indirect reuse, via the river, lake, and impoundment has, on the basis of existing facts, certainly not eliminated the problems which public health opinion ascribes to reclaimed water recycled to the public supply.

Public acceptance of the recycling of reclaimed water to the public supply, in admixture with the normal supply, can be a major contribution to wastewater reuse. A comprehensive educational effort is required (a) to make the public aware that their welfare is at stake—they must be convinced that there is a need for such a facility and that the potential costs are preferable to any alternative available; (b) to familiarize them with the technical details of the processes through education, demonstration, and visits to the purification facilities where they can observe and ask questions. To quote Jessie Rudnick (1971): "Lake Tahoe effluent meets USPHS drinking water standards, yet it is not returned to the Lake. Why? What keeps us from taking that logical next step of reusing the water directly? There appears to be a widespread belief that the public will not accept such a step. I believe that the public *will* accept such a step when given the right information." Experience with the Windhoek project proved Mrs. Rudnick's viewpoint conclusively.

Many countries are today moving along a critical path with respect to the availability of water resources for further socioeconomic development. Superimposed on this problem are the increasing proportions of wastewater discharged to rivers and lakes, which limits their usefulness. Because of this, water pollution control, wastewater reclamation in advanced treatment plants for reuse or return to rivers and lakes, and the upgrading of conventional water treatment plants must be regarded as an integral part of any country's water economy. The achievement of this objective requires, at national level, far-sighted research and development effort, planning, and financial input. The public has the right to demand this—after all, water is people.

ACKNOWLEDGMENTS

The author wishes to acknowledge the valuable information and help from India, United Kingdom, United States of America, Denmark, Sweden, Russia, Spain, Israel, France, Bulgaria, and South Africa.

REFERENCES

Cecil, L. K. (1967). Complete water re-use. *Chem. Eng. Progr., Symp. Ser.* **65**, No. 78, 258.

Downing, A. L. (1968). "Pollution Control and Related Research in the United Kingdom." Stationery Office, London.

Fish, H. (1967). Sewage effluent use in water supply for South-East England. *Effluent & Water Treat. J.* **7**, 645.

Gaffney, J. G., Kunin, R., and Downing, D. G. (1969). Ion exchange for industrial water. *Water Resour. Symp.* **3**.

Gloyna, E. F., and Eckenfelder, W. W. eds. (1969). "Water Quality Improvement by Physical and Chemical Processes," Water Resour. Symp. No. 3. University of Texas, Austin.

Haney, P. D. (1969). Water re-use for public supply. *J. Amer. Water Works Ass.* **61**, No. 2.

Henzen, M. R., Stander, G. J., and Van Vuuren, L. R. J. (1972). "The Current status of Technological Development in Water Reclamation," Workshop Sess. 6th Int. Conf. I.A.W.P.R. Pergamon, Oxford.

Metzler, D. F., Culp, R. L., Stoltenberg, H. A., Woodward, R. L., Walton, G., Chang, S. L., Clarke, N. A., Palmer, C. M., and Middleton, F. M. (1958). Emergency use of reclaimed water for potable supply at Chanute, Kan. *J. Amer. Water Works Ass.* **50**, 1021.

Nupen, E. M., and Stander, G. J. (1972). "The virus problem in the Windhoek Waste-water Reclamation Project. *Advan. Water Pollut. Res. Int. Conf. 6th I.A.W.P.R.*, Jerusalem, Israel, p. 133.

Rudnick, J. (1971). "Advanced Waste Treatment and Water Re-use Symposium." Dallas, Texas.

Sebastian, F. P. (1970). Waste-water reclamation and re-use. *Water Wastes Eng.* **7**, 46.

Stander, G. J., and Clayton, A. J. (1971). Planning and construction of waste-water reclamation schemes as an integral part of water supply. *Water Pollut. Contr.* **70**, No. 2, 228.

Stander, G. J., and Van Vuuren, L. R. J. (1969). The reclamation of potable water from waste water. *J. Water Pollut. Contr. Fed.* **41**, No. 3, Part I, 355.

Stephan, D. G., and Schaffer, R. G. (1970). Waste-water treatment and renovation status of process development. *J. Water Pollut. Contr. Fed.* **42**, Part I, 399.

Van Vuuren, L. R. J., Henzen, M. R., Stander, G. J., and Clayton, A. (1971). The full-scale relcamation of purified sewage effluent for the augmentation of the domestric supplies of the city of Windhoek. *Proc. Int. Conf. Water Pollut. Res. 5th, 1970* p. I-32/1–I-32/9.

Weinberger, L. W., and Stephan, D. G. (1968). Waste-water re-use: Has it arrived? *J. Water Pollut. Contr. Fed.* **40**, 529.

World Health Organization. (1971). "European Standards for Drinking Water." World Health Organ., Copenhagen.

6

Pressure-Driven Membrane Processes and Wastewater Renovation

Georges Belfort

I. INTRODUCTION

A. New Water Sources and Pollution Reduction

Several semiarid regions in the world, including Israel, South Africa, and Southern California, are actively searching for supplementary sources of water to help fulfill future demands. This quest includes new sources of water such as the renovation and reuse of wastewater and desalination of brackish and seawater.

New technological developments necessary for achieving a greater quantity of water and protecting the quality of various waters have begun to appear, particularly during the past two decades. Several new unit processes for water and wastewater treatment have recently been developed. These include the membrane processes as a group, which can be divided into pressure-driven (reverse osmosis and ultrafiltration) and electrically driven (electrodialysis and transport depletion) processes. The membrane separation processes are thought to be especially useful in water renovation because they allow separation of dissolved materials from one another or from a solvent, with no phase change.

B. Why Pressure-Driven Membrane Processes?

Several questions arise regarding pressure-driven membrane processes. Why and how were these processes developed in the first place? Why do we think they *will* play an important and unique role in the future with respect to wastewater treatment? What are the advantages and disadvantages of these processes within the spectrum of available and new unit processes? Finally, what is the state of the art of pressure-driven membrane processes as applied to wastewater treatment? The first and second questions will be discussed below, while the answers to the last two questions will be presented later in Sections III and IV.

From the outset, it should be made clear that widescale acceptance and usage of membrane processes for wastewater treatment are still only a future hope and, although several commercial applications already exist, most of the data discussed in this review are from experimental pilot plants and small-scale research studies.

Reverse osmosis and electrodialysis (and, later, transport depletion) have been developed during the past 20 yr for the purpose of removing salt from brackish waters with a total dissolved solids concentration of from about 1,000 to 10,000 ppm. Most, or all, of the dissolved solids in the brackish feedwaters were inorganic (ionic) in nature, with negligible dissolved organic species present. In the midsixties, various research laboratories involved in desalination research and development realized that these same processes, especially the pressure-driven ones, could be used both in municipal and industrial wastewater treatment as one element in the train of unit processes for recycling and/or treatment of water prior to disposal. The application of membrane processes to wastewater treatment is about 10 yrs old. A classification of the membrane processes according to their type of driving potential and their general behavior with respect to wastewaters is presented in Table I. Figure 1 presents the removal range of particle sizes for various separation processes. Note that electrodialysis and reverse osmosis cover essentially the same particle removal size range while ultrafiltration, using size as the primary parameter effecting separation, covers a particle size range of more than three orders of magnitude. The spectrum of substances to be removed from municipal, industrial, and wastewater streams can vary across the whole range of particle size shown in Fig. 1.

In this chapter, we will concern ourselves with the treatment by pressure-driven membrane processes of municipal wastewaters, industrial wastewaters,

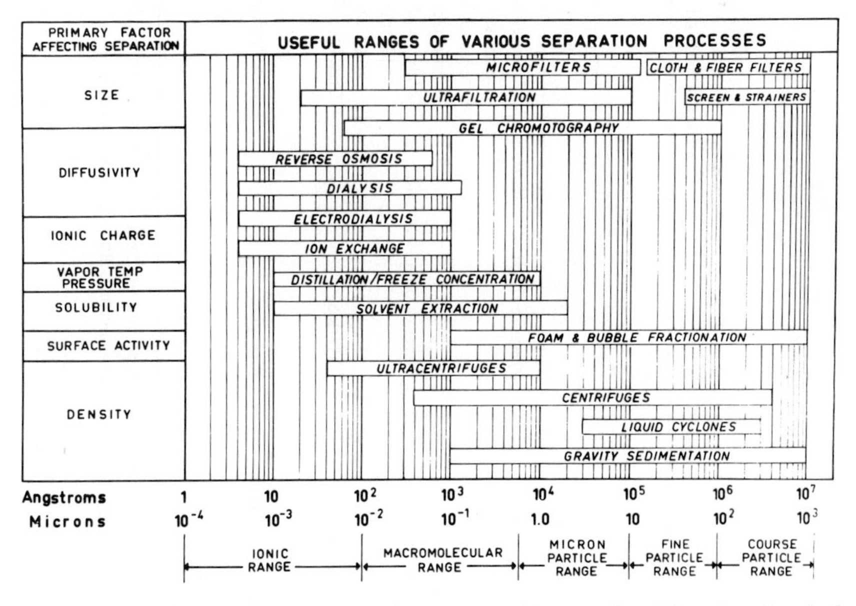

Fig. 1. Useful ranges of various separation processes. (Courtesy Dorr-Oliver, Inc., Stamford, Conn.)

Table I
Classification of Membrane Processes[a]

Process	Driving potential	Constituents removed from wastewater	Constituents remaining in product (besides water)	Possible size ranges of permeable species (Å)
Reverse osmosis	Pressure (high ~40 atm)	Water without dissolved and non dissolved inorganic and organic constituents	Little salt (due to membrane leakage) BO_3^-, NO_3^-, urea, low molecular weight organics[b]	4–300
Ultrafiltration	Pressure (low ~10 atm)	Water without dissolved and nondissolved organic constituents	All the salt and low molecular weight organics	$300–10^5$
Electrodialysis	Electrical	Dissolved inorganics	Little salt, all the organics (dissolved and nondissolved) including viruses, bacteria, etc.	4–300
Transport depletion	Electrical	Dissolved inorganics	More than a little salt, all the organics (dissolved and nondissolved) including viruses, bacteria, etc.	10–1000

[a] The feed is assumed to be a typical municipal secondary effluent.
[b] With the Loeb–Sourirajan asymmetric cellulose acetate membrane.

and polluted river waters. It should be noted that electrically driven processes have not proved attractive for treating effluents with substantial amounts of dissolved organic compounds. Early work on electrodialysis treatment of municipal secondary effluents indicated that virtually all the dissolved organics had to be removed from the feed prior to treatment for adequate performance (Smith and Eisenman, 1964, 1967). Thus, the chief function of the electrodialysis process was the removal of inorganic ions, which left bacteria, viruses, and neutral organics in the product stream. This could become a serious problem if you consider recycling for potable uses.

A slight improvement over electrodialysis is the transport depletion process. Although it is very similar to the electrodialysis process, it is marked by two important differences: (1) the troublesome anion exchange membranes used in electrodialysis are replaced by near-neutral membranes; (2) while conventional electrodialysis is a well-established process with existing plants operating on a brackish feedwater capacity of over one million gallons per day, transport depletion is still a small laboratory pilot plant curiosity without commercialization. In spite of this, there is some evidence that, where municipal effluents are concerned, transport depletion would perform better than electrodialysis (Lacey and Huffman, 1971). The objection with respect to bacteria, viruses, and neutral organics in the product stream also holds for the transport depletion process.

Because of the above considerations, the electrically driven processes will not be discussed further. The theory and operating experience of electrodialysis is well described in the literature (Wilson, 1960; Shaffer and Mintz, 1966), and data for transport depletion are also available (Southern Research Institute, 1967). Other interesting membrane separation processes which are in the developmental stage and which will not be discussed here include dialysis, piezodialysis, Donnan dialysis, gas permeation, and pervaporation (Lacey, 1972a).

Because multiple recycling results in a buildup of conservative constituents*, one very important question is whether the wastewater will be reused only once or whether it will be recycled many times. Dissolved inorganic ions, refractory organics, viruses, and some bacteria are examples of conservative constituents for normal biological secondary treatment. Thus, the removal of these elements may become necessary if, on recycle and buildup, they become detrimental to the intended reuse (Shuval and Gruener, 1973; World Health Organization, 1973). It is with the purpose of removing these conservative constituents that several advanced treatment techniques are being developed: These include activated carbon adsorption, ion exchange, chemical precipitation and clarification, and membrane processes.

*A conservative constituent is not removed from the water during normal treatment.

Table II
Approximate Treatment Efficiencies and Costs[a]

Wastewater constituent	Treatment process	% Removal	Cost, cents/1000 gal
Trace organic removal	Activated carbon adsorption	95	20–40
Soluble inorganic removal	Electrodialysis	90	>40
(heavy metals, radioactivity,	Ion exchange	90	>40
and salts)	Reverse osmosis	>90	>40
	Chemical precipitation	20–95	5–20

[a]The information in this table was obtained from Sawyer, 1972, Table I.

C. Economics of Pressure-Driven Membrane Processes

In Table II, we compare the approximate treatment efficiencies and costs for the aforementioned processes for the removal of trace organics and soluble inorganic constituents. The processes, grouped together for the removal of inorganic constituents, are not entirely comparable because the first three—electrodialysis, ion exchange, and reverse osmosis—are used to remove salt, while chemical precipitation has been used to remove specific ions such as the phosphate ion. Also, reverse osmosis is capable of removing organics, viruses,

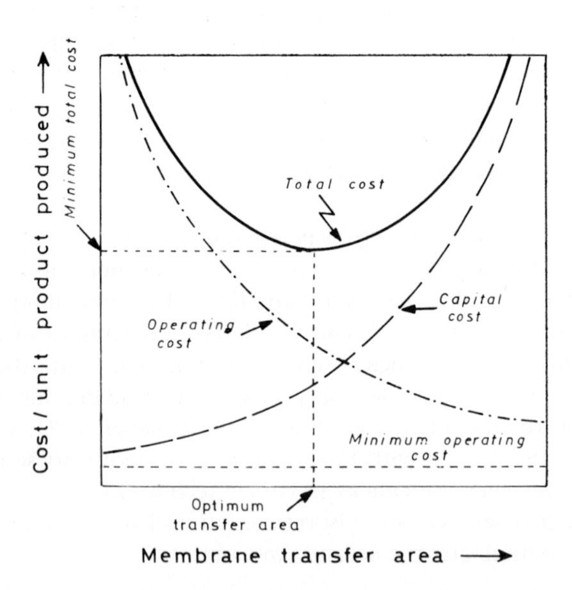

Fig. 2. Cost per unit product of treated wastewater as a function of system membrane-transfer area for a fixed production rate (after Probstein, 1973).

and bacteria (in addition to soluble inorganic ions) from the product, while electrodialysis, ion exchange, and chemical precipitation will not do this.

Total product water costs for membrane processes can be roughly divided into (1) capital costs or fixed charges based on capital recovery of initial investment and (2) operating costs for energy, labor, pretreatment, and membrane cleaning. These two group costs are shown diagrammatically in Fig. 2 as a function of membrane transfer area. The operating costs per unit volume of product will decrease with decreased driving force, which, in turn, will decrease with an increased membrane transfer area. The capital cost per unit product will increase with increased membrane transfer area. The total cost curve in Fig. 2 is, of course, the sum of the operating cost curve and the capital cost curve.

With respect to wastewater treatment technology, the traditional biological treatment processes have, as a percentage of the total cost, lower operating and maintenance costs, but higher capital investment, than the "advanced" physico-chemical treatment processes, such as the pressure-driven membrane processes discussed here. Because of this, municipalities in the United States, for example, which have various financing options such as very large grants-in-aid provided to the public sector for construction of treatment works, may be reluctant to burden themselves with high long-term operating and maintenance costs. Thus, it appears, in the United States, at least, that these advanced processes will first find acceptance in the industrial sector. Furthermore, only pressure-driven membrane processes of a capacity of a few million gallons per day (mgpd) and less are commercially available at present. For purposes of municipal water supply, these volumes are very small indeed. These smaller units will in the short term be most appropriate for the industrial and small municipal sectors (Channabasappa, 1969). In the long term, as larger plants become available, this picture may change.

II. PRINCIPLES OF PRESSURE-DRIVEN MEMBRANE PROCESSES

A. Definition

A membrane process, as discussed in the context of this chapter, can be defined as one whose purpose is to separate, using selective membranes, one (or more) component(s) from a two (or more) component system using a differential driving potential across the membrane. We thus begin with a feed solution, nominally called the wastewater stream, from which we would like to remove either (a) the unwanted pollutants such as dissolved inorganics, or (b) relatively clean water, and leave behind a more concentrated polluted water.

A differential driving potential across the membrane thickness is needed to attract or push the mobile component through the membrane. The choice of a driving force is, of course, a function of the type of membrane used. The driving force across the membrane may be the result of differences in concentration, as in dialysis; differences in electrical potential, as in electrodialysis and transport depletion; or differences in hydrostatic pressure, as in reverse osmosis and ultrafiltration. Several kinds of driving force may be, and usually are, operable simultaneously in any one process.

B. Ideal Minimum Energy

It is useful to discuss the ideal minimum energy needed to separate salt from water or vice versa for any given desalination process. It is known that the equilibrium vapor pressure of a salt solution is less than that of pure water under isothermal conditions. This is due to the fact that the activity of the water is lower in the solution than in pure water. Thus, if two large reservoirs, one containing the solution and the other the pure water, were connected and sealed from the outer environment, work energy would have to be supplied, say by a compressor, to prevent the movement of water vapor from the pure water reservoir to the solution reservoir. This energy is the ideal minimum energy and has been calculated at approximately 3 kWh/1000 gal of freshwater produced from seawater at standard temperature (Probstein, 1972). In practical terms, however, the actual energy consumed by an operating membrane process for desalting is usually several times higher than the ideal minimum energy. This is due both to the existence of energy losses or inefficiencies and to a finite flow rate and driving potential imposed on the system by economic requirements.

C. No Phase Change

One of the most important factors responsible for recent interest in membrane processes is the fact that they are able to separate dissolved species from one another or from a solvent with no phase change. Thus, the large energy requirements associated with the evaporation and crystallization processes are avoided in membrane processes. Since the energy costs are a major part of the operating costs, use of membrane processes may be an attractive alternative to vaporization or crystallization processes. In addition, many volatile low molecular weight organics may not be easily separated from water by vaporization due to the proximity of their boiling points.

D. Membranes

Although the membranes used for the pressure and electrically driven processes are functionally different, they do have similar features. They must separate two fluid-containing compartments without leakage and they must provide for differential transport rates through the membrane for different molecules, i.e., be permselective. Membranes of this kind can be visualized as consisting of many long-chain organic molecules randomly associated in a sort of spaghetti structure. The void spaces between the chains represent the interstitial volume in the membrane through which transferring species pass. Depending on the function and type of membrane, the long-chain polymers will (1) have long or short lengths, (2) be highly crystalline or cross-linked, (3) be homogeneous or heterogeneous, and (4) have neutral or highly charged functional groups (positive or negative) associated with (or grafted onto) the chains. Details describing reverse osmosis and ultrafiltration membranes will be presented later in Sections III,C and IV,B, respectively.

E. Transport Equations and Coefficients

Since we are interested in the relative motion of various components through a membrane, it would be convenient to be able to describe this motion quantitatively and thus be able to establish some basis for membrane performance. To attempt this, most researchers have invoked the theory of thermodynamics of irreversible processes (Prigogine, 1955; de Groot, 1959; de Groot and Mazur, 1962; Haase, 1969.) It is not our purpose here to develop this theory for membrane processes. Only the major results useful for our discussion will be presented, for further details, the reader will be referred to various references. Before proceeding, however, two additional points should be made. The first is that the theory of thermodynamics of irreversible processes is a phenomenological description of the relative motion of various components within the membrane, which is itself considered to be a "black box." This implies that the true microscopic mechanism of flow (and rejection) will not and cannot be explained by this theory. To the extent that this theory is combined with some "internal" membrane model, such as the solution-diffusion model in reverse osmosis, a mechanism can be inferred. The second point is that the theory of thermodynamics of irreversible process has been applied mainly to the pressure-driven as opposed to the other membrane processes.

Based on the theory of thermodynamics of irreversible processes, several approaches have been used to develop the basic transport equation, which relates the fluxes of solvents and solutes (J_i) with their respective driving forces (X_i) (Staverman, 1951; Spiegler and Kedem, 1966). These equations describe a coupling phenomenon that occurs between species when moving through the membrane. In general, for small deviations from equilibrium, one can write the following linear flux equations:

$$J_i = \sum_{j=1}^{m} L_{ij}X_j \qquad (i = 1,2,\ldots,m) \tag{1}$$

Onsager has shown theoretically, and others have verified experimentally, that the following symmetry exists for the phenomenological coefficients:

$$L_{ij} = L_{ji} \qquad (i,j = 1,2,\ldots,m) \tag{2}$$

Other restrictions on the coefficients are also operable and are due to entropy considerations. They include

$$L_{ii} > 0 \qquad (i = 1,2,\ldots,m) \tag{3}$$

and

$$L_{ii}L_{jj} - L_{ij}^2 > 0 \qquad (i,j = 1,2,\ldots,m,\ i \neq j) \tag{4}$$

The approach presented below uses the methods described by Haase (1969) to obtain the generalized equations for the isothermal heterogeneous (discontinuous) membrane system. Thereafter, a specific case, such as for the reverse osmosis process, is examined. Thus, we will merely define the system and present the results obtained using the procedure described below.

Here, we consider two liquid subsystems separated from each other by a semipermeable membrane. Let the two homogeneous subsystems of our heterogeneous system (see Table III) be designated as Phase ′ and Phase ″. According to this, we can attach a definite value for the pressure (P' or P''), for composition (molar concentrations C_k' or C_k''), and for the electrical potential (ϕ' or ϕ'') to each phase at constant temperature and at any arbitrary instant.

After performing a mass, energy, and entropy balance across the membrane, an explicit expression of the dissipation function (of entropy) is derived. Then the fluxes (J_i) and forces (X_i) acting on the membrane system are chosen (by

Table III
Heterogeneous (Discontinuous) System Consisting of Two Homogeneous Isotropic Subsystems (Phase′ and Phase ″)

Phase ′	Phase ″
Pressure P'	Pressure P''
Composition variable C_k'	Composition variable C_k''
Electrical potential ϕ'	Electrical potential ϕ''

observation) so that all these quantities are independent and disappear at equilibrium.

For this system, without external forces, the following kinds of forces result at constant temperature (Haase, 1969):

$$X_i = \text{grad}\ \mu_i = v_i\ \text{grad P} + \left(\frac{\partial \mu_i}{\partial C_j}\right)_{T,P} \text{grad}\ C_j + z_i F\ \text{grad}\ \phi \tag{5}$$

where v_i, partial molar volume of species i; μ_i, chemical potential of species i; z_i, electric charge on species i; F, Faraday's constant, grad E, refers to the gradient of a function E between phase $''$ and phase $'$.

Equation (5), together with some approximations, will be used later to develop the practical transport equations for the reverse osmosis process.

F. Fluid Mechanics

It is well to emphasize the central role played by the movement of bulk fluid over the surface of the membranes in all membrane processes. Mass and viscous boundary layers are either growing or are present at steady-state thickness in all these membrane processes. The mass boundary layer is due to the relative motion of various components through the membrane. In reverse osmosis and ultrafiltration, for example, since the water is forced through the membranes at a much higher rate than the solute molecules, a buildup of the solute species is observed at the solution-membrane interface. This phenomenon, known as concentration polarization, is schematically portrayed in Fig. 3. The viscous boundary layer, on the other hand, is a function of the water removed and the gross fluid mechanics of the system. For fully developed flow, the viscous boundary layer thickness, δ, is equal to half the channel

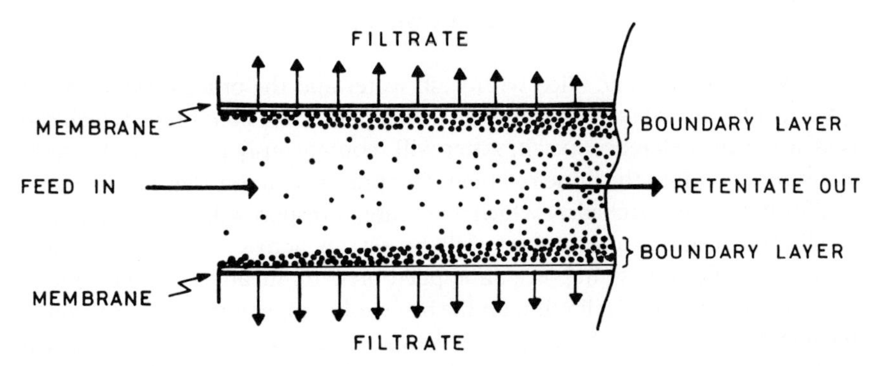

Fig. 3. Development of the mass or polarized boundary layer known as concentration polarization.

width or the radius of the tube. The importance of the hydrodynamic condition of the brine stream expresses itself through the shear force that is exerted at the membrane-solution interface. The higher this shear force, the easier it is for the solute molecules concentrated there to diffuse back into the bulk solution. It also plays a role in scouring the adsorbed particles which are responsible for membrane fouling (Thomas *et al.*, 1973; Copas and Middelman, 1973). Recently, this author has initiated a research study to investigate the relationship between the dissolved and nondissolved solute concentration profiles, especially at the membrane-solution interface (Belfort *et al.*, 1976).

III. REVERSE OSMOSIS

A. Introduction

Although it has been known for some time that, using pressure as a driving force, synthetic membranes such as those made from cellulosic derivatives are able to pass water in preference to salt, it was only in the early sixties that a practical membrane with reasonably high water fluxes and excellent salt rejection was developed (Loeb and Sourirajan, 1962). This was the major technological breakthrough that established reverse osmosis as a viable, attractive process with a great many potential applications. Details describing the historical and theoretical development and the engineering applications of this process are available in several texts (Merten, 1966; Sourirajan, 1970; Dresner and Johnson, 1975, Harris *et al.*, 1976).

B. The Concept

If two reservoirs, one filled with fresh water and the other filled with salt water (called brine), are separated by a semipermeable membrane which can pass water in preference to salt, water will spontaneously pass from the freshwater reservoir into the brine reservoir. This process will continue until osmotic equilibrium is approached, at which time water transport will terminate and an equilibrium osmotic pressure π° will be set up across the semipermeable membrane. If one now imposes an applied pressure greater than the osmotic pressure at osmotic equilibrium on the brine solution, water will pass through the membrane from the brine to the freshwater reservoir. This process is called reverse osmosis (or hyperfiltration). This discussion is schematically presented in Fig. 4.

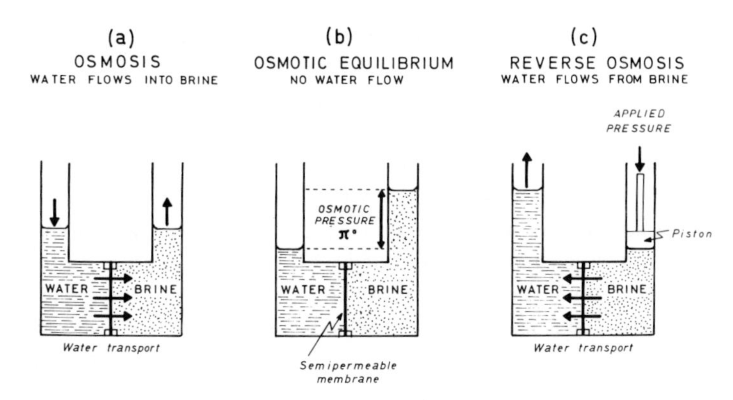

Fig. 4. Principles of normal and reverse osmosis.

Thus, to attain reasonable water fluxes [$J \sim 10$ gal/(ft^2, day)], the brine solution has to be pressurized well above the equilibrium osmotic pressure as estimated from the brine concentration. In practice, reverse osmosis systems are usually operated from about 4 to 20 times the equilibrium osmotic pressure $\pi°$, for sea and brackish water, respectively. For seawater, the operating pressure is of the order of 100 atm, while for brackish water and wastewater it is about 40 atm. Typical osmotic pressures can be obtained for dilute solutions from Van't Hoff's equation:

$$\pi° = \frac{n}{v_m} RT \tag{6}$$

where n is the number of moles of solute, v_m the molar volume of water, R the universal gas constant, and T the absolute temperature. For more concentrated solutions, the osmotic pressure coefficient ϕ is used to modify Eq. (6):

$$\pi° = \phi \frac{n}{v_m} RT \tag{7}$$

The osmotic pressure coefficients of many pure solutions are tabulated and available in the literature (Sourirajan, 1970). A useful rule of thumb for estimating the osmotic pressure of a natural water is 1 psi (pounds per square inch)/(100 mg/liter). Thus, for municipal wastewater containing about 1000 mg/liter dissolved salts, if we ignore for the moment the osmotic pressure exerted by the dissolved organics, we can estimate an approximate osmotic pressure of 10 psi and a corresponding applied pressure (multiply by 20) of about 200 psi (or 13.6 atm). Later, in Section III,G, we shall see that experimental reverse osmosis wastewater systems have been operated at even higher pressures than this, from 400 to 600 psi (27 to 41 atm).

C. Reverse Osmosis Membranes

A practical reverse osmosis membrane for water applications should possess several characteristics. First and foremost, it should be permeable to water in preference to all other components in the feed solution. Second, the rate of permeation of water per unit surface area (water flux) must be high enough to produce reasonable product volumes per unit time. A reasonable product volume per unit time can only be defined with respect to an economic analysis of the process. Third, the membrane must be durable, both physically and chemically, and have a reasonably extended life. Lifetimes for commercial reverse osmosis membranes using a brackish water feed are of the order of 1 to 3 years. Wastewater feeds with a high level of dissolved organics may naturally reduce the life of membranes. This will be discussed in more detail later, in Section III,H. Fourth, the membrane must be able to withstand substantial pressure gradients, either on its own or together with some porous backing or support material. Finally, the membrane should be easily cast into the configuration needed for use.

Since the development of a method for casting membrane films with high water fluxes and excellent salt rejections in the early sixties by Loeb and Sourirajan (1962), the "anisotropic" cellulose acetate membrane has been considered the leading commercial membrane. This "anisotropic" membrane has about 2.5 acetyl groups per monomer molecule and consists of a very thin dense skin, 0.15 to 0.25 μm in thickness, on top of a highly porous ($> 50\%$ void space), thick (> 100 μm) substructure. The asymmetric nature of a flat cellulose acetate membrane is illustrated by the electromicrograph in Fig. 5a. The desalination properties of the membrane are determined solely by the characteristics of the thin dense skin. Essentially, all the salt rejection takes place at the skin, while water flux rates through the membrane are limited by the permeability of the skin. The properties of the skin can be varied during the membrane preparation procedure, which has as its last phase an annealing step in water from 70–90°C. By accurately controlling the temperature during annealing, a range of different salt-rejecting membranes can be produced from a loose (70°C) to a very tight (90°C) membrane.* The water permeability of the membrane is inversely (although not linearly) related to the salt rejection. At present, typical average water fluxes are about 2.5 gal/(ft^2, day) per 100 psi of applied pressure for salt rejections of greater than 95%. Recent work has shown that the thin dense film (or skin) and the porous substructure need not be made from one material and that, by optimizing each layer and sandwiching them together, a membrane with superior performance characteristics can be

*In contrast to the Loeb and Sourirajan (1962) *wet* method for casting asymmetric reverse osmosis membranes, a *dry* method has recently been developed by Kesting (1973).

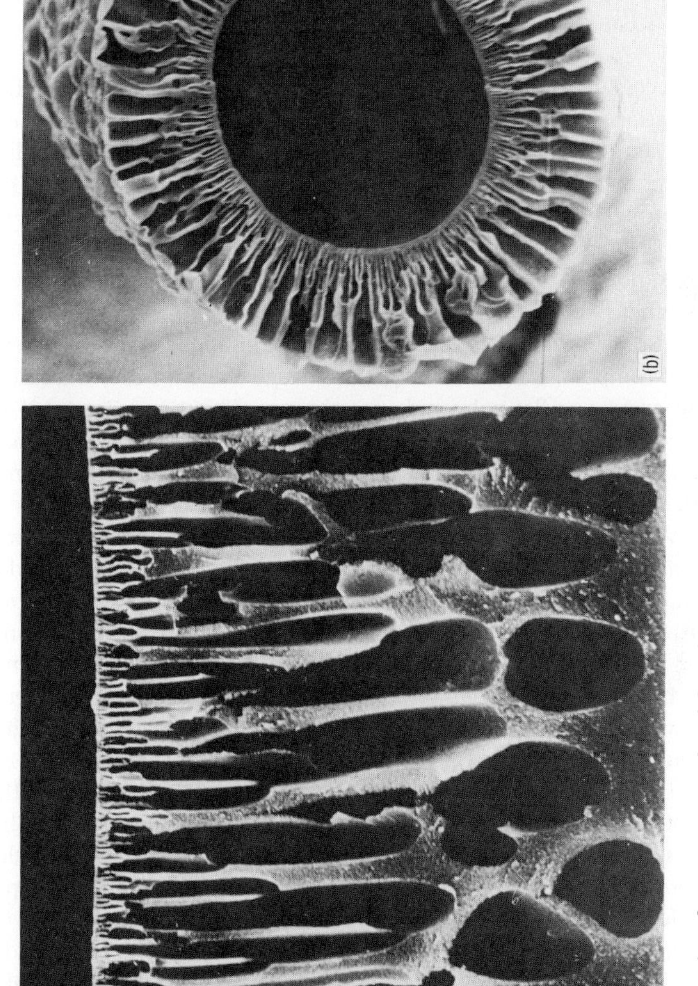

Fig. 5. Scanning electron micrographs of the substructure of asymmetric cellulose acetate membranes showing the dense skin and porous substructure. (a) Flat membrane (total thickness ~ 100 μm) (b) Hollow fiber membrane (inside diameter ~ 500 μm). (Courtesy of H. Strathmann, Forschungsinstitute, Berghof Gmb, West Germany.)

made (Riley *et al.*, 1971; Rozelle *et al.*, 1973). Thus, the permeability presumably can be increased without a decrease in rejection by reducing the dense film thickness. Of course, with very thin fllms (~ 1000 Å), mechanical strength of the membrane may suffer. These composite membranes are able to desalinate seawater (salt rejection ~ 99.5%) with excellent water fluxes of 25 gal/(ft^2, day) at 1500 psi applied pressure. Cellulose acetate membranes are commercially available in several configurations, namely as a tube with the dense film on the inside, as a flat sheet, or as a fiber with the dense film either on the inside or outside.

Another successful asymmetric membrane, developed mainly for the hollow fiber configuration, has been made from aromatic polyamide formulations or Nomex. See Fig. 5b for a scanning electron micrograph (SEM) of a hollow fiber membrane with the skin on the inside surface. Although the water permeabilities through these membranes are about an order of magnitude less than through the cellulose acetate membranes, their packing density (square feet of membrane surface per unit volume of permeator) is about an order of magnitude higher.

In a recent study, a new nonpolysaccharide membrane was shown to reject several model compounds found in a typical sewage effluent better than either the cellulose acetate or flat polyamide membranes (Chian and Fang, 1973).* This new membrane (referred to as "NS-1" by Rozelle *et al.*, 1973) consists of a polysulfone support film with a 6000-Å-thick coating of polyethylanimine reacted with diisocyanate.

In Table IV, we present a comparison of the cellulose acetate, aromatic polyamide, and NS-1 membranes. The rejection of salt is slightly better for cellulose acetate than for the aromatic polyamide and NS-1 membranes. Aside from degradation due to free chlorine, the latter membranes are chemically inert in water and are also less liable to biodegradation. This latter point is extremely important when treating wastewaters. Note also that some materials frequently present in wastewaters are poorly rejected by all three membranes. These include phenols, aldehydes, urea, methanol, and methyl acetate (Chian and Fang, 1973, Duvel and Helfgott, 1975). As will be seen later in Section III,H, an important part of reverse osmosis operation is membrane cleaning. Thus, the membranes must possess high stability and not degrade with time. Cellulose acetate is known to hydrolyze slowly with time. The rate of hydrolysis is dependent on the feed constituents and pH (Sachs and Zisner, 1972).

A wide variety of other membranes have also been developed (Lonsdale and Podall, 1972). These include the cellulose acetate-butyrate and cellulose acetate-methacrylate blends. These membranes have slightly different and,

*The model compounds were from the following groups: aromatics, acids, alcohols, aldehydes, ketones, amines, ethers, and esters.

Table IV
Comparison of Membrane Characteristics

Membrane material	pH range (continuous exposure)	Temperature limits (°F)	Operating pressure ψ	Materials causing membrane dissolution	Materials poorly rejected	Salt rejection (%)	Flux[a] gal/(feet^2, day)
Cellulose acetate (as tubes or flat sheets)	3–8	65–85	Up to 1500	Strong oxidizing agents, solvents, bacteria	Boric acid, phenols, de-detergents (ABS), carbon chloroform extract, ammonia, urea, methyl acetate	95–99.5	8.0
Aromatic polyamides (as fibers or flat sheets)	4–11	32–95	400 (fiber)	Strong oxidizing agents particularly free Cl_2	Aldehydes, phenols, methanol, methyl acetate	85–95	2.5
NS – 1 (as flat sheets)	1–13	—	Up to 1500	—	Aldehydes, esters, methyl acetate	>90	12.0

[a]Fluxes were obtained using a laboratory-made model sewage (Chian and Fang, 1973).

for some uses, more desirable properties than the standard asymmetric cellulose acetate membranes, but their overall performance has not yet warranted their replacing cellulose acetate membranes.

Another interesting recent development, especially with a view to wastewater applications, is the ability of certain materials, when deposited dynamically on a porous substructure or support, to reject salt at very high-water permeabilities (Johnson, 1972). A wide variety of additives have been found to form salt-rejecting dynamic membranes. These include materials such as humic and fulvic acid, Zr(IV) oxide and tannins, and polyelectrolytes (Sephadex CM-C-25) (Kraus, 1970). Fluxes of the order of 100 gal/ft^2/day with salt rejections greater than 50% have been reported for these dynamic membranes (Thomas *et al.*, 1973). It should be remembered that these low salt rejections may not be a disadvantage in municipal wastewater treatment since secondary effluents are usually in the range from 750 to 1500 ppm TDS (Sachs and Zisner, 1972). The two major disadvantages of dynamic membranes are their inherent instability and, because of the high permeation fluxes, the need for very high pumping rates of the feed stream to reduce concentration polarization or salt buildup at the membrane-solution interface.

In conclusion, commercially available membranes, such as the asymmetric cellulose acetate and the aromatic polyamide hollow fibers, are not perfectly suited to wastewater applications. The cellulose acetate membrane has problems with leakage of certain feed constituents into the product and is susceptible to biological attack. The aromatic polyamide hollow fiber membranes are both chemically and biologically stable toward typical municipal wastewaters, (except for highly chlorinated water), but because of their fiber design, the poor hydrodynamic condition of the feed solution results in a high potential for plugging and membrane fouling. This will be discussed in detail later. NS-1 (along with the newer NS-100, Fang and Chain, 1975) and dynamic polyelectrolyte membranes have yet to prove themselves in pilot plant operation for any considerable time, although all seem to be very promising. See Lonsdale (1974) for a review of recent reverse osmosis membrane advances.

D. Theoretical Considerations

There exist several theories whose main purpose is to describe and predict transport behavior of reverse osmosis membranes. The viscous flow theory assumes that all flow occurs through pores in the membrane, flow rate and permselectivity being governed by porosity, pore size distribution, and interactions (chemical and electrical) with the surface of the pores. This approach is widely used to describe transport through ultrafiltration membranes (Michaels, 1968), although it has also been invoked for special kinds of reverse osmosis membranes such as the porous glass system (Belfort, 1972).

With respect to the asymmetric Loeb-Sourirajan type membrane, the solution-diffusion theory has been successfully used (Merten, 1966). This theory explains the rejection phenomenon in terms of two steps. In the first step, the salt and water dissolve in the membrane film, while in the second step each molecular species is thought to move through the membrane by independent diffusion.

Returning now to the theory of thermodynamics of irreversible processes and neglecting the last term (grad $\phi = 0$) for the reverse osmosis process, Eq. (5) must be integrated across the thickness of the membrane. Using subscript 1 to designate the solvent (water) and subscript 2 to designate the solute in a two-component system, we get for the solvent

$$\Delta\mu_1 = \int \left(\frac{\partial\mu_1}{\partial C_1}\right)_{P,T} dC_1 + \int v_1 dP = \int \left(\frac{\partial\mu_1}{\partial C_2}\right)_{P,T} dC_2 + \int v_1 dP \tag{8}$$

We know that when $\Delta\mu_1 = 0$, we are left with the osmotic pressure difference $\Delta\pi$. Thus, for constant v_1,

$$v_1 \Delta\pi = -\int \left(\frac{\partial\mu_1}{\partial C_2}\right)_{P,T} dC_2 \tag{9}$$

and

$$\Delta\mu_1 = v_1(\Delta P - \Delta\pi) \tag{10}$$

For the solute,

$$\Delta\mu_2 = \int \left(\frac{\partial\mu_2}{\partial C_2}\right)_{P,T} dC_2 + \int v_2 dP \tag{11}$$

and for dilute solutions ($\mu_2 = \mu_2^\circ + RT \ln C_2$), and constant v_2, we get

$$\Delta\mu_2 = RT\Delta\ln C_2 + v_2\Delta P \tag{12}$$

where the second term on the right-hand side of Eq. (12) is negligible with respect to the first for the reverse osmosis process (Merten, 1966). Thus, we get

$$\Delta\mu_2 = RT\Delta\ln C_2 \approx \frac{RT}{C_2}\Delta C_2, \tag{13}$$

and can incorporate $\Delta\mu_i$ (or X_i) into Eq. (1) by neglecting the cross-coefficients which are fairly small for the asymmetric cellulose acetate membrane (Bennion and Rhee 1969):

for water

$$J_1 = K_1(\Delta P - \Delta\pi) \tag{14}$$

and for salt

$$J_2 = K_3\Delta C_2 = K_3 C'_2 R \tag{15}$$

where K_1 and K_3 are the water and salt permeability coefficients and related to the phenomenological coefficients, ΔC_2 is the difference in salt concentration between bulk streams of product and feed, C'_2 is the salt concentration of the bulk feed stream, and R is the coefficient of salt rejection defined by

$$R = 1 - C''_2/C'_2 \tag{16}$$

To achieve greater accuracy, the concentrations at the membrane surface interfaces (C'_{2m} and C''_{2m}) should be used in Eqs. (15) and (16) instead of the bulk stream concentrations. Bulk stream values are, however, easier to measure and are usually used.

Thus, using equilibrium and irreversible thermodynamics, and ignoring coupled flows, we have derived the flux equations. The resultant equations, assuming constant coefficients [Eq. (14) and Eq. (15)], are formulations of Ficks law of diffusion. Thus, K_1 has been described in terms of a diffusion coefficient, water concentration, partial molar volume of water, absolute temperature, and effective membrane thickness. K_3 has been described in terms of a diffusion coefficient, distribution coefficient, and effective membrane thickness (Merten, 1966).

E. Plant Equipment

A pressure-driven membrane separation plant consists of components shown in the simplified flow diagram in Fig. 6. Typically, the feed solution is first filtered and the pH adjusted to between 5 and 6 for the asymmetric cellulose acetate membranes. The feed is then pressurized (to say 600 psi for 3000 ppm TDS) and passed through the various membrane permeators or modules. Since the volume of the feed solution decreases with path length, fewer modules are needed for each successive stage. This can be seen in Fig. 6. The permeate of each stage is combined and stored for use. A back pressure regulator is used to reduce the pressure of the brine stream after the last stage. It has been suggested that this waste pressure be used to drive a turbine to produce electricity which could be used to run the motors of the pumps. Figure 6 is a once-through flow scheme. Many other alternatives are also used, including partial or total brine recycle to improve the system's recovery ratio. The recovery ratio is defined as the total volume of product obtained divided by the initial volume of feed. The recovery ratio is an economically significant performance parameter often neglected or minimized by manufacturers and researchers.

The filters, tanks, pumps, back pressure regulators, dampeners, and piping employed in these plants are conventional items commonly available to the chemical processing industry. The only special component is the membrane permeator or reverse osmosis module. A photograph of a commercial plant of the spiral wrap design is presented in Fig. 7.

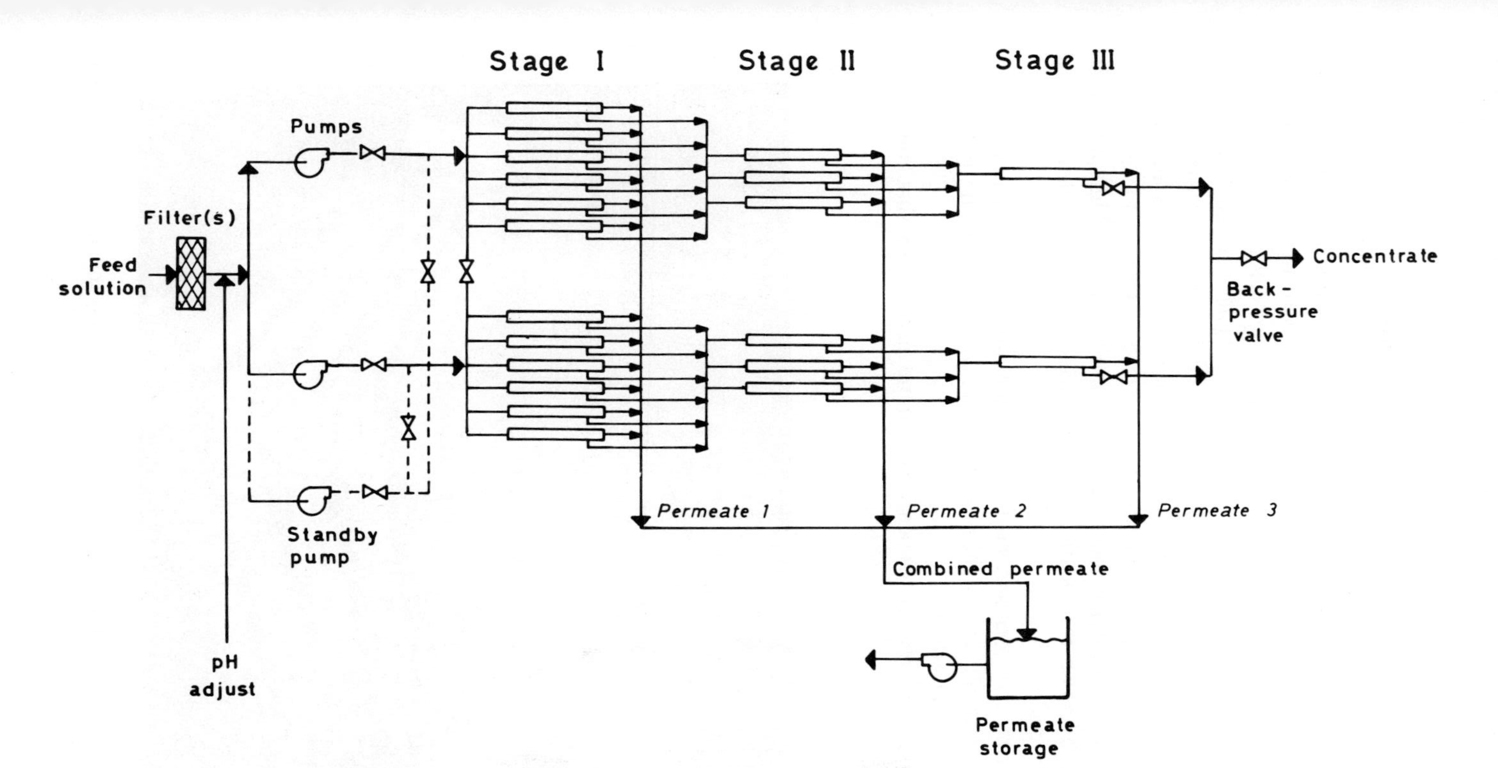

Fig. 6. Flow diagram of a pressure-driven membrane separator plant. (After Lacey, 1972a.)

F. Membrane Permeators

The main requirement of a reverse osmosis membrane permeator is that it house the membranes in such a way that the feed stream is sealed from the product stream. All other requirements are concerned with

a. *mechanical stability*, such as supporting a fragile membrane under very high differential pressures (200–1500 psi), preventing pressure leaks between the feed and product streams and between the feed stream and its surroundings (air), and avoiding large pressure drops in the feed and product streams,

b. *hydrodynamic considerations*, such as minimizing the buildup of both salt and fouling layers on the membrane surface, which might impede membrane performance, and

c. *economic considerations*, such as obtaining high membrane-packing density to reduce capital costs on the pressure vessels and designing the unit for ease of membrane replacement.

Several types of reverse osmosis membrane permeators that meet these

Fig. 7. Commercial spiral-wrap reverse osmosis plant. (Courtesy of R. L. Riley Fluid Systems Div., UOP. San Diego, California.)

requirements are now commercially available (Golomb and Besik, 1970). Based on the geometry of the membrane, they can be classified into five broad design categories: tubular, spiral wrap, hollow fiber, flat plate, and dynamic. Subclasses within each category are described in Table V, together with the manufacturer's address for each item.

A sketch of each of the major commercial reverse osmosis membrane permeators is shown in Figs. 8 and 9. Several performance and structural characteristics for the different permeators are also presented in Table VI. A brief explanation of the module designs is presented below with reference to Table VI.

The first thing to notice in Table VI is that the permeator with the lowest water output per unit volume (tubular with inside flow) is most easily cleaned, while the permeator with the highest water output per unit volume (fibers with brine flow on the outside) is the most difficult to clean. We are especially con-

Table V
Reverse Osmosis Membrane Permeators

Class	Designation	Description of available designs	Manufacturer[a]
1. Tubular	1a	Brine flow inside straight rigid support tube	A,B,C,D,E,F
	1b	Brine flow inside helical support tube	G
	1c	Brine flow inside straight squashed tube	E
	1d	Brine flow outside straight rigid support tube	H
	1e	Brine flow outside flexible rigid support tube	C
2. Spiral wrap	2a	Brine flow between alternate leaves of a spiral wrap	I,B
3. Fiber	3a	Brine flow outside flexible hollow fiber membranes	J,K
	3b	Brine flow inside flexible hollow fiber membranes	L
4. Flat plate	4a	Vertical filterpress design with brine flow radially between leaves	M
	4b	Same as 4a with the whole unit spining	E,N
5. Dynamic membrane	5a	A dynamic precoat membrane is laid down on a porous support	O

[a]The following letters are used to designate major manufacturers: A = Israel Desalination Engineering, Tel Baruch, Israel; B = Envirogenics Co., 9200 East Flair Dr., El Monte, Calif. 91734; C = Paterson Candy Int. Ltd., Laverstoke Mills, Whitchurch, Hamps., U.K.; D = Westinghouse Electric Corp., Beulah Rd., Churchill Boro, Pittsburgh, Penn. 15235; E = Kalle, D-6202 Wiesbaden-Biebrich, West Germany; F = Calgon Havens Systems, 8133 Aero Dr., San Diego, Calif. 92123; G = Philco-Ford, Fluid Processing, Ford. Rd., Newport Beach, Calif. 92660; H = Raypak, Inc., 31111 Agoura Rd., Westlake Village, Calif. 91361; I = Fluid Systems Div., UOP, 2980 N. Harbor Dr., San Diego, Calif. 92101; J = E.I. DuPont De Nemours & Co., Wilmington, Del. 19898; K = Dow Chemical Co., 2800 Mitchell Dr., Walnut Creek, Calif. 94598; L = Berghoff GmbH, 74 Tubingen, Luftnau, Berghoff, West Germany; M = De Danske Sukkerfabrikker, Langebrogade-5, DK-1001, Copenhagen, Denmark; N = Environmental Technology, Dresser, 1702 McGraw Rd., Santa Ana, Calif. 92705; O = Oak Ridge National Laboratory, Oak Ridge, Tenn. 37830.

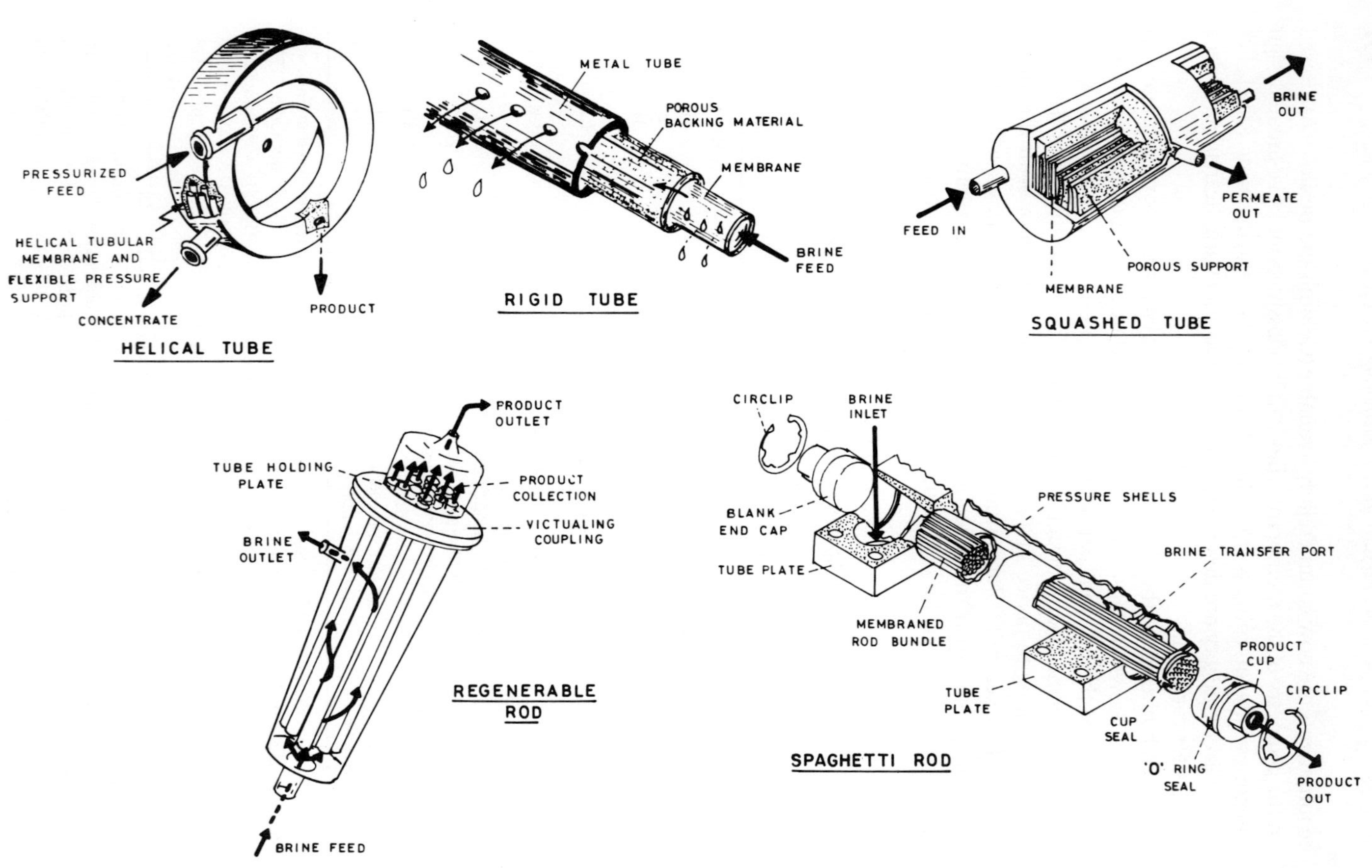

Fig. 8. Reverse osmosis membrane permeator designs.

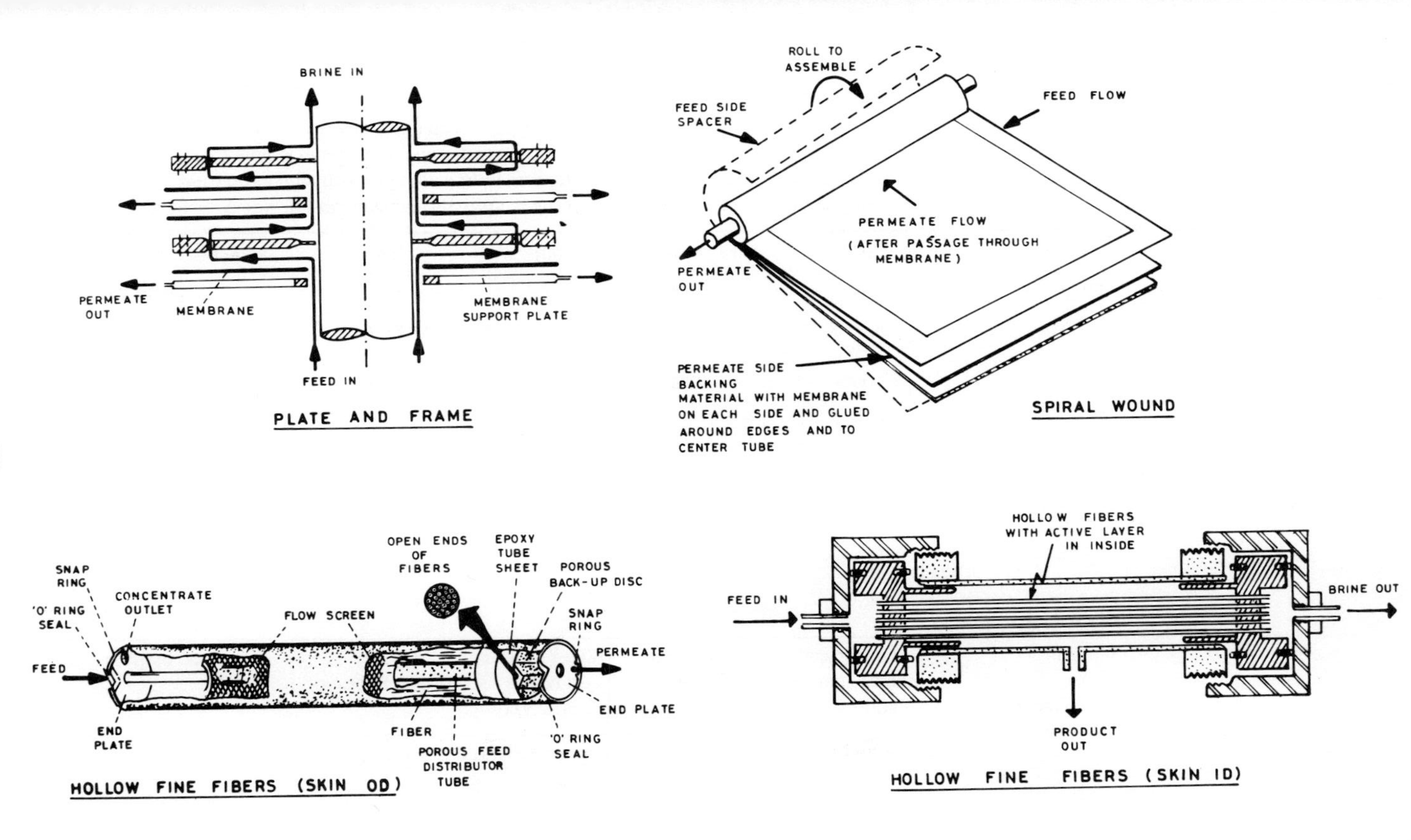

Fig. 9. Reverse osmosis membrane permeator designs.

Table VI
Comparison of Reverse Osmosis Membrane Permeators

Module design	Packing density (ft^2/ft^3)	Water flux at 600 psi gal/(ft^2, day)	Salt rejection	Water output per unit volume gal/(ft^3, day)	Flow channel size (in.)	Ease of cleaning
Tubular						
1a Brine flow inside tube	30–50	10	Good	300–500	0.5–1.0	Very good
1b Brine flow outside tube[a]	140	10	Good	1400	0.0–0.125[a]	Good
Spiral wrap	250	10	Good	2500	0.1	Fair
Fiber						
3a Brine flow inside fiber[b]	1000	5	Fair	5000	0.254	Fair
3b Brine flow outside fiber	5000–2500	1–3	Fair	5000–7500	0.002	Poor
Flat plate[c]	35	10	Good	350	0.01–0.02	Good
Dynamic membrane[d]	50	100	Poor	5000	~0.25	Good

[a]Data for spaghetti permeator obtained from Grover *et al.* (1973). The flow channel dimension can vary from zero (tubes touching) to about 0.125 in.
[b]Data for fiber with brine flow inside obtained from Strathman (1973). Maximum internal pressure for this unit is 28 atm (410 psi).
[c]Data for flat plate design obtained from Nielson (1972).
[d]Data for dynamic membrane design estimated from Thomas *et al.* (1973).

cerned with ease of cleaning when we have a turbid feed such as a typical wastewater. Unfortunately, it is the permeators, which are as yet virtually untried commercially, that offer the best of both worlds; i.e., (1) the tubular design with brine flow on the outside, (2) the fiber design with brine flow on the inside, and (3) the dynamic membrane concept. The flow channel size, in the penultimate column of Table VI, is presented as a measure of the cross-sectional area available for brine flow. In a crude calculation, a large value in the flow channel size should indicate little chance of fluid holdup due to blockage from floatables or suspended solids. A small value should suggest that a high degree of prefiltering is necessary.

1. HELICAL AND RIGID TUBE PERMEATORS (Fig. 8)

The membrane is cast as a tube on the inside of a porous support tube (e.g., paper or cloth) which is then placed inside a pressure vessel. The brine stream flows through the tube, while the product permeates the membrane radially. The pressure vessel may be a steel pipe with perforated holes (rigid design) or, if the support tube can withstand the pressure differential, a plastic or low-pressure housing (helical design) can be used to collect the product. Rods are sometimes placed inside the tube, along the center line, to increase fluid velocity and axial shear at the membrane-solution interface. Reynold's numbers as high as 20,000 have been used for this design. Tubular units are easily cleaned and much operating data exists for them. Their disadvantages include low water production per unit volume, high water holdup per unit area of membrane, and relatively expensive membranes (ca. $10–20/ft^2$).

2. SPIRAL-WRAP PERMEATORS (Fig. 9)

Several flat or planar membranes are sandwiched between porous plastic screen supports and then rolled up into a "swiss-roll." The edges of the membranes are sealed to each other and to the central perforated tube. The resultant spiral-wrap module is fitted into a tubular steel pressure vessel, such as a 4-in. nominal pipe. The pressurized feed solution is fed into the pipe so that it flows through the plastic mesh screens along the surface of the membranes. The product, which permeates the membranes, flows into the closed alternate compartments and spirals radially toward the weep holes in the central tube, from where it is removed. Advantages of this design include fairly high water output per unit volume, low membrane costs (ca. $3/ft^2$), low holdup per unit membrane area, and vast amount of operating data. The spiral-wrap design has probably been exposed to more hours of municipal effluent than any other design, aside perhaps from the rigid tubular design. However, it can be cleaned only chemically, not mechanically like the tubular design, and, due to the small dimensions of the flow channel (see Table VI), it has a high probability of plugging. This design is one of the leading contenders for large-scale municipal treatment (Ajax International Corp., 1973).

3. FIBER PERMEATORS (Fig. 9)

Several million hollow fibers almost as fine as human hair (100 to 200μm outside diameter) are bundled together in either a U-shape configuration (for brine flow on the outside) or in a straight configuration (for brine flow on the inside). The end of the fibers are epoxied into a tube sheet while making sure each fiber is not blocked. Thus, for the case where the brine flows at high pressure on the outside of the hollow fibers, the product permeates radially inward through the unsupported fiber. The product then moves inside the hollow fiber bore to the product collection chamber. In the other case, where the design is similar to a typical heat exchanger, the brine flows into the bore of the hollow fibers at one end and, after moving along the inside of the fiber, flows out of the other end of the unit. The product continually permeates radially outward through the fiber walls.

The shell-side feed hollow-fiber design (i.e., brine on the outside) is very compact, low in cost, has a low water holdup and, due to the compressive strength of the small diameter fibers, can withstand fairly high differential pressures (400 psi). It unfortunately plugs easily and is very hard to clean.

The inside-feed hollow-fiber design has the advantages of the shell-side feed design plus the added advantage of well-controlled hydrodynamics of the feed, which improves the possibility of cleaning. Because of its newness, meager operating data are available.

4. PLATE-AND-FRAME PERMEATORS (Fig. 9)

The original "plate-and-frame" unit was similar in principle to the filter press (Aerojet General Corp., 1964, 1966). This design became defunct in the late sixties because of many problems, the most important of which was the extreme difficulty and high expense of changing degraded membranes. Recently, a new commercial unit using the flat plate design has been made commercially available (Nielson, 1972; Madsen *et al.*, 1973). The design is similar to a stack of phonograph records. Alternate circular membranes and separating frames, which are also used for manifolding and sealing, are placed around a hollow central spindle through which the brine is fed via radial ports. This design does not need a cylindrical pressure vessel since each membrane is individually sealed by its neighboring separating frame. This is one of the main advances over the original unwieldy plate-and-frame design. General advantages of this new design include the low brine holdup per unit membrane area and the amenability of the design to desalting highly viscous solutions due to the thin channel size (0.01 to 0.02 in.). Its disadvantages include susceptibility to fouling, difficulties in cleaning and membrane replacement, and scanty operating data.

The other designs shown in Figs. 8 and 9 and presented in Tables V and VI are either not commercially available, such as the regenerable and the dynamic membrane designs, or have not made any impact on the market as yet, such as the squashed membrane or the flat-plate spinning unit. The concept that a membrane can be regenerated *in situ* has been successfully proven and reported in the literature (Belfort *et al.*, 1973).

G. Applications of Reverse Osmosis to Wastewater Renovation

The principal use of reverse osmosis to date is the desalination of brackish waters containing less than 10,000 ppm total dissolved solids. As an outgrowth of this application, many other applications have been pursued. Recent developments have made one-pass desalination of seawater (35,000 ppm) to drinking water (500 ppm) a real possibility (Johnson and McCutchan, 1973; Riley *et al.*, 1973). Reverse osmosis together with the ion exchange process has been successfully used to produce ultrapure water for the electronics industry (Haight, 1971; deBussy and Whitmore, 1972; Riedinger and Nusbaum, 1972). Evidently, water from the combined process is less expensive than from either process separately. In addition, the treatment of feedwater for boiler units has been attempted by reverse osmosis (Riedinger and Nusbaum, 1972; Leitner, 1973). The application of reverse osmosis to industrial processes with a high-value product is being tested. Examples of this include food processing such as the concentration of whey (McDonough and Mattingly, 1970; Madsen *et al.*, 1973; Goldsmith *et al.*, 1974a), orange juice (Merson and Morgan, 1968), maple sap (Willits *et al.*, 1967), and coffee (Underwood and Willits, 1969).

The remainder of this section will be devoted to the application of reverse osmosis to wastewater renovation and will be divided into three areas: industrial applications, municipal applications, and the treatment of polluted rivers.

1. INDUSTRIAL APPLICATIONS

Here we will be concerned mainly with the concentration of a given wastewater stream and the simultaneous production of improved quality water for reuse (Okey, 1972). The concentrate stream, in the liquid phase, may contain valuable constituents that could also be recycled and reused in the process, or could contain sludge or brine to be disposed of. Not much attention has been given to disposal or its potential costs, although deep-well disposal, landfilling, evaporation methods, and trucking to the ocean have been suggested. Two relevant monographs describing various applications of membrane processes are available (Flynn, 1970; Lacey and Loeb, 1972).

a. Pulp and Paper Industry. Considerable research has been conducted to test the applicability of reverse osmosis to concentrating dilute pulp wastewaters (Wiley *et al.*, 1972a, Anonymous, 1975). The idea is to concentrate these dilute streams (0.5 to 1% dissolved solids) to about 10% solids using reverse osmosis and then to further concentrate the solids, to say 50%, by evaporation, with eventual by-product recovery or disposal (Leitner, 1972). The overall objective is also to reduce direct discharge into rivers or lakes and to ensure that the biological oxygen demand (BOD), total dissolved solids (TDS), odor and color are "acceptable" to the receiving waters. The definition of acceptability is only now being defined by governmental authorities.

Average flux rates, using a rigid tubular reverse osmosis system, of about 2.5 to 5 gal/(ft^2, day), with 80% water recovery, and rejections of over 95%, were reported for a wastewater effluent from a high yield, sodium sulfite, chemi-mechanical pulping process (Wiley *et al.*, 1972b). Problems included high osmotic pressures because of the high salt (Na_2SO_4) content and membrane fouling due to hydration of polysaccharides. Operating costs for various capacity plants (125,000–1,000,000 gal/day) were estimated to vary from $0.82–1.48/1000 gal of product water. Based on these encouraging results, plans were made for a 300,000–400,000 gal/day reverse osmosis plant to be installed in Wisconsin.

Concentrates of high molecular weight lignosulfonates at 30% solids and reducing sugars at 20% total solids have been fractionated and concentrated from spent sulfite liquors using both ultrafiltration and reverse osmosis (Bansal and Wiley, 1975).

b. Mine-Drainage Pollution Control. Present methods for treating acid mine drainage, such as neutralization and aeration, are wasteful in their inability to produce reusable water for either industrial or municipal use. The residual contaminants are high in dissolved solids, hardness, and sulfate concentrations. The application of reverse osmosis to the removal of nearly all the dissolved solids and the production of a concentrate from which the valuable heavy metals can be recovered has been conducted and reviewed by Wilmoth and Hill (1972). The spiral-wrap, hollow fine fiber, and tubular configurations have been tested and compared. Table VII summarizes typical rejection by these configurations. Unfortunately, because of the maximum allowable standards on iron and manganese concentrations of 0.3 and 0.05 mg/liter, respectively, the product water in Table VII is not potable. Field tests have disclosed that the

Table VII

Typical Rejections by Reverse Osmosis Systems on Mine Drainage Effluents[a,b]

Systems	pH	Cond.[c]	Acidity	Ca	Mg	Total Fe	Fe II	Al	SO_4	Mn
Spiral wrap[d]										
Feed water	3.1	2070	460	260	170	77	64	12	1340	43
Product	4.4	17	38	0.4	0.3	0.4	0.3	0.2	0.9	0.5
Rejection (%)[e]	—	99.2	91.7	99.8	99.8	99.8	99.8	99.2	99.9	98.8
Tubular[d]										
Feed water	3.4	1050	250	125	92	78	61	12	660	14
Product	4.2	46	46	2.2	1.4	0.9	1.0	1.0	4.4	0.3
Rejection (%)[e]	—	95.6	81.6	98.2	98.5	98.8	98.4	91.7	99.3	97.8
Hollow fiber[d]										
Feed water	3.4	1020	210	150	115	110	71	15	940	14
Product	4.3	32	32	1.2	1.4	1.2	0.8	0.8	4.6	0.1
Rejection (%)[e]	—	96.9	84.8	99.2	98.8	98.9	98.9	94.7	99.5	99.1

[a]Source: Wilmoth and Hill, 1972.
[b]All units are mg/liter except pH and conductivity.
[c]Cond.—Specific conductance (microohms/cm).
[d]Recovery, 75%.
[e]Rejection = 100 (feed concentration − product concentration)/feed concentration.

two major causes of chemical fouling of reverse osmosis membranes are iron and calcium sulfate (Sleigh and Kremen, 1971; Mason and Gupta, 1972). Bacteria have been shown to concentrate on the membrane and to oxidize the ferrous ion and, thus, foul the membrane. Here, the limiting factor in high recovery reverse osmosis operation was calcium sulfate precipitation. Various flushing techniques were used to remove the fouling layers. These included high velocity flushes to dislodge the precipitate, acidified flush to render the salts soluble, and a BIZ flush (enzyme active laundry presoak) to remove the organics. Even ion exchange of the cations has been suggested as a pretreatment to reduce precipitation and allow for higher water recoveries (Pappano *et al.*, 1975).

The longest period of operation reported for mine-drainage feed waters on a single set of membranes was on the spiral modules that accumulated 4400 hours on approximately a −0.015 log-log slope* (Wilmoth and Hill, 1972). The longest period of operation on the hollow fiber modules was 2670 hours on a −0.037 log-log slope. For the tubular unit, the longest successful run was 807 hours of operation on a −0.04 slope. No loss of solute rejection was noted in any of these long-term tests. Current estimates of cost per 1000 gallons of product vary between $0.35 and $1.50, depending on such variables as plant size, water quality, and disposal techniques.

c. Plating and Metal Finishing Operations. Here, once again, we have the possibility of using reverse osmosis to treat rinse waters and wastes from plating and metal-finishing operations (Rozelle, 1971, Anonymous, 1975). The aim is to recover reusable water and to salvage valuable metals. Intensive laboratory-scale work has been conducted in Canada on electroplating effluents, especially on nickel-plating wastes (Golomb, 1972). Based on a small capacity plant of 2400 gal/day, processing costs for a nickel-plating rinse of about $3.00/1000 gal have been estimated (Leitner, 1973).

For treatment of cyanide bath wastes, which are generally at pH levels higher than 8, membranes other than cellulose acetate would have to be considered. (Allegrezza *et al.*, 1975). In addition, rejections for the cyanide ion from cyanide or cyanide-complex solutions have been very poor for all membranes tried (Hauk and Sourirajan, 1972; Leitner, 1973).

Hexavalent chromate rejection, on the other hand, has been found to be highly dependent on pH. At pH $\sim$ 7, the chromate rejection is greater than 99%, while for pH $\sim$ 2.5, the chromate rejection is less than 94% (Riedinger and Nusbaum, 1972). Increasing the pH may cause other compounds such as

*If one plots the log of the permeate flux versus the log of time, after a short initial nonlinear period, a linear negatively sloping plot is usually observed. This log-log slope designated as b has been used as an indicator of the severity of membrane fouling and a rough predictor of membrane life.

aluminum hydrate (from aluminum anodyzing) to precipitate out (at pH $\sim$ 4), causing reverse osmosis membrane-fouling problems.

d. Photographic Processing Industries. Chemical recovery and rinse water renovation have been studied using the tubular reverse osmosis process in photographic processing industries. One study reports that 80 to 90% of the final washwater can be renovated for reuse (Cohen, 1972). In addition, ferro- and ferricyanides can be concentrated and reused as a bleach replenisher and electrolytic recovery of the concentrated silver is possible (Mahoney *et al.*, 1970, reported by Cohen, 1972).

e. Food-Processing Plant Effluent. In the United States, two laws governing the operation of food-processing plants present obstacles to the use of the reverse osmosis process. The first prohibits the reuse of water for food manufacturing, although some secondary applications may be considered, such as boiler feed and cleanup. The second requires sterilization temperatures that cellulose acetate membranes (and probably most others) are not able to withstand.

In spite of these restrictions, reverse osmosis, electrodialysis, and ultrafiltration have all been applied to food industry wastes. Ultrafiltration and reverse osmosis have been used in series to concentrate whey from a dairy plant resulting in two saleable by-products, protein concentrate and lactose (Horton *et al.*, 1972). Pilot plant testing in a dairy research center, using a tubular unit, has also been conducted in Ireland (reported by Cohen, 1972). The new flat-plate thin-channel configuration has also been tested on a whey effluent from a dairy plant (Nielson, 1972; Madsen *et al.*, 1973). They point out the advantages of their thin channel design, emphasizing the hydrodynamic advantage over other units for highly viscous solutions.

Tubular modules have also been used to concentrate the effluents from a potato-starch factory (Porter *et al.*, 1970). The permeate is good for reuse, while the concentrate with some further treatment can produce soluble proteins and free amino acids. Reverse osmosis costs, however, will restrict the commercial feasibility of such treatment until low-cost membranes become available.

Because of recent legal pressure on industries to prevent their effluents from polluting the natural surface and groundwaters, the financing arrangements of municipalities (Section I,C), and, further, the disparity between municipal volume needs and the present capacities of membrane technology, we expect to see industry lead the way in using membrane processes. Other possible applications not, as yet, mentioned include the treatment of cooling tower make-up water, laundry wastes, petrochemical effluents, the removal of detergents from nuclear wastewater, the treatment of hospital wastes, the removal of pesticides, the reuse of textile dyeing wastes, and waste-cutting oils (Cohen, 1972; Johnson *et al.*, 1973; Cohen and Loeb, 1973; Markind *et al.*, 1973; Markind *et al.*, 1974;

Schmitt, 1974; Fang and Chian, 1974; Brandon *et al.*, 1975; Chian *et al.*, 1975; Gollan *et al.*, 1975; Minard *et al.*, 1975).

2. MUNICIPAL APPLICATIONS

Because of new stringent requirements concerning the quality of municipal effluents and because of the general need to consider wastewater reuse, advanced methods such as reverse osmosis have been extensively tested to determine their future role in various municipal treatment trains. Thus, the reverse osmosis process has been tested not only on raw sewage (Conn, 1971, W. H. Sprague, private communication), but also on secondary effluents with different types of pretreatment. At least two intensive studies comparing the performance of several reverse osmosis configurations on municipal wastes have been conducted in the United States (Smith *et al.*, 1970; Currie*, 1972). In addition, many studies on municipal wastewaters have been reported and are discussed in the following sections.

a. Initial Studies. The fact that cellulose acetate membranes could indeed concentrate organic as well as inorganic solutes from a municipal wastewater was established in the early sixties (Aerojet General Corp., 1965). Thereafter, a series of field tests was conducted to evaluate spiral-wound modules with diatomaceous earth-filtered secondary effluent (Bray and Merten, 1966). These tests showed effective removal of many secondary effluent constituents, but revealed very fast permeate flux decline (within hours). This, as will be evident later, is a major problem encountered in treating sewage with membrane processes. Considerable effort has gone into attempts to minimize this flux-decline phenomenon.

A series of pilot plant investigations of the spiral-wrap and the tubular configurations, as well as of the flat plate unit (Smith *et al.*, 1970), led to the following conclusions:

1. Cellulose acetate membranes were capable of rejecting more than 90% of the total dissolved solids, phosphates, particulate matter, total organic carbon, and chemical oxygen demand (COD). Ammonia nitrogen and nitrate nitrogen are only rejected from 80 to 90% and from 60 to 70%, respectively. The rejection performance exhibited by the three types of modules were similar.
2. pH control of the feed was successful in minimizing inorganic precipitation at recoveries as high as 80%.
3. The principal cause of flux-decline was attributed to organic fouling of

*The final project report (Boen and Johannsen, 1974) was issued during the preparation of this chapter.

the membranes. Although a correct identification of the class of material responsible for flux-decline was not determined, soluble, colloidal, and suspended species were all fundamentally involved in fouling.

4. Organic fouling was most successfully controlled by routinely depressurizing the unit and flushing it with an enzyme-active detergent.

5. The tubular unit, because of its well-defined hydrodynamics, seemed to show less tendency to plug.

b. Recent Studies. Several recent comprehensive investigations have been conducted on individual configurations at various sewage plants (Feuerstein and Bursztynsky, 1969; Bishop, 1970; Eden *et al.*, 1970; Fisher and Lowell, 1970; Hardwick, 1970; Cruver *et al.*, 1972; Grover and Delve, 1972; Cruver and Nusbaum, 1974; Loeb *et al.*, 1974) or on several configurations tested simultaneously under similar conditions of feed, pretreatment, and pressure (Currie, 1972). Results from a few of the most important studies are presented in Tables VIII and IX and Fig. 10.

In Table VIII, we have summarized the results of three different studies, each using a different membrane configuration or permeator design. Within each study, either the type of effluent (Cruver *et al.*, 1972) or the membrane formulation was varied (Feuerstein and Bursztynsky, 1969; Fisher and Lowell, 1970). The purpose of the first two studies (spiral-wrap and tubular) reported in Table VIII was to determine reliable flux-decline slopes for long-term tests on different types of municipal effluent. The purpose of the last study (flat plate) was to evaluate the effectiveness of newly prepared blended cellulose acetate membranes for wastewater treatment.

Figure 10 includes details of three typical lifetime tests (i.e., of 56-, 50-, and 77-day duration, see Table VIII) for product water flux versus time of operation. The following conclusions are drawn from the results shown in Table VIII and Fig. 10:

1. If the necessary pretreatment (sand filtering) and periodic membrane cleaning are rigorously adhered to, cellulose acetate membranes perform satisfactorily for extended periods at reasonable flux decline rates ($b \approx -0.10$) in both the spiral-wrap and tubular configurations.

2. Activated carbon pretreatment has been shown to be unnecessary for successful reverse osmosis operation on secondary effluent (see tests on 56- and 50-days duration in Table VIII and Fig. 10), although sand-filtered secondary effluent is necessary to maintain reasonable product flux.

3. Flux-decline slopes for adequately treated and cleaned secondary effluents are in the range -0.025 to -0.052, while for primary effluents a correspondingly higher range of -0.040 to -0.074 is to be expected.

4. Although not obvious from Table VIII, it was found experimentally that the higher the initial flux of the membrane, the worse the flux-decline slopes

Table VIII
Performance of Cellulose Acetate Membranes with Sewage Effluents[a]

Configuration and membrane	Type of[b] effluent	Cleaning during operation	Initial[c] flux gal/(ft^2, day)	Flux[d] decline slope	Extra-polated[e] average flux after 1 year (%)	Recovery (%)	Feed pressure lb/sq in.	Duration of test days
Spiral wrap and cellulose diacetate[f]	CFSE•	Daily 15 min tap water flush and 2% BIZ flush/wK	10.7	−0.025	79	80	320	56
	SFSE	Daily 15 min tap water flush and twice 2% BIZ flush p/wk	10.9	−0.037	71	80	320	50
	SFSE[g]	Same	10.0	−0.052	62	80	320–690	104
	CCPE	Same	15.8	−0.074	51	75–80	400	146
	CFCCPE	Same	15.1	−0.049	64	75–80	400	66
Tubular and cellulose diacetate[h]	CCSFPE	None	13.0	−0.056	60	80–95	700	77
Tubular and cellulose acetate and cellulose triacetate blend[h]	CCSFPE	None	25.0	−0.040	69	80	700	11

Flat plate and ds 2.55 blend (type 35C)[i]	Secondary effluent	None	46.0	−0.032	75	—	600	17
Flat plate and E-383-40 cellulose[i]	Secondary effluent	2 gm/liter BIZ solution at 600 psi for 30 min/day	50.0	−0.054	61	—	600	6
Flat plate and CAM-360 70°C[i]	Secondary effluent	Same as above	58.0	−0.049	64	—	600	6

[a]In general, all parameters had rejections above 85%, except for the NO_3^- ion which was rejected between 50 and 70%.

[b]The following abbreviations are used: CFSE, carbon-filtered secondary effluent; SFSE, sand-filtered secondary effluent; CCPE, chemically coagulated (alum) primary effluent; CFCCPE, carbon filtered, chemically coagulated, primary effluent; CCSFPC, chemically coagulated, sand-filtered, primary effluent.

[c]Multiply gal/(ft^2, day) by 0.0407 to get m^3/(m^2, day).

[d]Flux decline slope is defined by $b = -\Delta\log(\text{flux})/\Delta\log(\text{time})$, dimensionless.

[e]Expressed as a percent of the initial flux.

[f]Source: Cruver, Beckman, Bevage (1972).

[g]Constant product experiment at 10 gal/(ft^2, day).

[h]Source: Feuerstein and Bursztynsky (1969).

[i]Source: Fisher and Lowell (1970).

Table IX

Rejection of Various Constituents from Municipal Sewage Plant Secondary Effluents by Cellulose Acetate Membranes

	Feuerstein and Bursztynsky (1969)[a]	Eden *et al.* (1970)	Cruver, *et al.* (1972)[a]		Eastern Municipal Water District, Hemet, Calif.[b] Currie (1972)[c]						
Configuration	Tubular	Tubular	Spiral	Spiral	Tubular	Fiber	Spiral	Rod	Tubular	Fiber	Rod
Pretreatment Conditions	Alum	Sand filtered	Sand filtered	Carbon filtered	Sand filtered	Sand filtered	Sand filtered	Sand filtered	None	None	None
Duration of test, days	—	83	50	56	17	17	17	17	14	14	14
Pressure, psi	700	—	320	320	400	400	400	400	400	400	400
Recovery (%)	80	56	80	80	44	56	37	9	37	37	6
Rejections (%)											
a. COD	96	98	96	91	>99	91	96	99	97	83	96
b. Conductivity	93	92	95	—	—	—	—	80	—	—	92
c. Sulfate	—	97	>99	>99	>99	80	97	97	>99	77	99
d. Chloride	—	93	95	93	—	—	—	—	—	—	—
e. Nitrate, Nitrogen	41	81	—	—	80	71	55	77	82	63	73
f. Phosphate	98	98	>99	>99	98	80	90	96	>99	90	98
g. $Ca^{2+} + Mg^{2+}$	—	95	>99	>99	99	86	98	94	99	88	96
h. Ammonia-nitrogen	98	—	95	97	—	—	—	—	—	—	—
i. Turbidity (JTU)	—	62	—	—	92	88	94	89	—	—	—

[a]Product flux results appear in Table VIII.

[b]For 17 days all the reverse osmosis units received a feed of sand-filtered, postchlorinated secondary effluent; thereafter for 14 days the sand filtration was by-passed. The tubular, fiber, spiral, and rod (membrane on the outside of a porous stick) permeators were supplied by Universal, Du Pont, Gulf and Raypack, respectively, and the percent reduction for the product flux per unit pressure difference, as a result of discontinuing sand filtration, was 8.01, 43.03, 22.38, and 26.16, respectively.

[c]Also see Boen and Johannsen (1974).

became (Eden *et al.*, 1970; Nusbaum *et al.*, 1972a). This phenomenon is probably related to membrane fouling.

5. Rejections of various constituents for the tests in Table VIII were all above 85% except for NO_3^- ions.

Details from four different studies on the rejection of various constituents from municipal sewage plant secondary effluents by cellulose acetate membranes are presented in Table IX. Only nitrate-nitrogen consistently showed very poor rejections (between 41 and 82%). All the other compounds tested exhibited rejections of over 85%. For the spiral unit, sand-filtered secondary effluent showed essentially the same rejection percentages as did carbon-filtered secondary effluent (Cruver *et al.*, 1972). Results from the most recent comparative study (Currie, 1972) indicate that with or without any pretreatment of secondary effluent feed, the tubular and spiral units displayed the best performance. The fiber unit tended to have slightly lower rejections, while the rod unit (membrane cast on the outside of a porous ceramic stick) had very low percent recoveries.

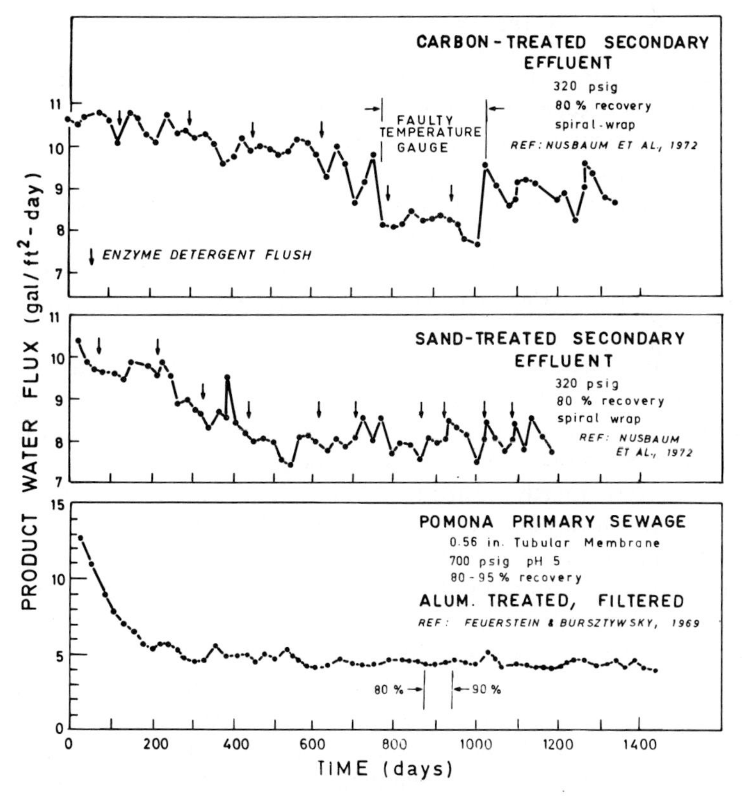

Fig. 10. Product water flux versus time for different pretreated municipal effluent feed waters.

The tubular, fiber, spiral, and rod designs displayed K_1 values [see Eq. (14)] of 1.4077, 0.1097, 0.5439, and 0.7922 $\times$ 10^{-5} gm/(cm^2, sec, atm), respectively.

Recently, Hamoda *et al.* (1973) and Matsuura and Sourirajan (1972) tested a newly developed porous cellulose acetate membrane (designated as Batch 316) for low-pressure applications. Hamoda *et al.* (1973) evaluated both the water and solute permeability for a series of organic compounds (glucose, sucrose, soluble starch, beef extract, glutamic acid, sodium stearate, and detergents) as a function of solute concentration. For these compounds, they obtained rejections greater than 88.5% (with most rejections above 95%) and fluxes between 35.6 and 42.7 gal/(ft^2, day).

Other investigators have used reverse osmosis to concentrate trace organic contaminants in drinking water (Deinzer *at al.*, 1974, 1975) and to evaluate the removal efficiencies and product fluxes of selected organic chemical species (sometimes called organic refractories) found in abundance in most treated sewage and other waste effluents (Bennet *et al.*, 1968; Edwards and Schubert, 1974).

3. TREATMENT OF POLLUTED RIVERS

Several feasibility studies have been conducted recently to determine whether reverse osmosis is capable of treating polluted river water (Kuiper *et al.*, 1973, 1974; J. D. Melbourne, private communication, 1973; D. G. Miller, private communication, 1973). All of these tests have been carried out with asymmetric cellulose acetate membranes in tubular configurations. The purpose of these studies were specifically to determine the effect of the polluted river water on process variables such as flux-decline rate, required pretreatment, effectiveness of various membrane-cleaning techniques, recovery ratios, and percent rejections.

One of the earliest studies concentrated on the effect of tangential brine velocities on the arrest of product flux-decline for a feed of untreated river water (Sheppard and Thomas, 1970). This study was the first to show that for a particular feed, from the Tennessee River, both the absolute flux and the flux-decline rate were directly related to the tangential brine velocity. It suggested that there exists a threshold velocity above which the product flux, J, remains fairly constant and also that the tighter the membrane, the lower the threshold velocity. This study also showed that at very high tangential velocities such as 24 ft/sec, the flux-decline parameter, b, was consistently low (-0.02 to -0.03) irrespective of feed composition. These and other studies have highlighted the relationship between the hydrodynamics of the brine stream and the reduction of the membrane flux-decline parameter, b (Sheppard and Thomas, 1971; Sheppard *et al.*, 1972). The next important question to be asked in this area is one of economics, namely, what pumping power will be required to maintain these high velocities and how does this relate to the optimum economic condition for plant operation?

Other concurrent studies using reverse osmosis on sand-filtered Rhine River water in the Netherlands and Thames and Trent river waters in the United Kingdom have recently been conducted. Detailed operation over 19 m for a 16 m^3/day reverse osmosis pilot plant using Rhine River water has been reported (Kuiper *et al.*, 1973). Besides some anomolous behavior at the end of the reported study, the main problem of membrane fouling and associated flux decline was adequately controlled by using chlorination, coagulation with iron, and rapid sand filtration as a pretreatment, together with membrane-cleaning procedures comprising daily depressurization, washing with acid, flushing, and mechanical cleaning. The average applied pressure was 40 atm and the average recovery was about 70%.

A 7 m^3/day (1850 gal/day) pilot plant was operated on sand-filtered Thames River water for 18 months at 27 to 34 atm (400 to 600 psi) (D. G. Miller, private communication). This study sought to establish the pretreatment techniques and/or membrane-cleaning methods required for the treatment of river waters. They showed that membrane fouling strongly correlated with both the turbidity of the feed waters and with the percent recovery of feed water. The study used both detergent membrane cleaning and foam ball swabbing and showed that both methods adequately arrested product flux-decline. Over the period of operation of 8000 hr, they were able to maintain permeabilities from 0.22 to 0.20 $m^3/(m^2$, atm, day).

Another pilot plant using Trent River water as a feed, was able to reduce TDS from 650–700 mg/liter to 20 mg/liter with adequate BOD removal (equivalent to 30 min through an activated carbon stage) (J. D. Melbourne, private communication, 1973). This tubular plant was operated with rough upflow sand filtration, pH adjustment to 6, and membrane cleaning by a solution and foam swab at regular intervals.

H. Control of Product Flux Decline

Clearly, when any conventional municipal effluent is treated with a pressure-driven membrane process, the product flux will decline with time eventually rendering the process as a whole uneconomical. Three phenomena have been shown to be responsible for product-flux decline in long-term performance of reverse osmosis membranes (Belfort, 1974b). They are (a) membrane *hydrolysis* resulting in an increase in both water and salt flux. (b) membrane *compaction* resulting in an initial flux decline with little effect on rejection (Bennion and Rhee, 1969); and (c) membrane *fouling* resulting in (1) a decrease of water flux and possibly salt flux, or (2) a fairly constant water flux with a decrease in rejection (Kuiper *et al.*, 1973, 1974).

As we will see below, feed pretreatment and membrane-cleaning methods have been developed to curtail product flux decline. These two processes are intimately linked and a determination of the economic optimum for the control of product flux-decline must include a consideration of both.

1. PRETREATMENT

Most researchers, using municipal effluents as a feed, have assumed that fewer problems might be encountered if effluents of the highest available quality were treated. Recent thinking, however, infers that this assumption may not be entirely valid (B. Sachs, private communication, 1973; Belfort, 1974b).

Several different pretreatment methods are shown in the flow diagram in Fig. 11. The most frequently used pretreatment methods for municipal feeds include clarification with chemical coagulants, such as alum, to remove large suspended solids and some dissolved organics and activated carbon filtration to remove small suspended solids and a large proportion of the dissolved organics.

Acidification reduces the rate of product-flux decline by increasing the solubility of inorganic precipitates such as $CaSO_4$, $CaCO_3$, or $Mg(OH)_2$ and is essential for minimizing the rate of hydrolysis of the cellulose acetate membranes. the pH is usually kept between 5 and 6 by using a mineral acid.

Cartridge filters are often used for final polishing of the feed solution with their removal size dependent on the permeator configuration.

Increasing the temperature of the feed stream has improved the efficiency of the electrodialysis process (Forgacs, 1967). Thus, it was thought that one might gain by increasing the temperature of the secondary effluent feed for the reverse osmosis process. Unfortunately, it was found that increased fouling at higher temperatures offset any potential advantage (Bailey *et al.*, 1973).

Ultrafiltration has been used with some success as a pretreatment process

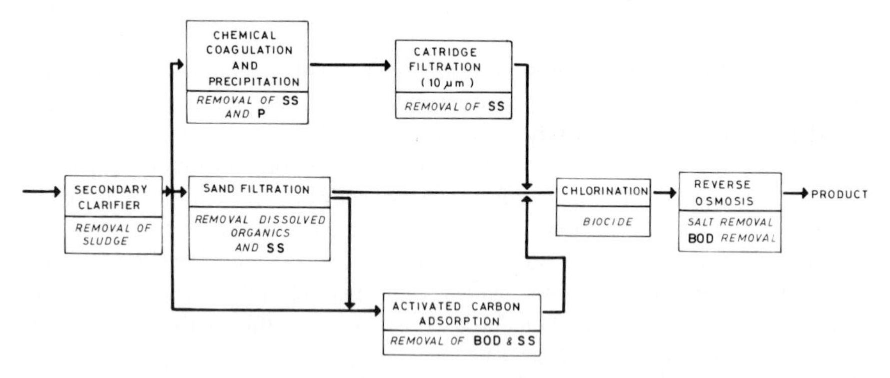

Fig. 11. Reverse osmosis pretreatment for a secondary effluent feed.

for reverse osmosis (Bailey *et al.*, 1973). Details will be presented later in the Section IV,E,2 on ultrafiltration.

There is evidence that pretreating with activated carbon filtration is in the long run (> 50 days) no better than sand filtration (Nusbaum *et al.*, 1972a,b) and that fouling also occurs even after the ultrafiltration pretreatment (Bailey *et al.*, 1973). What does this evidence lead us to conclude? It appears that some fouling material for the reverse osmosis membranes is either dissolved or in fine colloidal suspension, the particles certainly being smaller than the average pore size of the ultrafiltration, sand, and activated carbon prefilters.

The more treatment a raw sewage is subjected to, the more likely it will be to contain very small particles, i.e., the spread of the particle size distribution will widen with increased treatment. Thus, if very small particles ($\ll 1 \mu m$) contribute to fouling, the assumption stated at the beginning of this section—namely, that fewer problems will be encountered with an effluent which has a higher degree of treatment—may not necessarily hold. In fact, raw sewage with virtually no pretreatment (only minor pulverization of the feed to protect the centrifugal pump) has been successfully treated by reverse osmosis (Conn, 1971). Radically different membrane-cleaning techniques are used for this system. They will be discussed in the next section.

Before we discuss cleaning methods, it is appropriate to examine the most hazardous fouling constituents of domestic sewage. They include biological activity, dissolved organics (as measured by BOD, COD, or TOC)*, and suspended solids (SS). The exact contribution to fouling of each of these "constituents" is not known. As mentioned earlier, model compounds found in sewage have recently been tested for their rejections. In addition, a similar study, to evaluate potential fouling constituents, has recently been initiated (Belfort *et al.*, 1975b). Biological activity is, however, a particular hazard, since the cellulose acetate membrane is susceptible to biological degradation, while the sewage feed source is, of course, an ideal environment for microbial growth. Chlorination of the feed water prior to reverse osmosis treatment has been used to prevent biological attack. However, residual chlorine is itself dangerous to the membranes, especially to the aromatic polyamide membranes. Recent results from a study of the causes of biodegradation of tubular cellulose acetate membranes by agricultural run-off wastewater suggest a two-step mechanism (Richard and Cooper, 1975). The first step is the metabolism of the acetyl groups and thereafter the cellulutic microorganisms are able to act upon the remaining modified material. Quoting Richard and Cooper (1975): "Acidification to pH 4-6 with sulfuric acid or removal of dissolved oxygen with catalyzed solium sulfite (best method) or chlorination at 0.1–0.2 ppm total residual prevented

*BOD, COD and TOC are biological oxygen demand, chemical oxygen demand, and total organic carbon, respectively.

Table X
Membrane Cleaning Techniques for Reverse Osmosis

Technique	Method	Description	Investigator
Physical	Mechanical	Foam-ball swabbing	Loeb and Secover (1967), Kuiper *et al.* (1973)
	Hydrodynamical	Tangential velocity variation	Sheppard and Thomas (1970)
		Turbulence promoters	Thomas (1972)
	Reverse flow	Depressure and use forced or osmotic reverse flow of product	—
	Air/water flushing	Daily 15 min depressurized flush	Bray and Merten (1966)
	Sonication	Regular ultrasonic cleaning with wetting agent	R. Smith and W. Grube (1972, in Wilmoth and Hill, 1972)
Chemical	Additives to feed	pH control to reduce hydrolysis and scale deposit	—
		5 ml/gal of 5% sodium hyperchlorite at pH = 5	Fisher and Lowell (1970)
		Friction-reducing additives (poly-etheleneglycol) soil dispersants (sodium silicate)	Bailey *et al.* (1973)
	Flushing with additives at low pressure	Complexing agents (EDTA, Sodium hexametaphosphate)	Cruver *et al.* (1972)
		Oxidizing agents (citric acid)	Grover and Delve (1972)
			Bailey *et al.* (1973)
		Detergents (1% BIZ)	D. G. Miller (Private communication, 1973)
		Precoat (diatomaceous earth, activated carbon, and surface-active agent)	Conn (1971)
		High concentration of NaCl (18%)	Lacey and Huffman (1971)
	Membrane replacement	*In situ* membrane replacement	Belfort *et al.* (1973)
	Inorganic membranes	Encourage biogrowth to consume fouling film	Belfort (1974b)
	Active insoluble enzymes attached to membrane	Degradation of fouling film	Fisher and Lowell (1970)
	Polyelectrolyte membranes	Composite membranes or dynamic layer technique	Marcinkowsky *et al.* (1966)

bacterial attack of the membranes. Spongeball cleaning appeared not to appreciably reduce colonization of the membrane surface by bacteria nor subsequent biodeterioration of the colonized membranes."

2. MEMBRANE CLEANING TECHNIQUES

Several approaches have been used in attempts to reduce the rate of product-flux decline in systems treating municipal effluents (Belfort, 1974b). See Table X for a detailed summary of these techniques. What follows is a summary of the membrane-cleaning experience of several groups in reverse osmosis and wastewater treatment.

The early studies with the spiral-wrap and tubular modules were moderately successful in maintaining fluxes by daily air-water flushes at low pressure (Merten and Bray, 1966). Initial studies with citric acid and an anionic detergent solution proved unsuccessful in arresting flux decline in the flat plate unit (Smith *et al.*, 1970). Periodic cleaning (approximately once every 10 days) with an enzyme active presoak solution (BIZ) was successful in holding the flux fairly constant for tubular modules with a carbon-treated secondary effluent (Belfort *et al.*, 1973). Other studies on a similarly treated feed solution, but with the spiral-wrap modules, used daily air and water flushes and weekly flushes with a solution of 10,000 mg/liter of an enzyme presoak product (Nusbaum *et al.*, 1972a). These methods were also able to maintain a fairly low product-flux decline rate ($b = -0.05$). Cruver and associates (1972) also showed that sodium perborate, EDTA, and BIZ flushes each restored the flux to 80–85% of the initial value. Each additive gave approximately equal results. One of the problems with the application of these flushes is that their pH is usually harmfully high (>9) for the cellulose acetate membrane. Raw sewage is being treated successfully by reverse osmosis in San Diego using the Havens tubular units. The product-flux decline rate is arrested by precoating the membranes every 8 hr under pressure, with diatomaceous earth, powdered activated carbon, and a surface-active agent (Conn, 1971). The precoat is probably protecting the membranes for fouling. One disadvantage of this method may be the abrasiveness of the precoat, which may itself reduce the lifetime of the membranes. Most systems operating with tubular modules have used the foam-swab flushing technique especially for feeds with high suspended solids concentration. A disadvantage of this kind of flushing is that any abrasive material adhering to the swab could appreciably damage the membrane. Two reverse osmosis tubular pilot plant studies with river water as feed have successfully maintained fluxes by using daily depressurization, washing with HCl (pH $= 3$) and foam-ball flushing (Kuiper *et al.*, 1973) and by using detergent flushing and foam swabbing (D. G. Miller, personal communication, 1973).

From the typical chemical analysis of acid mine-drainage feeds presented earlier in Table VII, the two main causes of membrane fouling could be anticipated (Wilmoth and Hill, 1972). They are the bacterial conversion of Fe^{+2} to Fe^{+3}, which is precipitated on the membrane, and calcium sulfate inorganic scale precipitation. Disinfection by UV light or sudden decreases in pH to 2.5 delay bacterial growth for about 100 hr. Careful pH control is also used to minimize the $CaSO_4$ scale. The most successful cleaning method reported to date involves flushing with acidified water (pH = 2.5) and then leaving the unit idle for 1 week. Evidently, long periods of depressurization cause reverse product flow and the membrane to relax or destress, which leads to improved fluxes. Specific flushes such as an enzyme active presoak (BIZ), ammoniated citric acid, and sodium hydrosulfide were successful in the same study (Wilmoth and Hill, 1972), in removing organic, $CaSO_4$, and iron, respectively. Ultrasonic techniques have also been successful in cleaning the membranes (R. Smith and W. Grube, 1972, in Wilmoth and Hill, 1972).

Another promising technique for cleaning membranes has been pursued by Thomas and co-workers (1973). By increasing the tangential velocity in a tubular permeator, they were able to define a threshold velocity, above which flux decline was markedly smaller than at lower velocities. Their results were surprisingly similar for river water and primary treated sewage. This method has also been successfully adapted to the treatment of dilute pulp and paper effluents (Wiley *et al.*, 1972a).

Thus, from Table X, we see that the most common membrane-cleaning techniques for reverse osmosis are foam-ball swabbing and flushing with additives at low pressure. Several new approaches to membrane cleaning are listed in the same table, but from a practical point of view they all need further development. See Belfort (1974b) for further discussion of membrane-cleaning methods.

Thus, at present, membrane-cleaning methods are able, at the most, to reduce the flux-decline rate to a fairly small value ($b < 0.10$) for secondary effluents with some pretreatment. These methods can, however, be costly in downtime, expense of chemicals, and their degradative effect on membranes. Cheaper and more effective membrane-cleaning methods would reduce the operating cost of reverse osmosis and render it more competitive.

I. Economics of Reverse Osmosis

We have already discussed general aspects of the economics of pressure-driven membrane processes, in Section I,C of the Introduction. Although many cost estimates for brackish water treatment have been made (Harris *et al.*, 1969; LeGros *et al.*, 1970; Currie, 1972; Lacey, 1972b; Dresner and Johnson, 1975), cost estimates for wastewater treatment are not reliable, as insufficient large-scale experience is available.

Based on a million-gallons-per-day (mgd) capacity reverse osmosis plant, cost estimates from various sources range between 44.7 and 62.1 cents/kgal (Currie, 1972). Currie estimated that a completely equipped reverse osmosis plant at 80% recovery for a 1 mgd plant treating secondary effluent would cost about 84 cents/kgal. The reader is referred to the work by Currie (1972), Boen and Johannsen (1974) for details of these cost estimates.

A recent cost sensitivity analysis shows that reverse osmosis costs are much less sensitive to energy prices than thermal process (Arad and Glueckstern, 1975). For a 1-MGD plant and for 10% interest rate on capital investment with 30 year plant life and 80% load factor, desalting brackish water by reverse osmosis is about three times less expensive than using thermal processes on sea water. Desalting sea water by reverse osmosis, which is about to be commercialized, has a potential reduction of 10 to 20% over thermal process at prevailing high energy prices.

In comparing costs from various treatment plants, one must be aware that secondary effluent from different sewage plants may require completely different pretreatment and membrane-cleaning techniques. This would, in turn, result in a different product water cost. In addition, due to the recent large increase in the price of fuel, use of past cost estimates should be viewed with caution.

IV. ULTRAFILTRATION

A. Introduction and the Concept

Ultrafiltration, like reverse osmosis, is a pressure-driven membrane process using permselective membranes. The distinguishing features of the two processes are that reverse osmosis is a high-pressure process (400 to 1500 psi) whose membranes are capable of rejecting salt molecules, while ultrafiltration is a low-pressure process (5 to 100 psi) whose membranes are only capable of retaining molecules with a molecular weight of about 500 or higher.* Notice, however, from Fig. 1 that their particle size separation ranges overlap. This is a consequence of the fact that both reverse osmosis and ultrafiltration membranes can be "tailor-made" for larger or smaller particle rejections. The upper molecular weight cutoff for ultrafiltration is about 300,000 to 500,000. Above this molecular weight range another membrane process called microfiltration, which allows the passage of solvent and most solute molecules, but impedes the

*Retain and reject are synonymous terms, although they are used for ultrafiltration and reverse osmosis, respectively.

passage of large colloids and small particulate matter, is operative. The reason ultrafiltration is a low-pressure process as compared to reverse osmosis is that the large molecules being retained exert very little osmotic pressure. By their very nature, ultrafiltration membranes are "looser" and able to pass much higher fluxes of product water than reverse osmosis membranes. Thus, ultrafiltration membranes will not retain salt molecules or small low molecular weight organic molecules.

Because it is only the membranes and the differential pressure across them that differ, the plant equipment, membrane permeators, and associated operating problems are very similar for the reverse osmosis and ultrafiltration processes. Thus, rather than repeat what has been presented in the previous section on reverse osmosis, we will emphasize here the important differences between the two processes and refer to the previous sections for any similarities. For further information, see Michaels (1968).

B. Ultrafiltration Membranes

As in the case of reverse osmosis, the emergence of ultrafiltration as a practical industrial unit process is due in large measure to the development of the asymmetric polymer membrane (Loeb and Sourirajan, 1962). However, in contrast to reverse osmosis, where cellulose acetate and aromatic polyamides are predominantly used, a variety of synthetic polymers, such as polycarbonate resins, substituted olefins, and polyelectrolyte complexes have been employed for ultrafiltration membranes. Many of these anisotropic membranes can be handled dry and used with a variety of organic solvents. Temperature and pH resistance are often superior to those of conventional cellulose acetate membranes (Cruver, 1972).

The properties of typical commercial ultrafiltration membranes are summarized in Table XI. The spectrum of choice is wide and depends on the molecular size of the molecules to be concentrated. Molecular weight cutoff for 80 to 100% retention is used as an approximate measure of the rejecting characteristics of a particular membrane. Obviously, a molecular weight cut off characterization is at best nominal since retention depends on molecular shape and flexibility as well as size. Hydraulic permeability varies directly with molecular weight cutoff as is shown in Table XI.

C. Theoretical Considerations*

Ultrafiltration membranes that retain relatively large solute molecules (in excess of 500 molecular weight, or greater than 10 Å molecular diameter)

*Parts of this section rely heavily on Blatt *et al.* (1970).

Table XI
Properties of Typical Commercial Ultrafiltration Membranes[a]

Membrane designation	Manufacturer	Water flux at 100 psi gal/(ft^2, day)	Molecular weight[b] cutoff 80–100% retention
Diaflo XM 300	Amicon Corp., U.S.A.	1280 (55 psi)	300,000 (fibrinogen, bacteria, viruses, etc.)
Diaflo XM 100A	Amicon Corp., U.S.A.	—	100,000 (7S globulin)
Diaflo XM 50	Amicon Corp., U.S.A.	—	50,000 (hemoglobin)
Diaflo PM 30	Amicon Corp., U.S.A.	—	30,000 (pepsin)
Diaflo PM 10	Amicon Corp., U.S.A.	600	10,000 (cytochrome C)
Diaflo UM 10	Amicon Corp., U.S.A.	100	10,000 (dextran 10)
Diaflo UM 2	Amicon Corp., U.S.A.	70	1,000 (rafinose)
Diaflo UM 05	Amicon Corp., U.S.A.	35–9 (55 psi)	500 (sucrose)
HFA-100	Abcor, Inc., U.S.A.	25	10,000 (dextran 10)
HFA-300	Abcor, Inc., U.S.A.	500	70,000 (hemoglobin)
Pellicon	Millipore Corp., U.S.A.	10	1,000 (rafinose)
Cellulose Acetate	Fluid Systems Div., UOP, USA	30	1,000 (rafinose)

[a]Sources: Fig. 1 of Porter and Michaels (1971) and Table 7–5 of Cruver (1972).
[b]Rejections of 80–100% are for the compounds listed in parentheses.

appear to function as molecular sieves or screens in which viscous flow dominates the solvent motion through the micropores in the membrane and solute molecules are carried convectively with the solvent only through the pores large enough to accommodate them. For such membranes, the transport relationships are approximated by

for water
$$J_1 = \frac{\overline{K}_1 \Delta P}{t_m} \tag{17}$$

for solute
$$J_2 = C_2'(1 - R)J_1 \tag{18}$$

where $\overline{K}_1$ is the local hydraulic permeability coefficient of the membrane and t_m is the membrane thickness. The quantity $(1 - R)$ represents the fraction of the solvent flux carried by pores large enough to pass the solute. Because of continuity, we get

$$J_2 = J_1 C_2'' \tag{19}$$

and together with Eq. (18) we again get Eq. (16).

For such membranes, we expect the solvent flux to vary linearly with the hydraulic pressure difference and the rejection coefficient to be essentially constant and pressure independent. The quantity $\Delta\pi$ is absent in Eq. (17) as compared to Eq. (14) simply because for high molecular weight solutes the osmotic pressure of the retained solute is usually negligible compared to the applied hydraulic pressure.

As a consequence of using "open" membranes with high fluxes in ultrafiltration as opposed to "tight" membranes with low fluxes in reverse osmosis, concentration polarization or the buildup of solute molecules on the upstream (or brine) side of the membrane is much more acute. The result of concentrating macromolecules in solution is the formation of a porous gel-like layer on the surface of the membranes. It can be recognized as a "slime" or "cake" adhering to the membrane surface. In practice, as the gel layer grows, its properties become very important to the performance of the ultrafiltration unit (Forbes, 1972; Porter, 1974). The gel, in essence, is an added resistance through which the solvent must flow. Thus, to a good approximation, the gel-layer-membrane laminate can be treated as two hydraulic resistances in series, i.e.,

$$J_1 = \frac{\Delta P}{\left(\dfrac{t_m}{\overline{K}_1} + \dfrac{t_g}{\overline{K}_g}\right)} \tag{20}$$

where $\overline{K}_g$ and t_g are the local hydraulic permeability and thickness of the gel layer, respectively. $\overline{K}_g$ can be estimated by the Kozeny-Carmen relation for porous solids, viz.,

$$\overline{K}_g \simeq \frac{d^2\epsilon^3}{180\mu(1-\epsilon)^2} \tag{21}$$

where d is the diameter of the particles packed together to form the gel or "cake" and ϵ is the porosity of the gel. Assuming a 10 psi pressure difference across a 1-μm thick gel consisting of either 1-μm or 30 Å gel particles, and using $J_g = \overline{K}_g \Delta P / t_g$ and Eq. (21), one gets fluxes of $\sim 4 \times 10^5$ and ~ 3 gal/(ft^2, day), respectively, across the gel. Hence, it is evident that gel layers formed by relatively large size dispersions will have a negligible effect on ultrafiltration rate, whereas layers formed from small macromolecular solutes may reduce the flux rate markedly. There is indeed experimental evidence to support this (Baker and Strathman, 1970; Blatt *et al.*, 1970; Porter, 1974). In the real situation, $\overline{K}_g$ probably decreases with time due to the compressibility of the gel, while t_g, increases at first and then remains constant at steady-state operation.

D. Plant Equipment and Membrane Permeators

Almost all of the reverse osmosis equipment manufacturers offer ultrafiltration membranes. In addition, several other industrial companies offer only ultrafiltration membranes and/or equipment. Three manufacturers have attempted to reduce the concentration polarization phenomenon and associated gel layer buildup by offering units with very thin channels for feed flow (Strathman and Keilin, 1969; Blatt *et al.*, 1970; Nielson, 1972). By narrowing the brine channel height, the shear rate at the membrane surface is increased, and this, it is hoped, will significantly decrease concentration polarization. See Fig. 6 for the flow diagram of a typical ultrafiltration plant.

E. Applications of Ultrafiltration to Wastewater Renovation

Virtually any waste stream whose solids content consists of macromolecules and colloids can be renovated by ultrafiltration. If, in addition to reusing or recycling the permeate, the concentrated solids could be sold as a by-product, the cost of pollution abatement would be offset, in some cases. The by-product sales might even bring in a net profit. Of course, this would be true for any process stream, not necessarily only for wastewater.

1. INDUSTRIAL APPLICATIONS

As in the case of reverse osmosis, we are mainly concerned here with the treatment of wastewater streams and not with the use of ultrafiltration as a unit process within a flow sheet. Examples of industrial applications of ultrafiltration include the concentration and recovery of products such as viruses, pharmaceuticals, and enzymes from biological fermentations and the recovery and concentration of products from food-processing operations in the dairy, vegetable foods, beverage, meat and fish industries (Flynn, 1970; Porter and Michaels, 1971, 1972; Forbes, 1972; Madsen *et al.*, 1973). In the following section, we'll review some industrial applications of ultrafiltration to the purification of various wastewater effluents.

a. Fractionation of Whey from Cheese Production. A severe pollution problem facing the cheese industry is the disposal of cheese whey with its COD of about 55,000 ppm. The whey contains 50% of the milk solids, including most of the lactose, vitamins, and minerals, and 20% of the protein. Reverse osmosis has been used to remove most of the constituents in the whey including the dissolved minerals (McDonough and Mattingly, 1970), while ultrafiltration has been utilized to selectively fractionate the proteins from the unwanted salts, lactic acid, and lactose (deFilippi and Goldsmith, 1970; Porter *et al.*, 1971; Madsen *et al.*, 1973; Goldsmith *et al.*, 1974a; Porter, 1974; Fenten–May *et al.*, 1974) COD reductions from 65,600 to 800 ppm in a two-stage ultrafiltration system were reported by Madsen *et al.* (1973).

b. Recovery of Valueable Constituents from Electrocoating-Painting Operations. Automobile companies in the United States and in the United Kingdom are presently operating ultrafiltration plants to recover water-soluble primer paints used in the electropainting of car bodies (Forbes, 1970; Anonymous, 1971; Gross *et al.*, 1975). Typical water fluxes in thin channel units with electropaints containing 10% paint solids, range from 29–108 gal/(ft^2, day) at an average pressure of 57.5 psi, depending on the type of paint used (Porter *et al.*, 1971, Porter, 1974).

c. Other Industrial Applications. Ultrafiltration has also been used for the treatment of waste liquors in the production of paper (Bansal and Wiley 1974). Fluxes ranged from 9 to 26 gal/(ft^2, day) while BOD rejections varied from 44 to 56% (Porter *et al.*, 1971). Other applications include the recovery of polymer from latex manufacturing effluents, the concentration of soluble oil wastes; the recovery of metal colloids from nuclear power plants, photographic processing effluents, and electroplating wastes; the concentration of dyes from textile plant effluents, and fine particulate-activated carbon from tertiary wastewater (Goldsmith *et al.*, 1974b. Porter *et al.*, 1971; Cohen and Loeb, 1973).

2. MUNICIPAL APPLICATIONS

Ultrafiltration has been tested for potential use in the treatment of surface waters and municipal effluents. With municipal effluents it was used as a pretreatment for reverse osmosis (Bailey *et al.*, 1973). It has also been suggested as a possible method for dewatering sludges (Porter *et al.*, 1971).

a. Treatment of Surface Supplies. Ultrafiltration has been developed as a candidate process to treat surface water supplies to meet the U. S. Public Health Service and WHO standards (Smith and Di Gregorio, 1970).

b. Virus Concentration. A cellulose acetate hollow-fiber ultrafiltration permeator has been used to concentrate polio-1 virus from 5 liters of water in about 1 hr with 85% virus recovery for a 50-fold dehydration (Belfort *et al.*, 1975a). The purpose was to develop the first step of a system to concentrate and quantify virus in contaminated waters.

c. Pretreatment to Reverse Asmosis. In two studies ultrafiltration of sewage effluent was intended purely as a pretreatment to reverse osmosis (Gollan *et al.*, 1975; Bailey *et al.*, 1973). However, for one of the studies the quality of the ultrafiltrate was much better than expected and was suitable for many industrial uses (Bailey *et al.*, 1973). See Chapter 14, Section IV, D for details of this application of ultrafiltration.)

d. Unit Process for Municipal Treatment. If salt removal is not a requirement, as in the blending of renovated water with surface waters with a low mineral content, ultrafiltration could find application in lieu of reverse osmosis at the end of the municipal treatment train. Two such ultrafiltration systems, a 3,000 gal/day raw sewage facility and a 20,000 gal/day tertiary unit, indicate good BOD and suspended solids rejections (Okey, 1970).

New noncellulosic assymetric membranes with characteristics between reverse osmosis and ultrafiltration membranes are being tested for municipal effluent renovation (Sachs *et al.*, 1975). The membranes are said to be resistant to biological attack and stable in the pH ranges 1 to 14. Preliminary results after 600 hours gave a permeation flux of about 70 gfd and a b-slope of −0.098 at 7.5 atm (110 psi) and 35°C on oxidation pond effluent.

F. Economics of Ultrafiltration

As in reverse osmosis, cost estimates will vary considerably depending on the application and the required operating capacity. As mentioned above, if by-products in the concentrate stream can be reused or sold, the net product water cost will drop correspondingly. Since ultrafiltration operates at lower pressures than reverse osmosis, this may result in a slightly lower capital cost. However, capital costs are not usually the major component of the total costs for pressure-driven membrane processes where operating costs are usually a very large proportion of the total costs (usually greater than 50%). Porter and Michaels (1971) present a range of 20 to 500 cents/kgal for both ultrafiltration and reverse osmosis. According to them, the key cost factors are flux rate, membrane costs, and membrane durability. These three parameters plus the power requirements determine the operating conditions for the plant. The product flux rate is directly related to membrane fouling, degree of pretreatment, and membrane cleaning methods.

It is difficult at present to estimate realistically the cost of treating wastewaters by ultrafiltration. Not only is it too early in the development of the process, but the spectrum of applications has been wide and the studies have been mainly the size of a pilot plant. In addition, costs of most commercial plants have not been published. Thus, cost projections for ultrafiltration should be viewed with care and some skepticism.

V. CONCLUSIONS

We have attempted to answer the questions posed at the beginning of this chapter in the section "Why Pressure-Driven Membrane Processes?" and have presented the historical development of both reverse osmosis and ultrafiltration. We have marshaled the advantages and disadvantages of each process and have reviewed the present state of the art of the treatment of various wastewaters. The cost data and the cost projections are tenuous at best. Finally, it is inferred from the many applications and the increasing interest and activity in pressure-driven membrane processes that they are, at present, being taken very seriously for wastewater treatment.

Given this bright picture, why, we are prompted to ask, have these membrane processes not received wider acceptance in either industrial or municipal applications? One important reason is that they are structurally and operationally different from conventional municipal wastewater treatment processes, such as sedimentation, filtration, and even biological treatment (Belfort, 1974a). Second, the membrane processes have higher operating (and sometimes total) costs than the latter processes. In addition, a higher level of trained operator is needed for these pressure-driven membrane processes. Understandably, there exists a certain hesitancy in accepting these new processes. Yet, with respect to wastewater reclamation for potable use, reverse osmosis, with its capacity to reject viruses, bacteria, and dangerous organic and inorganic compounds, is certainly an attractive, viable process.

As for industrial applications, especially for pollution abatement, it is a lamentable, but undeniable, fact that most commercial polluters need to be "legally encouraged" to do something about their plant effluents. It is also a truism that, although laws exist in most developed countries, they are not adequately enforced. With environmental awareness on the increase and man coming to realize that he had better save his surface and groundwaters from industrial and agricultural degradation, prosecutions for illegal polluting are becoming more numerous, especially in the advanced industrial countries. In light of this, ultrafiltration and reverse osmosis will surely find wide acceptance for industrial uses.

DEDICATION

This chapter is dedicated to the memory of Professor Shai Winograd, who lost his life in the tragic "Yom Kippur" War of October 1973.

ACKNOWLEDGMENT

Thanks are due to Professor K. Sam Spiegler for his continual encouragement and advice; to Professor Sandford Budick for his editorial assistance; to Mr. John Convery and Dr. Sydney A. Hannah of the U.S. Environmental Protection Agency for promptly sending me all the EPA reports relevant to this study.

I would also like to thank Dr. Marlene Belfort for her indefatigable efforts to make this chapter shorter and clearer. Of course, the final responsibility for accuracy, clarity, and length is the author's.

REFERENCES

Aerojet General Corp. (1964). *U.S. Off. Saline Water, Res. Develop Progr. Rep.* **86**.

Aerojet General Corp. (1965). "Reverse osmosis as a treatment for wastewater" (Contract No. 86-63-277). *U.S., Pub. Health Serv., Publ.* **2962**.

Aerojet General Corp. (1966). *U.S. Off. Saline Water, Res. Develop. Progr. Rep.* **213**.

Ajax International Corp. (1973). "Sales News Flash 73-10-1," p. 1.

Allegrezza, A. E., Jr., Charpentier, J. M., Davis, R. B., and Coplan, M. J. (1975). "Hollow Fiber Reverse Osmosis Membranes" (Presented at 68th Annual AIChE Meeting, Los Angeles, 1975), Paper No. 34b.

Anonymous. (1971). Primer coat for Chevy's Vega applied by electrodeposition. *Ind. Finish. J. Paint Colour Rec.* (July issue) (cited in Cohen, 1972).

Anonymous. (1975). Membrane makes strong treatment bid. *Chem. Week* **September**, 31–32.

Arad, N., and Glueckstern, P. (1975). "Sensitivity of Desalted Water costs from Reverse Osmosis Plants to Changing Investment and Operating Costs Including Energy (Presented at the Israel National Council for Research and Development Symposium on Desalination, Oholo, Israel 1975). Nat. Counc. Res. Develop., Prime Minister's Office, Jerusalem, Israel.

Bailey, D. A., Jones, K., and Mitchell, C. (1973). "The Reclamation of Water from Sewage Effluents by Reverse Osmosis" (Presented to the Joint Meeting of the Scottish Branch of the IWPC, IPHE and IWE. Available from Dept. of the Environment, Water Pollut. Res. Lab., Stevenage, United Kingdom.

Baker, R. W., and Strathman, H. (1970). Ultrafiltration of macromolecular solutions with high flux membranes. *J. Appl. Polym. Sci.* **14**, 1197–1214.

Bansal, I. K., and Wiley, A. J. (1974). Fractionation of spent sulfite liquors using ultrafiltration cellulose acetate membranes. *Environ. Sci. Technol.* **8**, 1085–1090.

Bansal, I. K., and Wiley, A. J. (1975). Membrane processes for fractionation and concentration of spent sulfite liquors. *Tappi* **58**, 125–130.

Belfort, G. (1972). The role of water in porous glass desalination membranes. Ph.D. Thesis, University of California, Irvine.

Belfort, G. (1974a). Interfacing newly developed technology within present wastewater treatment trains. *J. Amer. Water Works Ass.* **1–3**, 29–35.

Belfort, G. (1974b). 'Cleaning of Reverse Osmosis Membranes in Wastewater Renovation (Presented at Joint AIChE-VTG Meeting, Munich, Germany, 1974).

Belfort, G., Littman, F., and Bishop, H. K. (1973). Membrane regeneration for wastewater reclamation using reverse osmosis. *Water Res.* **7**, 1547–1559.

Belfort, G., Rotem, Y., and Katzenelson, E. (1975a). Virus concentration using hollow fiber membranes. *Water Res.* **9**, 79–85.

Belfort, G., Alexandrowicz, G., and Marx, B. (1975b). "A Study of the Mechanism and Prevention of Membrane Fouling in the Application of Hyperfiltration (Reverse Osmosis) to Wastewater Treatment." Nat. Counc. Res. Develop., Prime Minister's Office, Jerusalem.

Belfort, G., Mahlab, D., and Ben-Yosef, N. (1976). Concentration polarization profile for dissolved species in batch hyperfiltration (reverse osmosis). *Proc. Int. Symp. on Fresh Water from Sea, 5th, 1976.* Vol. 4, pp. 249–258.

Bennet, P. J., Narayarian, S., and Hindin, E. (1968). Removal of organic refractories by reverse osmosis. *Eng. Bull. Purdue Univ., Eng. Ext. Ser.* **132**, 1000–1017.

Bennion, D. N., and Rhee, B. W. (1969). Mass transport of binary electrolytes in membranes. *Ind. Eng. Chem., Fundam.* **8**, 36.

Bishop, H. K. (1970). Use of improved membranes in testiary treatment by reverse osmosis. *Water Pollut. Contr. Res. Ser.* **17020 DHR12/70**.

Blatt, W. F., Dravid, A., Michaels, A. S., and Nelson, L. (1970). "Solute Polarization and Cake Formation in Membrane Ultrafiltration: Causes, Consequences and Control Techniques" *In* "Membrane Science and Technology" (J. E. Flinn, ed.), pp. 47–97. Plenum, New York.

Boen, D. F., and Johannsen, G. L. (1974). Reverse osmosis of treated and untreated secondary sewage effluent. *Environ. Protect. Technol. Ser.* **EPA-670/2-74-077**.

Brandon, C. A., El-Nashar, A., and Porter, J. J. (1975). "Reuse of Wastewater Renovated by Reverse Osmosis in Textile Dyeing" (Presented at 2nd National Conference on Complete Water Uses. Chicago, 1975).

Bray, D. T., and Merten, U. (1966). Reverse osmosis for water reclamation. General Atomic Division, General Dynamics and Los Angeles County Sanitation District. *J. Water Pollut. Centr. Fed.* **100**, No. 3, 315.

Channabasappa, K. C. (1969). Reverse osmosis process for water reuse application. *Chem. Eng. Progr., Symp. Ser.* **65**, 140–147.

Chian, E. S. K., and Fang, H. H. P. (1973). "Evaluation of New Reverse Osmosis Membranes for Separation of Toxic Compounds from Water. (Presented at 75th National AIChE Meeting, Detroit, 1973).

Chian, E. S. K., Bruce, W. N., and Fang, H. H. P. (1975). Removal of pesticides by reverse osmosis. *Environ. Sci. Technol.* **9**, 52–59.

Cohen, H. (1972). The use of pressure-driven membranes as a unit operation in the treatment of industrial waste streams. *In* "Utilization of Brackish Water" (G. A. Levite, ed.), pp. 63–69. Nat. Counc. Res. Develop., Prime Minister's Office, Jerusalem.

Cohen, H., and Loeb, S. (1973). Industrial Wastewater Treatment in Israel using Membrane Process," Rep. No. 132. Negev Institute for Arid Zone Research, Beer-Sheva, Israel.

Conn, W. M. (1971). "Raw Sewage Reverse Osmosis" (Presented at 69th Annual AIChE Meeting, Cincinnati). City of San Diego, Pt. Loma Sewage Treatment Plant, San Diego, California.

Copas, A. L., and Middleman, S. (1973). "The Use of Convective Promotion in Ultrafiltration of a Gel-Forming Solute" (Presented at 66th Annual AIChE Meeting, Philadelphia, 1973). Paper No. 53a.

Cruver, J. E. (1972). Membrane Processes. *In* "Physico-chemical Processes for Water Quality Control" (W. J. Weber, Jr., ed.), Chapter 7, pp. 307–362. Wiley (interscience), New York.

Cruver, J. E., and Nusbaum, I. (1974). Application of reverse osmosis to wastewater treatment. *J. Water Pollut. Contr. Fed.* **46**, No. 2, 301–311.

Cruver, J. E., Beckman, J. E., and Bevage, E. (1972). "Water Renovation of Municipal Effluents by Reverse Osmosis," EPA Proj. No. EPA17040EOR. Gulf Environmental Systems Co., San Diego, California.

Currie, R. J. (1972). "Study of Reutilization of Wastewater Recycle Through Groundwater" (prelim. copy). Eastern Municipal Water District, Hemet California (also cited in final project report of Boen and Johannsen, 1974).

deBussy, R. P., and Whitmore, H. B. (1972). Reverse osmosis: Some aspects of industrial water/waste treatment. *Nat. Eng.* (February issue).

deFilippi, R. P., and Goldsmith, R. L. (1970). *In* "Membrane Science and Technology" (J. E. Flinn, ed.), pp. 33–46. Plenum, New York.

de Groot, S. R. (1959). "Thermodynamics of Irreversible Processes." North-Holland Publ., Amsterdam.

de Groot, S. R., and Mazur, P. (1962). "Non-Equilibrium Thermodynamics." North-Holland Publ., Amsterdam.

Deinzer, M., Melton, R., Mitchell, D., Kopfler, F., and Coleman, E. (1974). "Trace organic contaminants in drinking water: Their concentration by reverse osmosis." Presented at Amer. Chem. Soc., 1974, Los Angeles.

Deinzer, M., Melton, R., and Mitchell, D. (1975). Trace organic contaminants in drinking water: their concentration by reverse osmosis. *Water Res.* **9**, 799–805.

Dresner, L., and Johnson, J. S., Jr. (1976). Hyperfiltration (reverse Osmosis). *In* "Principles of Desalination" (K. S. Spiegler and A. D. K. Laird, eds.), draft of chapter revised for 2nd ed. Academic Press, New York.

Duvel, W. A., Jr., and Helfgott, T. (1975). Removal of wastewater organics by reverse osmosis. *J. Water Pollut. Contr. Fed.* **47**, 57–65.

Eden, G. E., Jones, K., and Hodgson, T. D. (1970). Recent development in water reclamation. *Chem. Eng. (London)* Jan./Feb. Issue, CE24-CE29.

Edwards, V. H., and Schubert, P. F. (1974). Removal of 2,4-D andd other persistent organic molecules for water supplies by reverse osmosis. *J. Amer. Water Works Ass.* **October**, 610–616.

Fang, H. H. P., and Chian, E. S. K. (1974). "RO Treatment of Power Cooling Tower Blowdown for Reuse" (Presented at 67th Annual AIChE Meeting, Washington, D.C., 1974). Paper No. 40C.

Fang, H. H. P., and Chian, E. S. K. (1975). Removal of alcohols, amines and aliphatic acids in aqueous solutions by NS-100 membranes. *J. Appl. Polym. Sci.* **19**, 1347–1358.

Fenton-May, R. I., Hill, C. G., Jr., Amundson, C. H., and Auclair, P. D. (1974). *AIChE Symp. Ser.* **120**, 31–40.

Feuerstein, D. L., and Bursztynsky, T. A. (1969). Reverse osmosis renovation of municipal wastewater. *Water Pollut. Contr. Res. Ser.* **ORD-17040 FFO12/69**.

Fisher, B. S., and Lowell, J. R., Jr. (1970). New technology for treating wastewater by reverse osmosis. *Water Pollut. Contr. Res. Ser.* **17020 DUD09/70**.

Flinn, F., ed. (1970). "Membrane Science and Technology." Plenum, New York.

Forbes, F. (1970). *Chem. Process Eng.* (December issue).

Forbes, F. (1972). Considerations in the optimization of ultrafiltration. *Chem. Eng. (London)* January Issue, pp. 29–34.

Forgacs, C. (1967). *Proc. Int. Symp. Water Desalination, 1st 1965* paper No. SWD/83.

Goldsmith, R. L., deFilippi, R. P., and Hossain, S. (1974a). New membrane process applications. *AIChE Symp. Ser.* **120**, 7–14.

Goldsmith, R. L., Robert, D. A., and Burre, D. L. (1974b). Ultrafiltration of soluble oil wastes. *J. Water Pollut. Contr. Fed.* **46**, 2183–2192.

Gollan, A., Goldsmith, R., and Kleper, M. (1975). "Advanced Treatment of MUST Hospital Wastewaters" (Presented at 5th Intersocity Conference on Environmental Systems, San Francisco, 1975).

Golomb, A. (1972). An example of economic plating waste treatment. *Proc. Int. Conf. Water Pollut. Res., 6th, 1972* Paper 15/2/3/1.

Golomb, A., and Besik, F. (1970). RO- a review of the applications to waste treatment. *Ind. Water Eng.* **7**, 16.

Gross, M. C., Hastings, C. R., and Minard, P. G. (1975). "Electrodeposition Painting with Ultrafiltration" (Presented at 2nd National Conference on Complete Water Reuse, Chicago, 1975).

Grover, J. R., and Delve, M. H. (1972). Operating experience with a 23m^3/day reverse osmosis pilot plant. *Chem. Eng. (London)* January Issue, pp. 24–29.

Grover, J. R., Gaylor, R., and Delve, M. H. (1973). *Proc. Int. Symp. Fresh Water Sea, 4th, 1973* Vol. 4, pp. 159–169.

Haase, R. (1969). "Thermodynamics of Irreversible Processes." Addison-Wesley, Reading, Massachusetts.

Haight, A. G. (1971). "Demineralized Water through Reverse Osmosis and Ion Exchange" (Presented at American Association for Contamination Control meeting, Washington D.C.). Obtainable from E. I. DuPont de Nemours & Co., Wilmington Delaware.

Hamoda, M. F., Brodersen, K. T., and Souirajan, S. (1973). Organics removal by low pressure reverse osmosis, *J. Water Pollut. Contr. Fed.* **45**, 2146–2154.

Hardwick, W. H. (1970). Water renovation by reverse osmosis. *Chem. Ind. (London)* February Issue, pp. 297–301.

Harris, F. L., Humphreys, G. B., Isakari, H., and Reynolds, G. (1969). Engineering and economic evaluation study of reverse osmosis. *U.S. Off. Saline Water, Res. Develop. Progr.* **509**.

Harris, F. L., Humphreys, G. B., and Spiegler, K. S. (1976). Reverse osmosis (hyperfiltration) in water desalination. *In* "Membrane Separation Process" (P. Meares, ed.). In press.

Hauk, A. R., and Sourirajan, S. (1972). Reverse osmosis treatment of diluted nikel-plating solutions. *J. Water Pollut. Contr. Fed.* **44**, 1372–1382.

Horton, B. S., Goldsmith, R. L., and Zall, R. R. (1972). Membrane processing of cheese whey reaches commercial scale. *Food Technol.* **26**, 30–35.

Johnson, J. S., Jr. (1972). *In* "Reverse Osmosis Membranes Research" (H. K. Lonsdale and H. E. Podall, eds.), pp. 379–403. Plenum, New York.

Johnson, J. S., Jr., and McCutchan, J. W. (1973). "Desalination of Sea Water by Reverse Osmosis" (Presented at AIChE Meeting, Dallas, Texas, 1973). Obtainable from J. W. McCutchan, Univercity of California, Los Angeles.

Johnson, J. S., Jr., Minturn, R. E., Westmoreland, C. G., Csurny, J., Harrison, N., Noore, G. E., and Shor, A. J. (1973). "Filtration Techniques for Treatment of Aqueous Solutions," Annu. Progr. Rep., ORNL-4891. Chem. Div., Oak Ridge Nat. Lab., Oak Ridge, Tennessee.

Kesting, R. E. (1973). Concerning the micro-structure of dry-RO- membranes. *J. Appl. Polym. Sci.* **17**, 1771–1784.

Kraus, J. A. (1970). Application of hyperfiltration to treatment of municipal sewage effluents. *Water Pollut. Contr. Res. Ser.* **ORD 17030 EOH01/70**.

Kuiper, D., Bom, C. A., van Hezel, J. L., and Verdouw, J. (1973). *Proc. Int. Symp. Fresh Water Sea, 4th, 1973* Proc. No. 4.205.

Kuiper, D., van Hezel, J. L., and Bom, C. A. (1974). The use of reverse osmosis for the treatment of river Rhine water. Part II. *Desalination* **15**, 193–212.

Lacey, R. E. (1972a). Membrane separation process. *Chem. Eng. (London)* September 4 Issue, pp. 56–74.

Lacey, R. E. (1972b). The costs of reverse osmosis. *In* "Industrial Processing with Membranes" (R. E. Lacey and S. Loeb, eds.), Chapter 9, p. 179. Wiley (Interscience), New York.

Lacey, R. E., and Huffman, E. L. (1971). Demineralization of wastewater by the transport depletion process. *Water Pollut. Contr. Res. Ser.* **17040 EUN02/71**.

Lacey, R. E., and Loeb, S., eds. (1972). "Industrial Processing with Membrane." Wiley (Interscience), New York.

LeGros, P. G., Gustafson, C. E., Sheppard, B. P., and McIlhenny, W. F. (1970). *U.S. Off. Saline Water, Res. Develop. Progr. Rep.* **587**.

Leitner, G. F. (1972). Reverse osmosis for wastewater treatment: What? when? *Tappi* **55**, 258–261.

Leitner, G. F. (1973). Reverse osmosis for water recovery and reuse. *Chem. Eng. Progr.* **69**, 83–85.

Loeb, S., and Selover, E. (1967). Sixteen months of field experience in the Coalinga pilot plant. *Desalination* **2**, 63–68.

Loeb, S., and Sourirajan, S. (1962). Sea water demineralization by means of an osmotic membrane. *Advan. Chem. Ser.* **38**, 117.

Loeb, S., Levy, D., and Melamed, A. (1974). "Reclamation of Municipal Wastewater for Reuse," Final Report, NEG-ES-73-1/2 (Presented to the Israel National Council for Research and Development, Jerusalem, Israel. Ben Gurion University Research and Development Authority.

Lonsdale, H. K. (1974). Recent advances in reverse osmosis membranes. *Desalination* **13**, 317–332.

Lonsdale, H. K., and Podall, H. E., eds. (1972). "Reverse Osmosis Membrane Research." Plenum, New York.

McAllister, D. G., Jr. (1971). A Tour Guide Booklet of the Pomona Advanced Waste Treatment Facility, Ponoma, California.

McDonough, F. E., and Mattingly, W. A. (1970). *Food Technol.* **24**, 88.

Madsen, R. F., Olsen, O. J., Nielsen, I. K., and Nielson, W. K. (1973). Use of hyperfiltration and ultrafiltration with chemical and biochemical industries. *In* "Environmental Engineering, A Chemical Engineering Discipline" (G. Linder and K. Nyberg, eds.), pp. 320–330. Reidel Publ., Dordrecht, Netherlands.

Mahoney, J. G., Rowley, M. E., and West, L. E. (1970). *In* "Membrane Science and Technology." (J. E. Flinn, ed.), pp. 196–208. Plenum, New York.

Marcinskowsky, A. E., Kraus, K. A., Phillips, H. O., and Shor, A. J. (1966). Hyperfiltration studies. IV. Salt rejection by dynamically formed hydrous oxide membranes. *J. Amer. Chem. Soc.* **88**, 5744.

Markind, J., Minard, P. G., Neri, J. S., and Stana, R. R. (1973). Use of reverse osmosis for concentrating waste cutting oils. *Proc. Amer. Instit. Chem. Engr.—Canod. Soc. Chem. Engr., 4th Joint,* 1973.

Markind, J., Neri, J. S., and Stana, R. R. (1974). "Use of Reverse Osmosis for Concentrating Oil Coolants" (Presented at 78th National AIChE Meeting, Salt Lake City, Utah, 1974).

Mason, D. G., and Gupta, M. K. (1972). Ameanability of reverse osmosis concentration to activated sludge treatment. *Water Pollut. Contr. Res. Ser.* **14010 FOR03/72.**

Matsuura, T., and Sourirajan, S. (1972). Studies on reverse osmosis for water pollution control. *Water Res.* **6**, 1073–1086.

Merson, R. L., and Morgan, A. I., Jr. (1968). *Food Technol.* **22**, 631.

Merten, U. (1966). "Desalination by Reverse Osmosis." MIT Press, Cambridge, Massachusetts.

Merten, U., and Bray, D. T. (1967). Reverse osmosis for water reclamation. *Advan. Water Pollut. Res., Proc. Int. Conf., 3rd, 1966* Vol. 3, p. 000.

Michaels, A. S. (1968). Ultrafiltration *In* "Progress in Separation and Purification" (E. S. Perry, ed.), Vol. 1, pp. 297–334. Wiley (Interscience), New York.

Minard, P. G., Stana, R. R., and DeMeritt, E. (1975). Two years experience with a reverse osmosis radioactive laundry water concentrator. Presented at 2nd National Conference Complete Water-reuse, Chicago, 1975).

Nielson, W. K. (1972). "The Use of Ultrafiltration and Reverse Osmosis in the Food Industry and for Wastewaters from the Food Industry," Paper No. WKN/1h. Obtainable from DDS, Nakskov, Denmark.

Nusbaum, I., Cruver, J. E., Sr., and Kremen, S. S. (1972a). "Recent Progress in Reverse Osmosis Treatment of Municipal Wastewaters," Rep. No. Gulf-EN-A10994. Fluid System Div. UOP, San Diego, California.

Nusbaum, I., Cruver, J. E., and Sleigh, J. H., Jr. (1972b). Reverse osmosis—new solutions and new problems. *Chem. Eng. Progr.* **68**, 69–70.

Okey, R. W. (1970). The application of membranes to sewage and waste treatment. *Water Resour. Symp.* **3**, 327–338.

Okey, R. W. (1972). The treatment of industrial wastes by pressure driven membrane processes. *In* "Industrial Processing with Membranes" (R. E. Lacey and S. Loeb, eds.), Chapter 12, p. 249. Wiley (Interscience), New York.

Pappano, A. W., Blackshaw, G. L., and Chang, S.-Y. (1975). "Coupled Ion-Exchange Reverse Osmosis Treatment of Acid Mine Drainage" (Presented at 80th National AIChE Meeting, Boston, Massachussets, 1975), Paper No. 44d.

Porter, M. G. (1974). Ultrafiltration of colloidal suspensions. *AIChE Symp. Ser.* **120**, 21–30.

Porter, M. C., and Michaels, A. S. (1971). Membrane ultrafiltration. *Chem. Technol.* Part 1, 56–63; Part 2, 248–254; Part 3, 440–445; Part 4, 633–637.

Porter, M. C., and Michaels, A. S. (1972). Membrane ultrafiltration. *Chem. Technol.* Part 5, 56–61.

Porter, M. C., Schratter, P., and Rigopulos, P. N. (1971). Byproduct recovery by ultrafiltration. *Ind. Water Eng.* June/July Issue, pp. 18–24.

Porter, W. L., Siciliano, J., Krulik, S., and Heisler, E. G. (1970). Reverse osmosis: Application to potato–starch factory waste effluents. *In* "Membrane Science and Technology" (J. E. Flinn, ed.), pp. 220–230. Plenum, New York.

Prigogine, I. (1955). "Introduction to Thermodynamics of Irreversible Processes." Thomas, Springfield, Illinois.

Probstein, R. F. (1972). Desalination: some fluid mechanical properties. *Trans. ASME* **June**, 266–313.

Probstein, R. F. (1973). Desalination. *Amer. Sci.* **61**, No. 3, 280–293.

Richard, M. G., and Cooper, R. C. (1975). "Prevention of Biodegradation and slime Formation in Tubular Reverse Osmosis Units" (Presented at the Annual Conference of the National Water Supply Improvement Association, Key Largo, Florida, 1975).

Riedinger, A. B., and Nusbaum, I. (1972). Reverse osmosis applied to wastewater resure. *Amer. Soc. Mech. Eng., Publ.* **72-PID-8**.

Riley, R. L., Lonsdale, H. K., and Lyons, C. R. (1971). Composite membranes for sea water desalination by reverse osmosis. *J. Appl. Polym. Sci.* **15**, 1267–1276.

Riley, R. L., Hightower, G. R., Lyons, C. R., and Tagami, M. (1973). Thin film composite membranes for single-stage seawater desalination by reverse osmosis. *Proc. Symp. Fresh Water Sea, 4th, 1973* Vol. 4, pp. 333–347.

Rozelle, L. T. (1971). *Water Pollut. Contr. Res. Ser.* **12010 DRH 11/71**.

Rozelle, L. T., Cadotte, J. E., Nelson, B. R., and Kopp, C. U. (1973). Ultrathin membranes for treatment of waste effluents by reverse osmosis. *Polym. Symp.* **22**, 223–239.

Sachs, B., and Zisner, E. (1972). Reverse osmosis for wastewater reclamation. *In* "Utilization of Brackish Water" (G. A. Levite, ed.), pp. 70–80. Nat. Counc. Res. Develop., Prime Minister's Office, Jerusalem, Israel.

Sachs, B., Shelef, G., and Ronen, M. (1975). "Renovation of Municipal Effluents by Sewage Untrafiltration." Dept. of Membrane Processes, Israel Desalination Engineering, Tel Aviv, Israel.

Sawyer, G. A. (1972). New trends in wastewater treatment and recycle. *Chem. Eng.* Sept. 4 Issue, pp. 120–128.

Schmitt, R. P. (1974). Reverse osmosis and future army water supply. *Amer. Soc. Mech. Eng.*, publ. **74-ENAs-6**.

Shaffer, L. H., and Mintz, M. S. (1966). Electrodialysis. *In* "Principles of Desalination" (K. S. Spiegler, ed.), Chapter 6, pp. 199–289. Academic Press, New York.

Sheppard, J. D., and Thomas, D. G. (1970). Effect of high axial velocity on performance of cellulose acetate hyperfiltration membranes. *Desalination* **8**, 1–12.

Sheppard, J. D., and Thomas, D. G. (1971). *AIChE J.* **17**, 910–915.

Sheppard, J. D., Thomas, D. G., and Channabasappa, K. C. (1972). Membrane fouling. Part IV. *Desalination* **11**, 385–398.

Shuval, H. I., and Gruener, N. (1973). Health considerations in renovating water for domestic reuse. *Enivron. Sci. Technol.* **7**, 600–604.

Sleigh, J. H., and Kremen, S. S. (1971). Acid mine waste treatment using reverse osmosis. *Water Pollut. Contr. Res. Ser.* **14010 DYG08/71**.

Smith, C. V., Jr., and Di Gregorio, D. (1970). Ultrafiltration water treatment. *In* "Membrane Science and Technology" (J. E. Flinn, ed.), pp. 209–217. Plenum, New York.

Smith, J. D., and Eisenman, J. L. (1964). *Eng. Bull. Purdue Univ., Eng. Ext. Ser.* **117**, 738–760.

Smith, J. D., and Eisenman, J. L. (1967). *Fed. Water Pollut. Contr. Admin. (U.S.)* [*Publ.*] **WP-20-AWTR-18**.

Smith, J. M., Masse, A. N., and Miele, R. P. (1970). Renovation of municipal wastewater by reverse osmosis. *Water Pollut. Contr. Res. Ser.* **17040 05/70**.

Smith, R., and Grube, W. (1972). In Wilmoth and Hill (1972).

Sourirajan, S. (1970). "Reverse Osmosis." Academic Press, New York.

Southern Research Institute. (1967). Development of the transport depletion processes. *U.S. Off. Saline Water, Res. Develop. Progr. Rep.* **439**.

Spiegler, K. S., and Kedem, O. (1966). Thermodynamics of hyperfiltration (reverse osmosis) criteria for efficient membranes. *Desalination* **1**, 311.

Staverman, A. J. (1951). The theory of measurement of osmotic pressure. *Rec. Trav. Chim. Pays-Bas Belg.* **70**, 344.

Strathman, H. (1973). *In* "International Symposium on Membranes and Wastewater Treatment" (G. Belfort, organizer). Hebrew University, Jerusalem, Israel.

Strathman, H., and Keilin, B. (1969). Control of concentration polarization in reverse osmosis desalination of water. *Desalination* **6**, 179–201.

Thomas, D. G. (1972). *Membrane Dig.* **1**, 71–201.

Thomas, D. G., Gallaher, R. B., and Johnson, J. S., Jr. (1973). Hydrodynamic flux control for wastewater application of hyperfiltration system. *Environ. Protect. Technol. Ser.* **EPA-R2-73-228**.

Underwood, J. C., and Willits, C. O. (1969). *Food Technol.* **23**, 787.

Wiley, A. J., Dubrey, G. A., and Bansul, I. K. (1972a). Reverse osmosis concentration of dilute in pulp and paper effluents. *Water Pollut. Contr. Res. Ser.* **12040 EEL02/72**.

Wiley, A. J., Scharpf, K., Bansul, I., and Arps, D. (1972b). Reverse osmosis concentration of spent liquor solids in press liquors from high density pulps. *Tappi* **55**, 1671–1675.

Willits, C. O., Underwood, J. C., and Merten, U. (1967). *Food Technol.* **21**, 24.

Wilmoth, R. C., and Hill, R. D. (1972). "Mine Drainage Pollution Control by Reverse Osmosis" (Presented at American Institute of Mining, Metallurgical and Petroleum Engineers, 1972). Obtainable from R. C. Wilmoth, EPA, Box 555, Riversville, West Virginia.

Wilson, J. R. (1960). "Demineralization by Electrodialysis." Butterworth, London.

World Health Organization. (1973). Reuse of effleunts; methods of wastewater treatment and health safeguarding. *World Health Organ., Tech. Rep. Ser.* **517**.

7

Alternative Water Reuse Systems: A Cost-Effectiveness Approach*

Lucien Duckstein and Chester C. Kisiel

I. INTRODUCTION

The objective of this chapter is to compare alternative systems for reuse of reclaimed wastewaters by applying the cost-effectiveness approach—which is briefly reviewed first.

* Published with the kind permission of the American Society of Civil Engineers.

Early applications of cost-effectiveness analysis began in military and aerospace evaluations (Albert, 1963; Hitch and McKean, 1963). One of the advantages of this technique is that the word "effectiveness" has a powerful meaning in itself; it does not necessarily require an evaluation of every economic factor in monetary units. The word conveys specific information in water resources development wherein many economic and social effects can be or cannot be quantified. Reduction in lives lost can be quantified, at least in an expected (average) value sense as undertaken by insurance companies, but cannot be comprehensively valued in any market sense or in a legal sense (for example, in civil suits for loss of life or limb) because of the amorphous character of social values.

Application of this approach to public and civilian systems have been widespread in recent years. It should be noted that the terms output measures, evaluation criteria, measures of effectiveness, program effects, or project effects (either beneficial or adverse) are synonymous. Hatry (1970) suggests general measures of effectiveness for nondefense public programs, such as improvement of health, of highways, etc., Drobny *et al.* (1971) applied the cost-effectiveness approach to the analysis of waste management. De Neufville (1970) attempted to utilize this technique in designing a portion of the New York City water supply system. Kisiel and Duckstein (1972) follow a "standardized" cost-effectiveness approach to guide hydrologic modelers in the problem of hydrologic model choice with consideration of model effectiveness and of cost of model implementation (including costs of overestimation and underestimation by the model). Ko and Duckstein (1972) set forth the evaluation procedures for the analysis of wastewater reuses. Furthermore, the proposed guidelines for water- and land-use planning in the United States, as set forth by the U.S. Water Resources Council (1971), are within the framework of cost-effectiveness analysis as shown by Duckstein and Kisiel (1972).

Other applications include the following topics: municipal waste disposal schemes (Popovich *et al.*, 1973), development of the lower Mekong River (Chaemsaithong *et al.*, 1974), and long-range planning in the Central Tisza River Basin in Hungary (David and Duckstein, 1976). The flexibility of the methodology has been demonstrated by Bokhari (1975) who uses a cost-effectiveness approach to perform an expost study of the irrigation system development in Pakistan.

That there exists a need for a framework, broader than economic analysis, for the analysis of environmental problems has been recognized by the Sanitary Engineering Research Committee of the American Society of Civil Engineers (1972), Case (1972), and McGauhey (1972). The ASCE Committee points to the need for a ". . . broad evaluation of the facts, alternatives, costs, and environmental contacts . . ." for the problem of transporting raw or digested sludge to sea for locations on and beyond the continental shelf. Case (1972) proposes that water and wastewater be treated as a single good and that they should be subject to the same socioeconomic concepts. In fact, he challenges the solution syndrome (every problem has a single solution) and recommends that water problems be identified in ways that permit multiple solutions (the essence of our methodology). He gives elements for three steps of cost-effectiveness as applied to water problems: social goals, measures of effectiveness, and specifications. Similarly, McGauhey (1972) gives elements for organizing the water reuse problem: feasible technology, redefinition of water quality criteria, ecological relationships needed to evaluate existing and proposed water quality management schemes, and the environmental goals of society, All of this serves to emphasize the growing complexity of methods for evaluating alternative environmental systems. The multiple solutions, so derived, depend (a) on how the problem is defined, (b) on how much data is available for solving the problem, and (c) on how much money can be spent for trying to "solve" the problem. Cost-effectiveness analysis gives a unified framework, to be illustrated in the next section, for evaluation of these last three factors through sensitivity and uncertainty analysis.

The authors' experience in applying a cost-effectiveness approach to various problems in hydrology and water resources, from systems design (Chaemsaithong *et al.*, 1973, 1974; David and Duckstein, 1976) to salinity control (Duckstein and Kisiel, 1972) and model choice (Kisiel and Duckstein, 1972), has led to defining the following steps, which are guideposts and do not imply by any means that this is the ultimate methodology:

(a) Define the desired goals or objectives that the systems are to fulfill; this may be done in broad terms as in a word statement

(b) Translate the goals or objectives into sets of engineering, economic, social, and environmental specifications, which may be of a quantitative, qualitative, or even subjective nature; identification of standards and other constraints is included in this step

(c) Establish system evaluation criteria or measures of effectiveness (MOE) that relate system capabilities to specifications

(d) Determine if systems are to be designed on a fixed cost or fixed effectiveness basis

(e) Develop alternative systems to reach goals within the set of feasible technologies and institutions

(f) Determine by analytical or computational means the capabilities of the alternative systems in terms of MOE

(g) Generate an array of alternative systems versus MOE

(h) Analyze merits of systems by ranking (not weighting) MOEs

(i) Perform a sensitivity analysis on all the above steps in order to introduce a necessary feedback into the approach

(j) Document the hypotheses, rationale, model choice, data sources, and analyses underlying the above nine steps.

It may be noted that this approach differs from Kazanowski's procedure (1968, 1972) used in our earlier work (as noted in previous paragraphs) in the sense that alternative systems are determined as late as possible in the analysis; in this manner, the MOEs can be defined without reference to specific systems. The definition of MOEs independently of alternative systems is also recommended in the rigorous systems design methodology of Zapata *et al.* (1973). Further discussion about this approach will be pursued after illustration of the aforementioned methodology by means of the case problem of Tucson.

II. CASE STUDY OF TUCSON

City planners have to cope with two parallel problems in the area of water resources: to forecast and satisfy a growing demand for water and to dispose of a corresponding amount of wastewater. The latter problem is compounded by the presence of urban runoff. These problems become more acute for metropolitan areas located in a semiarid region, as is the case of Tucson, Arizona. A general sketch of water flow in the Tucson basin is shown in Fig. 1, including natural and human elements of the hydrologic cycle. A detailed description of the geographic area under study may be found in Cluff *et al.* (1971). The seriousness of the water problem in this area is related to a tenfold increase of the population in the past 30 years. Furthermore, as there are only intermittent streams in the region (30 days or less of flow per year), the City of Tucson procures its entire water supply by pumping groundwater. Other users in the region are competing for this diminishing groundwater resource: farmers pumping for irrigation, mining industries pumping for ore processing and milling, and other private water companies pumping for small subdivisions. This combined pumping has resulted in a continuous decline of the aquifer at a rate of 60 to 90 cm/year on the average, with some rates of decline above 150 cm (Cluff *et al.*, 1971).

III. APPLICATION OF METHODOLOGY

A. Define Desired Objectives

No meaningful analysis of an engineering problem may be undertaken without a clear statement of desired goals. This is not to say that goals are always

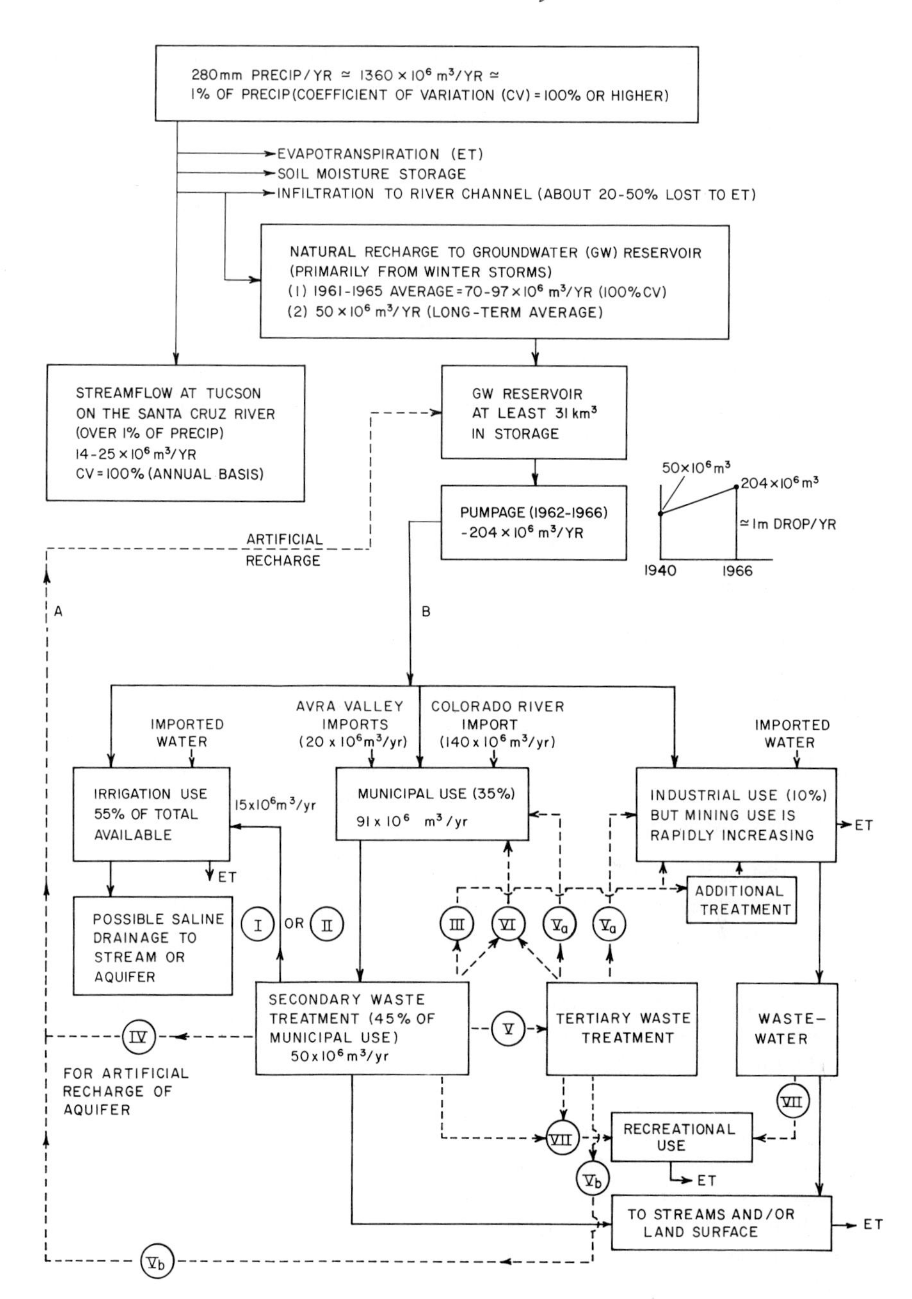

Fig. 1a, b. Water sources and sinks in the Tucson Basin. Figures are as of 1970 except where noted otherwise. Dashed lines indicate that source is not developed as yet or is identified as one of the alternative systems (I–VII) evaluated in this paper. Match marks indicated for 1a and 1b.

easily identifiable; in the case study, considerable interaction between city planners and analysts was necessary until the following objectives of wastewater reuse could be defined (Ko and Duckstein, 1972):

1. To alleviate the groundwater shortage in the basin
2. To alleviate the potential pollution hazard to groundwater from unused wastewater
3. To improve the efficiency of water resources utilization both quantitatively and qualitatively.
4. To develop a methodology applicable to other reuse schemes in similar semiarid urbanized areas

In the next step, these objectives are transformed into specifications written as precisely as possible.

B. Define Specifications

As stated earlier, specifications may be of an engineering, economic, social, or environmental nature; the point is that all acceptable alternative systems must satisfy or closely approach the set of specifications. Note that fulfilling specifications is only a necessary condition for a system to be an acceptable or feasible alternative. Also, a distinction should be made between adhering to strict standards, such as health requirements, and specifying "aspiration levels," such as water taste or hardness, within legal standards. Specifications in this case study are as follows, by approximate order of importance:

1. Meet the water quantity demand of farmers and mining industry
2. Meet the water quantity and quality requirements for the city
3. Meet Arizona State Health Department requirements concerning the reuse of wastewater
4. Stay within cost guidelines, such as (a) for the city, cost of the conveyance system should be less than the benefit accruing from the exchange scheme, or (b) for parties in the exchange scheme, costs are shared proportionally to benefits
5. Use all the sewage effluent that is produced by the city and Pima County secondary treatment plant. These specifications are not precisely detailed here as required for a comprehensive report on the cost-effectiveness analysis; inasmuch as our focus is on illustration of the methodology the interested reader may get specifics from the Annual Reports of the City of Tucson (1971), DeCook (1970), and Cluff *et al.* (1971).

The next step of the approach establishes how the degree of attainment of the goals will be measured.

C. Establish System Evaluation Criteria

Great care should be taken to define criteria or MOE, including costs and benefits, as the ranking of alternative systems ultimately depends on this choice. It is appropriate to keep in mind Kazanowski's remark (1968) that the number of criteria should be kept small lest the analyst loses clarity. MOEs are as follows:

1. *Differential benefit.* The differential or latent benefit is the difference between benefit accruing from the future exchange system and the current wastewater disposal system under a set of well-defined economic hypotheses to be given later. In other words, the benefits contributed by the present water supply schemes of farmers and mining companies are not entered into the calculations even though such absolute numbers are important (but difficult to calculate because all benefits and costs are not measurable). Only the marginal benefits resulting from a change in economic state are used. Such marginal benefits include (a) the nutrient value of water reused, at least \$4.85/1000 m^3; (b) the value of more efficient utilization of water (see DeCook, 1970); (c) the value of improved water quality (less dissolved salts, etc.) (see de Neufville, 1970).

2. *Differential cost.* This is defined in a similar way as the differential benefit and includes the following elements: (a) change in pumping cost between the two states; (b) marginal construction cost of conveyance system; (c) additional operations and maintenance costs; (d) additional treatment costs; (e) storage costs of additional excess effluent.

It may be noted that the sum of MOE 1 and 2 is akin to the change in collective utility (dU) between present economic state I and future exchange scheme II (Duckstein and Dupnick, 1971; Duckstein and Kisiel, 1971).

$$dU \Big|_{I}^{II} = \sum_i p(i) dQ(i)$$

where $i = 1, 2, \ldots,$ I goods and services (water, wastewater, fertilizer, construction, operation, maintenance)

$p(i)$ = price of i

$dQ(i)$ = differential change in consumption or production of good or service i.

3. *Environmental impact factors.* The Environmental Protection Agency has proposed a classification of factors that may be used as a guide to list possible impacts of alternative exchange schemes upon the environment:

(a) Air quality—no effect of consequence is foreseeable.

(b) Water quality—physical, chemical, and biological characteristics may be modified.

(c) Biochemical systems—soil quality characteristics may be modified by an exchange scheme; for example, salinity and hydraulic conductivity are expected to increase if effluent is used to irrigate. This MOE will be included under item 1 or 2, since it may be evaluated in monetary terms.

(d) Quality of life—tailing or holding ponds of secondary treated water may create a nuisance to nearby residents; on the other hand, recreation may be provided under the form of artificial lakes filled with tertiary treated water.

Other MOEs applicable to reuse schemes may be listed as follows:

4. *Contribution to future water supply.* The rate of aquifer depletion in the metropolitan area would decrease by the net amount exchanged, which depends upon the percentage of wastewater return and the proportion lost in conveyance.

5. *Water quality characteristics.* For human consumption or recreation, MOEs are color, taste, odor, hardness, turbidity, viruses, coliforms, and nitrate contents.

6. *Penalties for not satisfying given water quality standards.* In order to evaluate such penalties, a measure of the deviation from the standard must first be defined, then the effect of this deviation on the alternative system must be evaluated. For example, if the total dissolved salts are 1,2, . . . , mg/liter above the standard, what is the effect on agricultural yield, or how much additional treatment cost is necessary? It may be that only a simulation model of the polluted environmental system may give some evaluation of this question.

7. *Percentage of total sewage effluent utilized by respective alternative systems.*

8. *Human factors.* (a) Public acceptance and political feasibility; (b) health hazard (viruses, bacteria, worms); (c) reaction of news media; (d) risk of law suits.

The criteria under item (8) are interrelated; there are many public projects that have not been implemented because of public opposition, which may have been triggered by a newspaper article concerning the project. It is realized that these criteria are most important; if an alternative is not politically acceptable, it is a vain effort to analyze it further, except as an information source in the event future political conditions become more favorable. As the criteria under item (8) could constitute the subject of a separate research effort, only lumped considerations on risk and uncertainty pertaining to human factors will be given here.

D. Select Fixed Cost or Fixed Effectiveness

An alternative with maximum effectiveness and minimum cost belongs to Utopia; either the budget is given and an alternative system with good overall MOEs is sought (a "satisficing" system) (Monarchi *et al.*, 1973), or MOE are fixed, some rigidly, some within a tolerance (aspiration level), and the remaining MOEs are very loose. Then with the latter set of MOEs, a lower cost system is sought, not necessarily the cheapest, especially in public works. As the budget that may be allocated for an exchange scheme depends on changing social values (for example, conservation of resources or saving the environment) and economics (price of water), a fixed-effectiveness approach is appropriate in the present investigation.

E. Develop Alternative Systems

Gavis (1973) has given an overview of the alternatives for wastewater reuse for the purposes of a comprehensive water resources study by the U.S. National Water Commission.

Seven distinct alternative systems are proposed. The first four are specifically geared to the case study of Tucson; the last three, while still potentially applicable to Tucson, are alternatives of a more general nature.

1. SYSTEM I

The secondary treated sewage effluent is exchanged for the groundwater that is presently used for irrigation in a 115 km^2 area, by using a 2.80 m^3/second sewage effluent delivery system with 9.5 km of 168 cm gravity pipe conduit and 32 km of concrete-lined canals. The capacity of this system is sufficient to handle 55 km^3 of effluent per year, which is the estimated production in 1995. Further, additional wells must be drilled and a collecting system must be constructed in order to be able to utilize an existing city pipeline to return up to 1.32 m^3/second of fresh groundwater. Under this alternative, the farmlands belong either to individuals, to a group of farms, or to the state; a contract signed between the City of Tucson and the farmers specifies the conditions of the exchange.

2. SYSTEM II

While Systems I and II are physically equivalent, an important legal feature distinguishes these two systems: the city buys a minimum of 53 km^2 of farmland, of which 41 km^2 are irrigable. The city then either leases land back to individual farmers or uses the land as it sees fit. By owning the land, the city has control of the water according to court decisions in Arizona and is not tied down by a contract to deliver a certain amount of water which it may wish to utilize for other schemes later.

3. SYSTEM III

The sewage effluent is exchanged for the groundwater that is presently used as a water supply for mining and milling. This system requires building 34 km of pressurized pipelines; an existing pipeline may be used to return fresh groundwater. The underlying assumptions are that all parties to the scheme can effectively utilize the sewage effluent.

4. SYSTEM IV

The secondary treated sewage effluent is discharged into Rillito Creek, so that the natural sands and gravel in the valley may provide purification and storage. The future pumped water in the vicinity of riverbeds is then a mixture of fresh and used water: a chemical model of that part of the basin is necessary and is being built so that salt concentrations in the resulting mixture can be forecast (Kisiel *et al.*, 1972; Supkow *et al.*, 1973). A 5-km segment of Rillito Creek with an average width of 50 m is used as a recharge area. The estimated percolation ratio varies between 0.60 m/day and 6 m/day, which corresponds to a recharge range of 36×10^6 to 36×10^7 m^3/year. These figures may have to be decreased because of possible losses to evapotranspiration caused by deep-rooted plants.

5. SYSTEM V

The sewage effluent receives advanced or tertiary treatment. Two marginally different cases may be distinguished: (Va) the tertiary treated water is recirculated into the city system and used conjunctively with fresh water for domestic, industrial, commercial, and municipal purposes. (Vb) the tertiary treated water is recharged into the aquifer partly with the Santa Cruz River and partly with the same section of Rillito Creek as in System IV. The merits of recharging tertiary treated water rather than secondary treated water will be discussed later, as well as the difference between systems Va and Vb.

6. SYSTEM VI

The secondary treated effluent is distributed to industrial or commercial users using a dual distribution system for the City. While such a system may be impractical to implement in an existing city, its construction at the time a water distribution network is being built may be advantageous.

7. SYSTEM VII

The secondary treated effluent is further treated to recreational standards and used in artificial lakes for boating, swimming, and fishing.

F. Determine Capabilities of Alternative Systems

1. ADVANTAGES—SYSTEM I

The merits of System I include the following:

(a) *Plant nutrient value.* According to DeCook (1970), the estimated value of the nutrient content of secondary sewage effluent, primarily nitrogen and phosphorus, was $4.00/1000 m^3 in 1969. An updated value would be $6.00/1000 m^3. The exact value of the nutrient depends upon cropping patterns and mixture of fresh water to effluent applied.

(b) *Avoidance of advanced treatment costs.* Currently, the excess effluent is discharged into the channel of the Santa Cruz River. At the same time, groundwater in wells near the effluent disposal grounds has recently shown a substantial increase in nitrate content. Thus, the danger of polluting groundwater seems to exist, to the point where the city is being sued for allegedly endangering the water supply of nearby users. The implementation of an exchange scheme may thus avoid potential groundwater pollution. This benefit is at least equivalent to the cost of additional treatment needed to eliminate the nitrate-producing element; this figure could be increased by the amount saved by avoiding lawsuits. Such advanced treatment cost is estimated at $9.30/1000 m^3 for a 1.97 m^3/sec plant (Smith and McMichael, 1969).

(c) *Other miscellaneous benefits.* The permeability of soil irrigated by sewage effluent is expected to increase (Gray, 1968). On the other hand, soil salinity is also expected to increase. Further study is necessary to ascertain long-range effects of these two phenomena; in the meantime, it will be assumed that their added effect is negligible.

2. DISADVANTAGES—SYSTEM I

(a) Only the contracted amount of water can be exchanged; thus, at contract expiration, there is no guarantee of renewal under similar terms. In fact, farmers may ask a very high price for water at contract renewal time since the price of a diminishing resource increases indefinitely.

(b) The use of effluent cannot be controlled once it is turned over to users. In particular, public health problems such as groundwater pollution may be caused by careless practices at certain locations.

(c) Scheduling the delivery of effluent is a difficult task in the face of multiple ownership of farmland. For operational purposes, a water district with well-defined operation rules would have to be set up.

(d) Should another exchange scheme become attractive during the lifetime of the contract, there would be considerable difficulty in switching over to that system; the flexibility of System I is poor.

3. SYSTEM II

If the City of Tucson owns the land, long-range integrated plans for groundwater supply and sewage disposal can be established. System II thus presents all the benefits and none of the disadvantages listed for System I, except that capital funds must be provided for land purchase.

The city currently trucks the solid waste resulting from the sewage treatment to various parks for use as a soil conditioner. The plant nutrition value of sludge is minimal. For one of several parks alone, e.g., Randolph Park, the annual transportation cost was $33,200 in 1969. If exchange System II were implemented, the solid waste could be transported through the effluent pipeline and discharged on irrigated fields, which would both save trucking costs and provide a continuous method of disposing of solid waste. The benefit corresponding to this feature is estimated at $1.56/1000 m^3/year.

As stated above, the origin of funds to purchase the farmland must be examined. A 6% bond could provide the capital necessary for such a venture. The estimated price of land is relatively small: approximately $6.5 million for 52,500 m^2; after purchase, about $48/1000 m^2 of income could be generated for the city for leasing land to farmers. Another possibility would be to retire part of the land from agriculture and utilize it for other purposes while retaining the water rights.

4. SYSTEM III

Only the second advantage of System I would be preserved, namely, avoidance of groundwater pollution, because the mines produce almost no discharge (although this advantage is not clearly established in view of a current lawsuit involving the city, farmers, and mines on allocation of groundwater withdrawal rates). This assumes proper internal management of water by the mines, which is reasonable in view of the present situation. However, the advantage of avoiding groundwater pollution would be nullified by the cost of bringing effluent quality up to mining standards, especially flotation of ore, which necessitates low nitrate and phosphate content. Next, this system would have the same drawbacks as System I; as groundwater becomes a scarcer resource, the mines are likely to compete with the city for land purchase. Also, the copper mines have already purchased part of the farmland south of the city to avoid further legal problems. The possibility thus exists that the mines may purchase land elsewhere in the vicinity to gain access to water rights.

5. SYSTEM IV

A very high amount of effluent can be reutilized under this system. The city estimates that 100% of the projected 1995 effluent can be absorbed. However, the problem of increase in salt content of groundwater must be investigated using the chemical model mentioned earlier; also, a solid waste-disposal system must be provided. The problem of future water supply shortage is partially reduced, but not solved.

6. SYSTEM V

If water becomes really scarce, then complete recycling using tertiary treatment may become the answer, irrespective of cost. However, the fundamental problems of complete destruction of viruses and toxic organisms in the recycled water must be solved with a reliability very close to one (Okun, 1968; Berg, 1971). Hence, two variations of System V may be proposed with the following features:

(a) The tertiary treated effluent is directly mixed with fresh water and sent into the city water-distribution network. Public reluctance to accept this solution may prevent its implementation for many years. The public may have a point in refusing to accept a system where the slightest "failure" may cause an epidemic of viral and chemical origin. In support of these public positions are the policy statements on water reuse by the Environmental Protection Agency (McDermott, 1972) and by the American Water Works Association (1971),

(b) The tertiary treated effluent is recharged into the ground, at a location where percolation is fairly rapid. The water is then pumped back with the fresh groundwater. This gives a residence time in the ground for the treated effluent, which facilitates control and potentially reduces health hazards. Naturally, for System Vb, cost of pumping and of transport losses must be added to the cost of System Va.

Generally speaking, recycling of tertiary treated effluent is only at the experimental stage throughout the world. Other chapters in this volume provide a state-of-the-art review of technological health and economic aspects of this question.

7. SYSTEM VI

Although a dual system for water supply and waste disposal may appear impractical in the City of Tucson, which is quite developed (population of 350,000 in 1973), it is proper to mention that such a system may become meritorious in the future. Singer (1969) mentioned that such a system may not be as costly as initially thought. Several cities, such as Paris (France) and Györ (Hungary), have a dual supply system: one for potable water and one for industrial water. The latter is taken out of a river, the Seine and the Danube, respectively, and receives very little treatment. The industrial water could instead be secondary treated effluent: it could be used for parks and gardens, cooling towers, etc., in addition to regular industrial consumption. In planning water supply systems for new cities, a dual supply system should be considered as an alternative. Another aspect is a dual disposal system: one for fecal waste, high in viruses and bacteria, the other one for other domestic or industrial waste. Many variations may be considered here. With their advantages and disadvantages, they constitute a separate study in themselves.

8. SYSTEM VII

This system may be used to enhance recreational capabilities of a region; if a freshwater reservoir is presently being used for recreation, that fresh water can be exchanged for the treated effluent. In a semiarid climate, evaporation and seepage losses may be very large; however, such a system is mentioned for the sake of completeness as a possible alternative, in the spirit of the fourth goal of this study—to develop a methodology applicable to other exchange schemes.

At this point, it must be noted that there is not a one-to-one correspondence between this presentation of merits of alternative systems and the column headings in Table I. This was done for the purpose of simplifying the presentation; only essential MOEs are represented in Table I. For a complete example of a cost-effectiveness array, see Chaemsaithong (1973).

G. Generate Matrix of System versus Criteria

The results of analysis are displayed in Table I in terms of cost and effectiveness criteria. Table II gives details on the calculation of two of the cost criteria, latent benefit and latent cost, as defined previously in this chapter. It is important to recognize that, under our hypotheses (given in Section J), the benefits and costs are the estimated differential or latent values (*not* absolute values) generated by the implementation of an exchange system, i.e., a marginal transformation from economic state I, no exchange, to economic state II, with exchange. Other reasonable cost criteria, not presented here, include cost of analysis and the cost of uncertainties in the system. The latter costs (including nonmeasurable utilities) are frequently coupled with risk in a subtle way in the mind of the decision maker. The former costs (of analysis or thinking) loom larger as the pressures grow for more involved studies. The key question is: How much meaningful analysis can be realized with limits on time, budgets, manpower capability, and available data (frequently we cannot afford to wait for more data)?

Within the category of effectiveness criteria, the enumeration of all the environmental impact factors would have evoked on the part of the matrix user a form of mental paralysis or indigestion. Thus, for purposes of illustration,

Table I.
Cost-Effectiveness Criteria [a]

	Cost criteria, in dollars per thousand years			Effectiveness criteria						
System (1)	Total benefit (2)	Latent benefit (3)	Latent cost [b] (4)	Possibility of damage to environment (5)	System contribution to future water in 1.23×10^6 m³/yr (6)	Quality of exchange ground-water (7)	Effluent utilized, as a percentage (8)	Flexibility for deviation (9)	Risk of losses (10)	Public acceptance at proposal time (11)
I	+679.5	+877.5	\$198 [b] 202 [c] 325 [d]	Small	45	Excellent	64/1995	Rigid	Moderate	Excellent
II	+1,902.2	+2,327.5	425 [b] — [c] 550 [d]	Extremely small	89.5	Excellent	100	Rigid	Extremely small	Good
III	−219	46.0	265 [b] — [c] 460 [d]	Very small	45	Good	64/1995	Very rigid	Moderate	Excellent
IV	+471.0	+518.0	46.6 [b] — [c] 81.5 [d]	Small	29–292	Good	30–300	Rigid	Fair	Average
V	—	—	(a)568 [b] (b)680 [b]	Extremely small	(a) 25–250 (b) 20–200	Excellent	20–250	Good	(a) Moderate (b) Small	(a) Very bad (b) Excellent
VI	—	—	—	Moderate	20–60	—	—	Good	Moderate	Good
VII	—	—	—	Moderate	0	—	100	Poor	Fair	Average

[a] From Ko and Duckstein (1972).
[b] Latent cost at zero discount rate.
[c] Latent cost at $3\frac{1}{4}$% discount rate at 40 payments.
[d] Latent cost at 6% discount rate and 40 payments.

Table II
Cost Estimation[a]

Structures (1)	Estimated annual cost, in dollars (2)
(a) For System I	
61 km of canal (0.85 m^3/sec)	340,000
16 km of canal (0.28 m^3/sec)	218,000
Right-of-way of canal	23,000
Canal structures	118,000
Holding pond and land	710,000
Pipelines (groundwater)	373,000
Engineering costs and contingencies	422,000
Subtotal	2,204,600
At 40 payments, no interest	55,000
Pumping costs	25,200
Operation and maintenance	118,000
Total per year	198,200
(b) For System II[b]	
32.2 km concrete-lined canal	1,500,000
2.0 m concrete pipe	2,300,000
11.25 km concrete-lined lateral	45,000
3.06 km concrete-lined lateral	360,000
Pipelines (groundwater)	373,000
Pumps	60,000
Right-of-way	29,500
Reservoir	500,000
Subtotal + 33%	6,872,700
At 40 payments, no interest	172,000
pumping cost	45,000
Operations and maintenance	48,000
Total per year	265,000
(c) For System III	
Pipeline cost	8,722,000
Reservoirs and pumping	1,000,000
Drivers	503,000
Fuel	375,000
Total	10,600,700
At 40 payments, no interest, per year	265,000

Table II (continued)

Structures (1)	Estimated annual cost, in dollars (2)
(d) For System IV	
Well sites	16,000
Wells	250,000
Pipeline (groundwater)	373,000
Pipeline (effluent)	214,000
River dikes	10,000
Riverbed	100,000
Booster station	20,000
Subtotal	983,000
Operations and maintenance	880,000
Total	1,863,000
At 40 payments, no interest, per year	46,600

[a]From Ko and Duckstein (1972).
[b]Not including cost of land acquisition.

we as analysts have exercised subjective judgment in relating system performance to specification. Thus, for example, in considering the possibility of environmental damage (column 5 in Table I) and the quality of exchange groundwater (column 7 in Table I), we have telescoped a vector set of environmental quality criteria into single judgmental words like small or extremely small; the damage is taken to mean the possibility of polluting the environment when an exchange system is implemented. In fact, this forecast should be done with the aid of a simulation model of the alternative system and its environmental impacts.

Concerning column 6, the system contribution to future water supply is in terms of the water volume saved for future groundwater supply. In column 9, the flexibility for deviation from system specifications indicates the adaptability of each system to another quality of exchanged effluent. Column 10 gives a judgmental evaluation of the possibility of the risk of social (including public health) and economic loss; readers may disagree with our judgments, but the results will vary with our respective personal values assigned to the various attributes of the systems. Column 11 reflects the presumed reaction of the public to various alternatives, as learned from experience elsewhere.

H. Analyze Merits of the Systems

At this juncture of the analysis, the analyst must decide if it is within his prerogative to exercise value judgments about the relative merits of each system, or if it is within the domain of the client, user, or decision maker. If the latter is decided, then the remainder of analysis (steps I and J) is pursued. If the former is decided, then the analyst must decide between the use of ranking and weighting methods. We recommend ranking of systems, rather than weighting, in view of the great difficulties in subjectively assigning a consistent and unambiguous quantitative measure to each MOE for each system. Ranking, in a sense, is the recommended procedure by the U.S. Water Resources Council (1971), whereas weighting has been espoused for multi objective problems by Major (1969), based on the principles given by Maass (1962). Major essentially reduces a set of noncommensurate and incomparable objectives into a single index (through a univariate objective function) by assigning weights to each objective. Weights have also been used by Drobny *et al.* (1971) on criteria for evaluating waste management systems. Kazanowski (1968), in writing of the quantification fallacy or syndrome, considers weighting a major fallacy in the use of cost-effectiveness techniques in the aerospace, defense, and public sectors. Freeman (1969) explicitly challenges the use of a priori weights in water resources planning. The same criticism can be assigned to the classical operations research approach (Monarchi *et al.*, 1973) and to the more traditional benefit-cost analyses (Howe, 1971) because of the tendency to include only the quantifiable elements in both benefits and costs and to submerge the subjective elements. In our judgment, it would be preferable, in order to facilitate implementation of the information generated by cost-effectiveness analysis, to ensure the decision-maker's participation in the process of ranking. In particular, his awareness of who is paying and who is benefitting from a given system should enter into the decision-making process.

Inspection of Table I suggests that, if we as analysts were to confine ourselves to the first four systems, System II stands out as best with respect to all criteria, except latent cost (higher than the cost for the other systems) and lack of flexibility. Yet, the total benefits and latent benefits are of such magnitude that the higher latent costs of System II are small in comparison. The method of comparison of System II with the last three (tertiary treatment, dual systems, and recreational use) is not so readily apparent because of inherent intangibles and uncertainties. Tertiary treatment can be costly and still is, in many respects, in the developmental stage (Middleton and Stenberg, 1972). It is too expensive a product to be fed back into irrigation systems, recreational systems, and even, perhaps, back into the aquifer as recharge water but, if the treatment is tailored to industrial needs, it might find high economic value there. But it is uncertain if the market is ready. Tertiary treated water would be logical as an input to dual water systems with the following two provisos: (1) construction of dual distribution systems is economically feasible for the community (in fact, it is not and there is no incentive as yet arising from possible future water shortages or much lower future quality of groundwater) and (2) the tertiary process is able to cope with the virus problem (in fact, it is not to the satisfaction of all concerned). In view of these relative merits of the seven systems, it does seem reasonable to consider combinations of the alternatives (II, V, VII). System VII is included on the possible merits of the public service view of subsidizing development of recreational use of high-cost tertiary treated water. As more feasible technology develops, a re-evaluation of all alternatives is in order. For example, the key question in viral control involves a difficult trade off between public health criteria and public expenditures; that is, should we respond to uncertainty in virus control by building dual systems? This possibility and others can be encompassed in the next step: sensitivity analysis.

I. Sensitivity Analysis

In a sense, this step should precede Step H, but it seems reasonable to expect that the analysts have consciously noted the assumptions in all previous steps and are now prepared to decide on appropriate factors to perturb. This is not a simple task in general because, at this point, the cost of the entire cost-effectiveness analysis may jump by orders of magnitude. We can only suggest the use of judgment in choosing uncertainty factors for sensitivity studies.

Among the factors that might be singled out for sensitivity analysis, we include (a) effluent price versus net total benefits, (b) discount rate versus latent cost, (c) planning horizon versus latent cost, (d) amount of irrigated land versus net total benefits, (e) effect of adjustments in major components of delivery systems on ranking of systems (we exclude here the effect of population growth on sizing of components), (f) effect of additional land value on net total benefits, (g) forecasts of future population, waste loads, and water usage; see Berthouex and Polkowski (1970) for an excellent review of the uncertainty in estimates of these forecasts and for recommendations about using decision theory for evaluating the effects of poor forecasts, (h) effect of municipal water pricing policy and effluent tax on future water consumption and waste loads; see Duckstein and Kisiel (1971) for a methodology that compares two different water-pricing structures (regressive and progressive), (i) effect of improved technology, and (j) uncertainties in goals, cost estimates, and chosen criteria (Kazanowski, 1972).

Items (d) (f) are operations research problems because it is possible to define a single dimensional objective function (cost or profit) subject to well-defined constraints on prices, available land, labor costs, etc. For illustrative purposes, items (a), (b), and (c) are the only sensitivity factors considered herein.

1. EFFLUENT PRICE VERSUS NET TOTAL BENEFITS

Tentatively, let the effluent price be set at \$0.81/1000 m³ to the farmers and mining companies. But the price may range from a negative value (a consequence of government subsidy) to the price of groundwater pumping (\$11.50/1000 m³). Figure 2 shows the results of the sensitivity analysis for Systems I, III, and IV. It is interesting to note that System III has higher net benefits after about \$7.30/1000 m³. System IV shows no change in net benefits because of the implicit assumption made throughout the study that the future value of water is the same as the present one. While such a hypothesis reflects the current thinking of water planners, the possibility of using a social discount factor to indicate the value of preserving a dwindling resource should be considered, and would clearly modify the curves in Fig. 2.

2. DISCOUNT RATE VERSUS LATENT COST

In Table I, no discount rate was used to obtain estimates of latent costs. Even though it is common to assume a 6% discount rate, farmers, under System I, could qualify for small project reclamation loans from the U.S. Bureau of Reclamation depending on the amount of land owned by the applicant. The

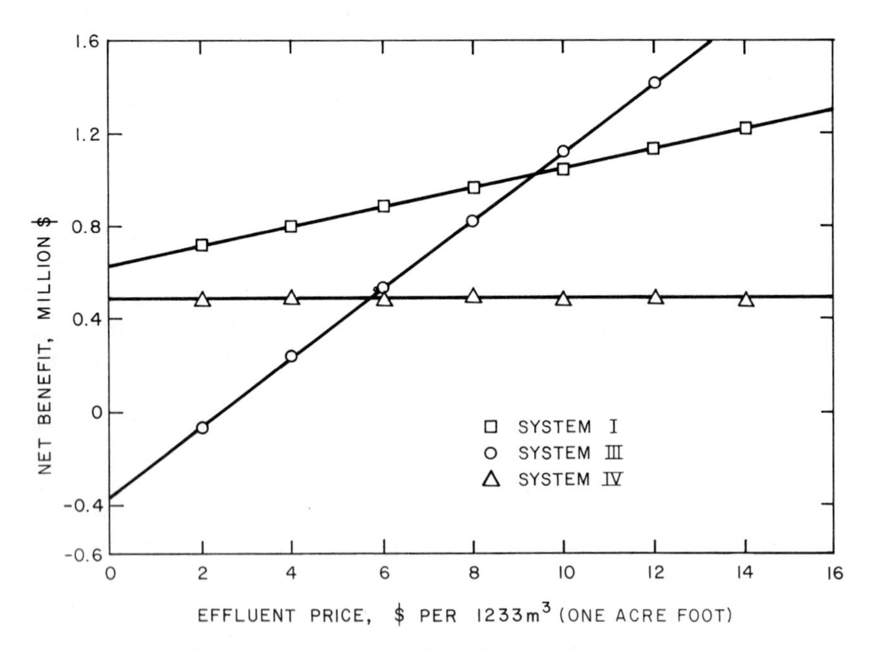

Fig. 2. Effluent price versus net total benefit. From Ko and Duckstein (1972).

discount rate for single ownership is zero for 64.75 hectares or less and is 3 1/4% for excess land area; but the cutoff point is 129.5 hectares for joint ownership by married couples. Hence, the latent costs in Table I are based on 0, 3 1/4 and 6% discount rates.

3. PLANNING HORIZON VERSUS LATENT COSTS

To determine latent cost, as a function of size of delivery system and, in turn, the planning horizon, requires a comparison of the forecast of total available effluent in the future and of the current capacity of the waste-disposal system. For the City of Tucson, effluent forecasts are 1.76 m^3/second in 1975, 2.28 m^3-second in 1985, 3.08 m^3/second in 1995 and 4.38 m^3/second in 2005. From Fig. 3, we observe that at a 25-year planning horizon the latent cost begin to level off. These results can change substantially if the uncertainty analysis recommended by Berthouex and Polkowski (1970) is undertaken.

J. Documentation of Assumptions and Conditions of Analysis

The factors identified as candidates for sensitivity analysis suggest a set of assumptions that can be quite extensive and detailed. Some of our assumptions have been discussed in Section H and will not be repeated here. For exemplifying

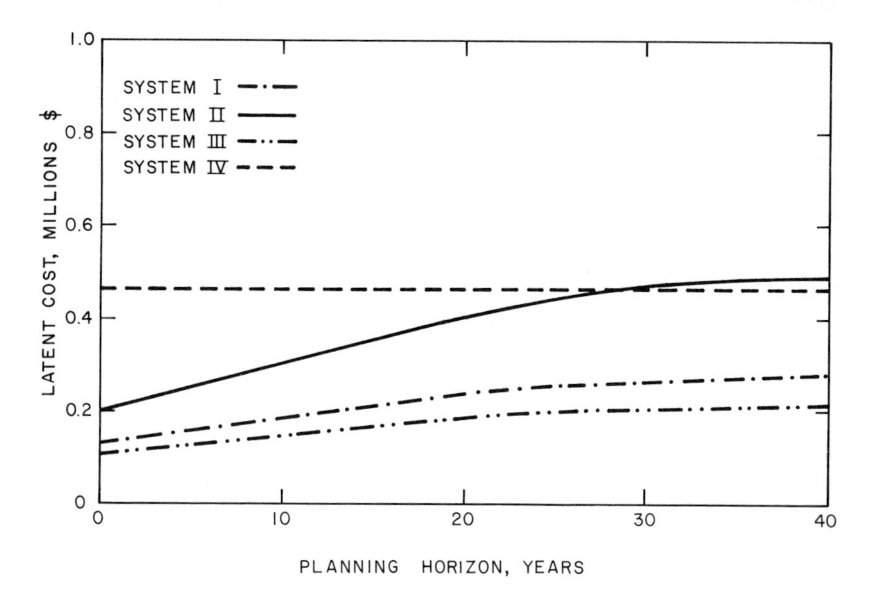

Fig. 3. Planning horizon versus latent cost. From Ko and Duckstein (1972).

our orientation, we focus only on the hypotheses made to compute latent benefits and latent costs. These are (a) the change from state I to state II effects only a marginal transformation (this is the primary axiom of the collective utility approach); (b) a fair market exists for exchange of goods and services, that is, each firm or sector pays the same price for the same goods or services (e.g., \$0.81/1000 m^3 of effluent); (c) each firm or sector is a price taker instead of a price maker, because there is no quantity discount; and (d) no additional water resources can be imported into the economic area under consideration.

Hypotheses (a) and (b) are related in the sense that large-scale imports, such as the proposed Central Arizona Project (CAP) designed to bring Colorado River water, can induce structural (major) transformations in the economy of the region. This imported water can have an important impact on each of the seven systems evaluated herein [see Fig. 1 for interaction of imports and these systems; also the work of Engineering Science, Inc. (1973) includes the CAP in its analysis]. Concerning (b), it is not clear that a fair market for water exists in the region, except perhaps as induced by water litigation. Concerning (c), all sectors or firms are price takers in the sense that the water is waiting to be pumped (under existing legal constraints); on the other hand, the City of Tucson does give quantity discounts to its consumers through a regressive pricing policy. Certainly, a more definitive study should consider the effects of these "complications."

An increasingly important facet to document is the budget, manpower, time, and computer used in the analysis. It should be clear from our earlier discussion on uncertainties that a form of infinite regress can set in, that is, some analysts, undertaking the cost-effectiveness approach, may never be satisfied. But all "good" things (like analysis for its own sake) must come to an end. Hence, the vector cost of thinking should be made a matter of record so as to promote a deeper perspective on the requisite scale of analysis for the increasingly complex environmental problems.

IV. SUMMARY AND CONCLUSIONS

We have sought to demonstrate by example the merits of a clearly defined cost-effectiveness approach to an environmental management problem. Because the environment of interest and the impinging impacts are frequently ill-defined and uncertain in many regions of the world, the need for more realistic analysis procedures is clearly evident as emphasized by us and other cited works in Section I. We hope that the approach, illustrated by our case study of alternate wastewater reuse schemes for Tucson, Arizona, gives insight valuable for informed decisions on similar problems to be made elsewhere. Even though, in our analysis, Systems I to IV are specific to Tucson because of the topography of Avra Valley in relation to the adjacent Tucson Basin and the presence of copper mines nearby, Systems V to VII may be considered for almost any medium sized town.

We see no limits on the applicability of the methodology, one that is not sacrosanct, but continually evolving. So-called weaknesses of the cost-effectiveness technique, like uncertainties in goals, evaluation criteria, alternative systems, and cost functions, are in fact reflections on the human condition, on our limited past data, and on our inability to forecast future natural and man-induced events.

As noted by Ko and Duckstein (1972) and Dávid and Duckstein (1976), a step by step approach to cost-effectiveness studies has the following merits:

(a) It specifies a common vocabulary and framework between analysts of diverse persuasions and disciplines and decision makers.

(b) It forces us to identify both quantitative and nonquantifiable criteria without recourse to the dangerous practice of a priori weighting.

(c) It encourages the constructive use of sober judgment through subjective ranking of criteria in an explicit manner (assumptions must be stated so that contemporaries and future users can judge for themselves).

(d) It supplies an adaptive procedure for adjusting, updating, and feeding back information at every step.

(e) It places in proper perspective (Steps e and f) the plethora of physical models, operations research models, and computer simulation models; many alternatives can be discarded before detailed studies or exhaustive simulation runs are made.

(f) It provides for sensitivity analysis to account for uncertainty in data and in models. For example, Bayesian decision theory (Davis *et al.*, 1972) gives a framework for such analysis and provides additional criteria for Step C.

REFERENCES

Albert, B. S. (1963). Cost-effectiveness of naval air weapons systems. *Oper. Res.* **11**, 173–189.

American Water Works Association. (1971). American water works association policy statement on use of reclaimed waste waters as a possible water-supply source. *J. Amer. Water Works Ass.* **63**, 490.

Berg, G. (1971). Integrated approach to problem of viruses in water. *J. Sanit. Eng. Div., Amer. Soc. Civil Eng.* **97**, 867–882.

Berthouex, P. M., and Polkowski, L. B. (1970). Design capacities to accomodate forecast uncertainties. *J. Sanit. Eng. Div., Amer. Soc. Civil Eng.* **96**, 1183–1210.

Bokhari, S. M. H. (1975). Ex post evaluation of river basin in Pakistan. Proceedings, United Nations

Interregional Seminar on River Basin Development, 1975, Budapest, Hungary. Water Resources Branch Center for Natural Resources, Energy and Transport, United Nations, New York.

Chaemsaithong, K. (1973). Design of water resources systems in developing countries: the lower Mekong basin. Natural Resources Systems Tech. Rep. Ser. No. 19, June 1973.

Chaemsaithong, K., Duckstein, L., and Kisiel, C. C. (1973). Hydrologic and social inputs to cost-effectiveness design of a water resources system. *Int. Ass. Hydraul. Res. Congr., 15th, 1973* pp. 274–285.

Chaemsaithong, K., Duckstein, L., and Kisiel, C. C. (1974a). Alternative water resource systems in the lower Mekong. *J. Hydr. Div. Amer. Soc. Civil Eng.* **100**, 461–475.

Case, F. E. (1972). Economics of water quality and wastewater control. *J. Sanit. Eng. Div., Amer. Soc. Civil Eng.* **98**, 427–434.

City of Tucson. (1971). "Annual Report, 1969–1970. Water & Sewage Systems, City of Tucson, Tucson, Arizona.

Cluff, C. B., DeCook, K. J., and Matlock, W. G. (1971). Technical and institutional aspects of sewage effluent-irrigation water exchange, Tucson region. *Water Resour. Bull.* **7**, 726–739.

Dávid, L. and Duckstein, L., (1976). Multi-criterion ranking of alternative long-range water resources systems. *Water Resour. Bull.* (In press.)

Davis, D., Kisiel, C., and Duckstein, L., (1972). Bayesian decision theory applied to design in hydrology. *Water Resour. Res.* **8**, No. 1, 33–41.

DeCook, K. J. (1970). Economic feasibility of selective adjustments in the use of salvageable waters in the Tucson region, Arizona. Thesis presented to the University of Arizona, Tucson, in fulfillment of the requirements for the degree Doctor of Philosophy.

de Neufville, R. (1970). Cost-effectiveness analysis of civil engineering systems: New York City's primary water supply. *Oper. Res.* **18**, 785–804.

Drobny, N. L., Qasim, S. R., and Valentine, B. W. (1971). Cost-effectiveness analysis of waste management systems. *J. Environ. Syst.* **1**, 189–210.

Duckstein, L., and Dupnick, E. (1971). Collective utility and cost-effectiveness in natural resources management. *40th Nat. Meet. Oper. Res. Soc. Amer.* 22. pp. (available as reprint 71–25, Systems and Ind. Engr. Univ. of Ariz., Tucson).

Duckstein, L., and Kisiel, C. C. (1971). Collective utility: a systems approach to water pricing policy. *Proc. Int. Symp. Math. Models Hydrol. 1971* p. 881–888.

Duckstein, L., and Kisiel, C. C. (1972). Cost-effectiveness approach: an example of water quality control in a river basin. *In* "Public Response to the Proposed Principles and Standards for Planning Water and Related Land Resources and Draft Environmental Statement," pp. 3455–3475. U.S. Water Resour. Counc. Washington, D.C.

Engineering Science, Inc. (1973). "Final Report on Urban Systems Engineering Study of Tucson's Water Resource and Solid Waste Disposal System" (to the City of Tucson and the United States Department of Housing and Urban Development). Eng. Sci. Inc., Tucson, Arizona.

Freeman, A. M., III, (1969). "Project design and evaluation with multiple objective in the analysis and evaluation of public expenditures," a compendium of papers submitted to the subcommittee on Economy in Government, p. 57. United States Congress Joint Economic Committee, Washington, D.C., 1969.

Gavis, J. (1973). "Waste Water Reuse." U.S. Nat. Water Comm., US Gov. Printing Office, Washington, D.C.

Gray, J. F. (1968). "Practical Irrigation with Sewage Effluent, Municipal Sewage Effluent for Irrigation." Agr. Eng. Dep. Louisiana Polytechnic Institute, Ruston.

Hatry, H. P. (1970). Measuring the effectiveness of non-defense public programs *Oper. Res.* **18**, 772–784.

Hitch, C. J., and McKean, R. N. (1963). "The Economics of Defense in the Nuclear Age." Harvard Univ. Press, Cambridge, Massachusetts.

Howe, C. W. (1971). "Benefit-cost analysis for water systems planning," Water Resources Monograph 2. American Geophysical Union, Publication Progress, Inc., Baltimore, Maryland.

Kazanowski, A. (1968). A standardized approach to cost-effectiveness evaluation. *In* "Cost-Effectiveness: The Economic Evaluation of Engineering Systems" (J. English, ed.), pp. 113–156. Wiley, New York.

Kazanowski, A. (1972). Treatment of some of the uncertainties encountered in the conduct of hydrologic cost-effectiveness evaluation, *Proc. Int. Symp. Uncertainties Hydrol. Water Resour. Syst. 1972* pp. 771–785.

Kisiel, C. C., and Duckstein, L. (1972). Economics of hydrologic modelling: a cost-effectiveness approach, *Proc. Int. Symp. Model. Water Resour. Syst. 1972* pp. 319–330.

Kisiel, C. C., Supkow, D. J., and Tetley, W. (1972). Role of digital computer models of aquifers in water resources planning: case study in Tucson, Arizona, *Proc. Int. Symp. Planning Water Resour., 1972* Vol II pp. 35–55.

Ko, S., and Duckstein, L. (1972). Cost-effectiveness analysis on waste water reuses. *J. Sanit. Eng. Div., Amer. Soc. Civil Eng.* **98,** Proc. Paper 9434, 869–881.

Maass, A., et al, (1969). Benefit-cost ratios for projects in multiple objective investment program. *Water Resour. Res.* **5,** No. 6, 1174–1178.

Major, D. C. (1969) Benefit-cost ratios for projects in multiple objective investment program. *Water Resour. Res.,* **5,** No. 6, 1174–1178.

McDermott, J. H. (1972). Waste Water Reuse. *J. Amer. Water Works Ass.* **64,** 627–630.

McGauhey, P. H. (1972). Clean water—an environmental challenge. *J. Sanit. Eng. Div., Amer. Soc. Civil Eng.* **98,** 375–378.

Middleton, F. M., and Stenburg, R. L. (1972). Research needs for advanced waste treatment. *J. Sanit. Eng. Div., Amer. Soc. Civil Eng.* **98,** 515–528.

Monarchi, D., Kisiel, C. C., and Duckstein, L. (1973). Interactive multi-objective programming in water resources: a case study. *Water Resour. Res.* **9,** no. 3, 837–850.

Okun, D. A. (1968). The hierarchy of water quality. *Environ. Sci. Technol.* **2,** 672–675.

Popovich, M., Duckstein, L., and Kisiel, C., (1973). On the mismatch between data and models of hydrologic and water resources systems. *J. Envir. Eng. Div., Amer. Soc. Civil Eng.* **99,** 1075–1088.

Sanitary Engineering Research Committee of the American Society of Civil Engineers. (1972). Research needs in sanitary engineering. *J. Sanit. Eng. Div., Amer. Soc. Civil Eng.* **98,** 299–304.

Singer, S. F. (1969). Dual water supplies. *Environ. Sci. Technol.* **3,** 197–198.

Smith, R., and McMichael, W. F. (1969). "Cost and Performance Estimate for Tertiary Wastewater Treating Processes," Rep. No. TWRC-9. Robert A. Taft Water Resour. Cent. Cincinnati, Ohio.

Supkow, D. J., Kisiel, C. C., and Simpson, E. (1973). "Digital Modeling of the Tucson Basin," Progr. Rep. for the City of Tucson from Jan. 1 to Mar. 31, 1973, University of Arizona, Tucson, Arizona.

U.S. Water Resources Council. (1971). Proposed principles and standards for planning water and related land resources. *Fed. Regist.* **36,** (24, 144–24, 194.)

Zapata, R. N., Wymore, A. W., and Cross, B. K. (1973). A systems engineering formulation of the open pit mine problem, *Proc. Symp. Comput. Appl. Miner. Ind. 11th, 1973.* pp. 279–298.

Part II

Experience and Practice around the World

8

Water Reuse in California

Henry J. Ongerth and William F. Jopling

I. INTRODUCTION

In many ways, since its early days as a state, California has been in the vanguard of wastewater reclamation. This has been due to many interacting factors, including natural incentives related to weather, land use, water availability, and the geographical features of the state, as well as technical capability and a great interest in conservation and innovative practices.

California is an immense area of over 100 million acres, extending in a several-hundred-mile-wide band along the western coast of the United States for almost 900 miles. Instead of the usual four seasons, most of the state experiences just two. From May through October, the weather is dry and warm. Aside from the high mountain areas, there are virtually no summer storms or showers. During the summer growing season, even normal pasture crops require irrigation and the larger communities must have provisions for storing and importing water. Annual precipitation ranges from less than 5 in. in the desert southeastern portion of the state to more than 60 in. along the northern coastal areas and the windward side of the Sierra Nevada mountain range. The mean precipitation is 23 in. annually or in excess of 200 million acre-ft. It has been estimated that 70 million acre-ft of this is runoff. During the winter wet weather period, this runoff must be contained and stored for the summer needs.

Present net water use in California is approximately 31 million acre-ft. Expectations are that the total net water use will approach 35 million acre-ft by 1990. Over 80% of the use is for irrigated agriculture—domestic water constitutes only 10% of the total use.

Over two-thirds of the available water supply is located in the northern half of the state and over two-thirds of the water use needs are in the southern half. Consequently, this has led to major water storage and transfer facilities such as the Central Valley Project, the Colorado River aqueduct system, and the State Water Project which provide water to the areas of need.

In the water deficient areas of southern California, water cost along with water availability becomes a definite factor in the encouragement of wastewater reclamation. For example, the cost of imported water to irrigate a golf course may exceed $20,000 annually. At such cost, many of the less essential uses cannot compete on the freshwater market.

II. STATUS OF WASTEWATER RECLAMATION

There are over 850 community sewerage systems serving a population of 19.4 million persons in California. The total volume of municipal and industrial wastewater is about 2.7 million acre-ft annually or 2.3 billion gallons per day.

Almost 70% of the wastewater is generated in the coastal areas and discharged into saline waters.

Roughly two-thirds of the wastewater flow which is presently wasted to the ocean or bays is of suitable mineral quality for reuse. The other third contains dissolved solids in excess of 1500 mg/liter and would require demineralization. Full reclamation of all suitable wastewaters which are now wasted to the saline waters would add 1.2 million acre-ft of water to the annual supply. While this is an immense amount of water, it constitutes only 4% of the total water demand.

It may be possible eventually to reclaim close to 80% of the total domestic wastewater flow in California; however, at present only 7% (192,000 acre-ft/year) of the wastewater flow is reused through planned reclamation operations.

Reuse of domestic wastewater is most intensively practiced in the lower portion of the great Central Valley of California, although in this area direct reuse of the wastewater for crop irrigation is also in many cases the most practical method of waste disposal. This is a major agricultural zone of the valley and effluent from 52 communities having a combined flow of 43 million gal/day is used for crop irrigation. (Almost all other communities in the area dispose of wastes to land, but do not practice deliberate reuse.) It should be noted that the irrigated crops are pasture, grain, and fiber crops which do not require extensive treatment under California regulations. Approximately 70% of the volume of wastewater generated in the Tulare Basin (the southern portion of the Central Valley) is reused for irrigation of crops.

On a *volume* basis, over 90% of the wastewater reused throughout California is used for irrigation of crops, public areas, and institutional grounds. About 5% is used for industrial cooling and a very small percentage is used for recreational purposes.

The potential high volume uses of reclaimed water are agricultural use, groundwater recharge, and industrial use; however, the operations that have had the greatest growth rate in recent years have been the small volume uses such as irrigation of golf courses, cemeteries, parks, and filling of recreational or ornamental lakes.

A 1975 inventory of water reclamation operations in California was conducted by the Department of Health and revealed a total of 200 reclamation facilities in operation. The facilities provide reclaimed water for the following uses:

Nonfood crops (includes fodder, fiber, and seed crops)	142
Landscape irrigation (includes parks, golf courses, and freeway landscape)	42
Food crops (orchard and vineyard)	32
Planned groundwater recharge	7
Ornamental lakes	5
Industrial use	8
Recreational lakes	5
Wildlife habitat	3

Other uses such as groundwater recharge by injection, green-belt irrigation, fire protection, road compaction, and even rocket test-pad cooling have employed reclaimed water to a limited extent. In some instances, a reclamation facility produced reclaimed water for more than one use; consequently the number of uses given above exceed 200. Other inventories have identified approximately 70 additional operations that are either in the planning or construction stage of development or are land disposal systems that may be considered as possible reclamation operations.

III. HEALTH CONCERNS

From the earliest days of reuse in California, there has been recognition and concern for the possibility of disease transmission from use of renovated wastewaters. Statutes and regulations were developed in 1918 to prevent infectious disease transmissions from crops irrigated with wastewater.

In recent years, as use of reclaimed water has advanced to new areas and as public exposure to renovated wastewater has increased, attention has been directed to three additional areas of health concern. These are (1) lack of reliability of wastewater treatment plants, (2) possible risks from long-term ingestion of stable organic components of wastewater that reach domestic water supply sources, and (3) public attitudes to wastewater reclamation practices.

A. Reliability

In 1964, the Department of Health conducted a study of the health aspects of sewage collection and disposal in the great Central Valley. The results showed that 56% of the plants had experienced serious equipment outages during the preceding year. Chlorination equipment was reported out of service by 18 plants, with the outage varying from 1 hr to an entire year. Thirty-three percent of the plants reported the necessity of bypassing untreated sewage from periods ranging from 6 hr to an incredible 300 days.

Repeatedly, the department has found that 30–60% of the sewage treatment plants have serious failures sometime during the year—at small plants in remote mountain areas and at plants in the urban San Francisco Bay area. This performance is unacceptable at reclamation systems where there is significant public exposure to the reclaimed water.

In 1970, a review was made of plant reliability at 45 reclamation plants. Particular attention was directed to personnel and administration, chlorination, and monitoring, all of which have a bearing on reliability of the reclamation process. Forty-two percent of the plants reported equipment failures during the previous 12 months, which underlines the need for effective reliability features. Most of the treatment facilities were attended 8 hr or less per day, and 11 plants were attended for only a few hours. Of the 35 plants which were not continuously manned, only 3 had remote sounding alarms. Only 2 plants that produced reclaimed water for landscape irrigation and for recreational lakes met the daily bacteriological and monitoring program as deemed essential by the Department of Health and as specified in its standards for such operations. For only 20% of the plants was bacteriological sampling done often enough to provide an acceptable documentation of the quality of reclaimed water produced.

The physical features for reliability of chlorination at reclamation operations also showed significant deficiencies. A majority of the chlorination facilities lacked the basic features for an adequate disinfection system. Only 15 of 42 plants had the basic essentials—standby cylinder, scales, and manifold—needed for an uninterrupted chlorination process. A significant number of operations depended on the unfailing operation of a single chlorinator for disinfection. Sixty percent of the reclamation facilities had no standby power source and half of these were also without emergency disposal provisions. The findings of the survey revealed that general deficiencies existed in the areas of operation, chlorination, and monitoring with consequent reduction of reliability of the water reclamation systems.

As a consequence of these findings, the State Department of Health sponsored a statutory change in California calling for the establishment of reliability requirements for wastewater reclamation facilities. These will be presented later.

B. Stable Organics

Currently a number of projects are proposed for major groundwater replenishment with reclaimed wastewater. A list of possible future recharge operations is presented in Table I. In some instances, the groundwater basins at equilibrium would be almost 100% reclaimed water. Other proposals, still in very preliminary form, would introduce reclaimed sewage into existing domestic water impoundments and even (in the extreme) utilize treated sewage directly as a portion of the raw water supply with or without dilution.

The advanced wastewater treatment plant at South Lake Tahoe produces an effluent containing 3–25 mg/liter of COD and 1–6 mg/liter of TOC; at the Orange County injection study, organics passed through the groundwater formation to withdrawal wells. This pilot study has led to questions about the possible long-term effects (e.g., chronic, carcinogenic, mutagenic, or teratogenic) from prolonged ingestion of organic residues originating in domestic sewage and other wastewaters reaching domestic water sources. These questions must be answered before sewage-originated organics can be accepted in domestic water sources; California has urged that these unknown areas be studied.

As a consequence, the California State Department of Health has adopted a policy with relation to groundwater recharge projects where renovated wastewater provides a supplement to the domestic water supply. In summary, the policy states that reclamation projects which result in significant augmentation

Table I
Proposed Groundwater Recharge Projects[a]

Project of area	Flow[b] (Acre-ft/Yr)
Watsonville	8960
Paso Robles	1200
King City	700
Pacific Grove area	6330
Eastside	1300
Camarillo	2150
Sepulveda	24,640[c]
San Jose Creek	47,040[c]
Whittier Narrows	12,320[c]
Las Virgenes	448[c]
Sacramento metropolitan area	27,280
Redding	6210
West Sacramento	3810
Mokelumne River area	2170
Modesto (Ceres)	12,640
Madera	4010
Oakdale	1450
El Nido (Gustine)	3020
Edison-Maricopa	220
Kern River Delta	2180
Visalia-Hanford	8470
Tulare	3960
Delano-Earlimart	3030
Shafton-Wasco	2230
Porterville	2130
Fresno	34,220
Surprise Valley	—
Honey Lake	—
Apple Valley-Desert Knolls	—
Victor Valley	4480
Barstow	5040
Upper Coachella Valley	2850
March AFB Perris Valley Sun City San Jacinto-Hemet	25,112
	259,610[c]

[a]The proposed projects are those identified in the Water Quality Control Plant Reports prepared for the California State Water Resources Control Board.
[b]The flow figures are the average daily flows for 1973 at the existing treatment facility in order to provide a rough idea of the amount of recharge. No recharge flows were generally given in the Reports.
[c]Year 2020 flows.

of a domestic water supply will not be accepted until there is adequate proof that public health will not be adversely affected, now or in the future.

The State of California has brought together a consulting panel of health and water experts to direct their knowledge and judgment on the problems associated with groundwater recharge. The panel will recommend a program of research that will provide information to assist the Department of Health to establish regulations for groundwater recharge that will assure public health protection. At present, the consulting panel has identified specific research needs in the areas of wastewater treatment, epidemiology, toxicology, soil studies, monitoring, and characterization of organic substances.

C. Public Attitudes

The trend in reuse in California has been toward uses where there is considerable public exposure such as the irrigation of golf courses, parks, and freeway landscape, and the filling of ornamental and recreational lakes. In 1955, when there were only a half-dozen California golf courses irrigated with reclaimed water, the practice was received with somewhat mixed and reserved response—it was looked upon as an undesirable necessity. Recognizing the impact of public attitudes, the State Department of Health engaged a research psychologist to perform a carefully designed public attitude survey to study public attitudes toward uses of water reclaimed from community sewage. Pairs of communities were located in the Tahoe Basin, the San Francisco Bay Area, Ventura County, Orange County, and San Diego County. Five specific water uses representing different degrees of personal contact were selected from each of five usage categories: domestic, general, recreational, commercial, and food production. Public response to the 25 uses showed that about 50% of the respondents were opposed to highest contact uses of reclaimed water with opposition dropping to negligible levels for middle and lower contact uses (see Table II). The pattern of opposition was extremely consistent within each of the five usage categories. Reasons for opposition to use of reclaimed water were based primarily upon psychological repugnance and concern over purity (Table III). Opposition was found to be more highly related to respondents' educational level and beliefs about technical purification than to north-south location in California or the presence of a reclamation facility in their community.

IV. CALIFORNIA REGULATIONS

The earliest reference related to a public health viewpoint of water quality requirements for the reuse of wastewater in California appears in the Monthly

Table II
Percentage of Respondents Opposed to 25 Uses of Reclaimed Water

Use	Northern (n = 386)	Southern (n = 586)	Total (n = 972)
1. Drinking water	55.0	57.3	56.4
2. Food preparation in restaurants	53.4	57.7	56.0
3. Cooking in the home	52.5	55.8	54.5
4. Preparation of canned vegetables	52.5	55.1	54.1
5. Bathing in the home	37.8	39.2	38.7
6. Swimming	24.8	23.0	23.7
7. Pumping down special wells	26.1	21.4	23.2
8. Home laundry	21.1	23.9	22.8
9. Commercial laundry	19.4	23.5	21.9
10. Irrigation of dairy pasture	15.6	13.1	14.1
11. Irrigation of vegetable crops	15.6	13.0	14.0
12. Spreading on sandy areas	13.2	13.3	13.3
13. Vineyard irrigation	14.0	12.1	12.9
14. Orchard irrigation	10.7	9.7	10.1
15. Hay or alfalfa irrigation	8.3	7.0	7.5
16. Pleasure boating	9.1	6.1	7.3
17. Commercial air-conditioning	7.8	5.6	6.5
18. Electronic plant process water	6.0	4.1	4.9
19. Home toilet flushing	3.9	3.7	3.8
20. Golf course hazard lakes	4.4	2.2	3.1
21. Residential lawn irrigation	2.3	2.9	2.7
22. Irrigation of recreation parks	2.8	2.4	2.6
23. Golf course irrigation	1.8	1.5	1.6
24. Irrigation of freeway greenbelts	1.8	0.9	1.2
25. Road construction	1.6	0.3	0.8

Table III
Reasons for Opposition to Uses of Reclaimed Water

Reason	Percentage stating listed reason		
	Northern (n = 386)	Southern (n = 586)	Total (n = 972)
1. Psychologically repugnant	29.0	29.4	29.2
2. Lack of purity	27.2	17.7	21.5
3. Can cause disease	9.3	10.1	9.8
4. Bodily contact undesirable	9.8	6.8	8.0
5. Undesirable chemicals added	6.2	4.4	5.1
6. Taste and odor problems	3.1	4.4	3.9
7. Cost of treatment unreasonable	1.6	0.3	0.8

Bulletin, California State Board of Health, February, 1906: "Oxnard is installing a septic tank system of sewage disposal, with an outlet in the ocean. Why not use it for irrigation and save the valuable fertilizing properties in solution, and at the same time completely purify the water? The combination of the septic tank and irrigation seems the most rational, cheap, and effective system for this State." Thus, the first California quality requirement for wastewater reuse was sewage passed through a septic tank.

Although the typhoid fever hazard of eating uncooked vegetables that had been fertilized with human excrement had been recognized, no mention of hazard from sewage irrigation of crops was made until 1907. In the Board's April 1907 Bulletin, local health authorities were asked to "watch irrigation practices" and not allow the use of "sewage in concentrated form and sewage-polluted water . . . to fertilize and irrigate vegetables which are eaten raw, and strawberries."

The ability of septic tanks to "purify" sewage was being questioned, although it was the highest degree of treatment then provided for community sewage in California. It was reported in 1910 that 35 communities "use their sewage for farm irrigation, eleven without previous treatment, twenty-four after septic tank treatment." Application of effluent to "sewer farms" was accomplished by broad irrigation, but no mention was made of the type of crops irrigated. The part that sewage irrigation played in the typhoid incidence plaguing the state at that time cannot be postulated. Deaths from typhoid were then reported in order of 20–50/month.

The California State Board of Public Health adopted its initial "Regulation Governing Use of Sewage for Irrigation Purposes" April 6, 1918. The regulations prohibited the use of "raw sewage, septic or Imhoff tank effluents, or similar sewages or water polluted by such sewage" for the irrigation of tomatoes, celery, lettuce, berries, and other garden truck produce eaten raw by human beings. Garden truck produce of the type that is cooked before being eaten could be irrigated if the application of effluent was not made within 30 days of harvest. The regulations provided several exemptions such as permitting irrigation of melons if the sewage did not come in contact with the vine or product and irrigation of tree-bearing fruit or nuts if windfalls or products lying on the ground were not harvested for human consumption.

The 1918 regulations were revised in 1933 to exempt restriction of effluents for the irrigation of garden truck produce eaten raw if the effluents were well oxidized, nonputrescible, and reliably disinfected or filtered and always met a California standard expressed in terms of specified allowable number of positive 10 and 0.1 ml portions in 20 consecutive samples—a bacterial standard approximately the same as the then current drinking water standards. Reliability of disinfection was emphasized in that two or more chlorinators, weighing scales, reserve supply of chlorine, twice daily coliform analyses, and records were required.

It was noted that the revisions were made because of an expressed interest by Los Angeles Chamber of Commerce and others in nearby communities to conserve water, to provide employment for fieldworkers in contemplated truck gardens, and to save the beaches.

A meeting was held May 4, 1933, by Dr. Giles S. Porter, state health officer with the Conservation Committee of the Los Angeles Chamber of Commerce and several local health officers to consider revision of the sewer farm regulations. Dr. J. L. Pomeroy, Los Angeles County health officer, is quoted in part from the notes of the meeting: "Dr. Pomeroy was in sympathy with the idea of the use of high grade sewage water, but asked what about the cost of supervision and these twice daily tests?—He is in sympathy with Mr. A. K. Warren's statement of the need for water in the southern part of the State, yet plants get out of whack and one worries over the factors of safety.—Had to destroy ten acres of vegetables on account of typhoid in Pasadena where there were eight cases among the employees, probably contracted originally from the vegetables.—He thought the regulations were a bouquet to the technocracy of sanitary engineering." The regulations continued in effect until passage of the Water Pollution Act of 1949 eliminated the permit system that constituted the statutory basis for the regulation.

For the next 20 years, the State Department of Health continued to apply these regulations though they no longer were legally effective. This hiatus in legal control was ended in 1967 when, in the course of review of California Water Pollution Control legislation, a legislative committee reported upon wastewater reclamation as follows:

> The Committee believes that its most important objective in recommending legislation relating to the use of reclaimed waste waters is the absolute protection of the public health. In order to assure this protection the Committee *recommends that the State Department of Public Health be required to establish "statewide contamination standards", which would represent the maximum concentration levels of various constituents of reclaimed water which are permissible for each type of use.* Standards would be established for such uses as swimming, fishing, irrigation, etc. These contamination standards will provide a guideline similar to Federal Public Health Service Drinking Water Standards, which can be applied prior to the initiation of waste water recalamation projects. The establishment of such standards will provide a *preventive control* over the use of reclaimed waters.
>
> We further *recommend that enforcement by accomplished: 1) by specifically applying the summary abatement authority of the State Department of Public Health to waste water reclamation activities. This represents a restatement and reinforcement of the basic responsibility of the State Department and local health agencies; and, in addition, 2) by requiring that the Regional Water Quality Control Boards establish waste discharge requirements for reuse of reclaimed waste waters used directly or otherwise. Such discharge requirements must be in conformity with the 'statewide contamination standards' for the particular use contemplated.*

These recommendations were followed explicitly and became the State policy with regard to wastewater reclamation. They are contained in Sections 13510 to 13512 and Sections 13520 and 13521 of the California Water Code as follows:

> 13510. It is hereby declared that the people of the state have a primary interest in the development of facilities to reclaim water containing waste to supplement existing surface and underground water supplies and to assist in meeting the future water requirements of the state.
> 13511. The Legislature finds and declares that a substantial portion of the future water requirements of this state may be economically met by beneficial use of reclaimed water. The Legislature further finds and declares that the utilization of reclaimed water by local communities for domestic, agricultural, industrial, recreational, and fish and wildlife purposes will contribute to the peace, health, safety and welfare of the people of the state. Use of reclaimed water constitutes the development of "new basic water supplies" as that term is used in Chapter 5 (commencing with Section 12880) of Part 6 of Division 6.
> 13512. It is the intention of the Legislature that the state undertake all possible steps to encourage development of water reclamation facilities so that reclaimed water may be made available to help meet the growing water requirements of the state.
> 13520. As used in this article "contamination standards" means the levels of constituents of water, which will be safe for any direct reuse purpose.
> 13521. The State Department of Public Health shall establish statewide contamination standards for each varying type of direct use of reclaimed waste waters where such use involves the protection of public health.

As a result, the State Board of Public Health enacted regulations entitled "Statewide Standards for the Safe Direct Use of Reclaimed Waste Water for Irrigation and Recreational Impoundments." These prescribe levels of wastewater constituents which will assure that the practice of directly using reclaimed wastewater for the specified purposes does not impose undue risks to public health. These establish limits for crop irrigation, landscape irrigation, and recreational impoundments and are summarized in Table IV.

In 1969, at the request of the Department of Health, Sections 13520 and 13521 of the Water Code were revised as follows:

> 13520. As used in this article "reclamation criteria" are the levels of constituents of reclaimed water, and means for assurance of reliability under the design concept which will result in reclaimed water safe from the standpoint of public health, for the uses to be made.
> 13521. The State Department of Health shall establish statewide reclamation criteria for each varying type of use of reclaimed water where such use involves the protection of public health.

As was previously pointed out, the Department of Health was concerned with the record of poor reliability associated with the performance of wastewater treatment facilities. This modification in the state law authorized the Department to establish regulations on treatment reliability. New regulations were adopted in 1975, in furtherance of Sections 13520 and 13521 which are quite comprehensive and which are included in the appendix of this chapter because they may be useful in guiding others in this area.

Table IV
Summary of Statewide Standards for the Safe Direct Use of Reclaimed Wastewater for Irrigation and Recreational Impoundments

Use of reclaimed wastewater	Description of minimum required Wastewater Characteristics: Primary[a]	Secondary and disinfected	Secondary coagulated, filtered[b] and disinfected	Coliform MPN/100 ml median (daily sampling)
Irrigation				
Fodder crops	X			No requirement
Fiber crops	X			No requirement
Seed crops	X			No requirement
Produce eaten raw, surface irrigated		X		2.2
Produce eaten raw, spray irrigated			X	2.2
Processed produce, surface irrigated	X			No requirement
Processed produce, spray irrigated		X		23
Landscapes, parks, etc.		X		23
Creation of impoundments				
Lakes (aesthetic enjoyment only)		X		23
Restricted recreational lakes		X		2.2
Nonrestricted recreational lakes			X	2.2

[a]Effluent not containing more than 1.0 ml/liter/hr settleable solids.
[b]Effluent not containing more than 10 Turbidity Units.

V. DEVELOPMENT OF WASTEWATER RECLAMATION

It is not possible to present a brief chronology of the milestones in water reclamation in California without being aware of the many important points in development that must be omitted. In particular, such a chronology must omit the men who were able to look ahead of their time and were able to overcome opposition and inertia to push forward the frontiers of reclamation. Here, briefly, are the milestones:

1890 Sewer farms in use at California communities
1918 First regulations covering agricultural use of sewage effluents
1932 Golden Gate Park initiates reclamation for filling ornamental lakes and landscape irrigation from a specially constructed reclamation plant

1943 Reclaimed water first used at military installations for landscape irrigation of recreational areas
1961 Santee develops recreational lakes for fishing and boating, and ultimately studies an experimental swimming operation
1963 County Sanitation Districts of Los Angeles County prepares the first comprehensive regional plan for reclamation
1964 Orange County Water District undertakes a pilot project of groundwater injection utilizing reclaimed water
1965 South Lake Tahoe Public Utility District constructs the initial advanced waste treatment system in California
1967 State Legislature sets positive policy on water reclamation and regulations developed for the quality of reclaimed water for specific uses
1968 Pomona formally initiates the water utility concept to reclaimed water
1973 Orange County Water District constructs a 30-mgd reclamation-desalination plant for saline water intrusion control.
1975 Regulations were adopted covering quality reguirements and the means for assurance of reliability at reclamation operations

A. Agricultural Use

The use of domestic wastewater for agricultural irrigation developed in the nineteenth century along with the development of sewage collection systems. Farms using raw sewage were established in England, Australia, Germany, France, and Italy during the period 1870–1890. By 1900, "Sewer Farms" were numerous in the new and old world.

Agricultural use of sewage effluent has been the major reuse practice on both a volume basis and with regard to the number of operations in California and, since the turn of the century, California has used sewage effluent for the irrigation of fodder, fiber, and seed crops. The March 1910 *Monthly Bulletin of the State Board of Health* recommended: "In California where water is so valuable for irrigation, the utilization of sewage for broad irrigation should be carefully considered. Some of our cities which operate sewer-farms realize from $500 to $5,000 a year (from these operations)." In 1935, sewage effluent from 62 municipalities in California was used to irrigate cultivated and "wild" crops and there are 153 installations where crop irrigation is now a major reuse activity. It has been estimated that more than 20,000 acres of agricultural lands are irrigated all or in part with reclaimed water. The largest two operations are at Bakersfield and Fresno. The City of Bakersfield has used sewage effluent for irrigation since 1912. At the present time approximately 2400 acres of alfalfa, cotton, barley, sugar beets, and pasture are irrigated. Fresno irrigates 3500 acres of the same types of crops. The two operations use close to 30,000 acre-ft of reclaimed wastewater.

It would appear that for agricultural use of reclaimed water to expand much beyond the present use major interbasin projects will be required. The trend line for agricultural use in recent years has been fairly flat and a new concept is needed to advance this use. Several considerations would tend to support a major interbasin reclaimed water project to supply agricultural areas:

1. The increased cost of waste disposal
2. The availability of irrigable lands and the apparent world need for increased food production
3. The increased awareness of the need for freshwater conservation.

Three areas of California are likely prospective sources of reclaimed water supply for interbasin projects: Monterey Bay area, Los Angeles area, and the San Francisco Bay area. A major study of potential uses of the municipal wastewaters generated in the San Francisco Bay area has concluded that the reclaimed wastewater could be used as a substitute irrigation supply in the San Joaquin Valley, one of the major agricultural area of the state, provided adequate drainage possibilities were available.

B. Landscape Irrigation

One of the direct uses of water reclaimed from sewage in which many communities are currently interested is the watering of parks, golf courses, and other public areas. The concept is particularly attractive because the peak demand comes at the time of year when stream flows are at their lowest and are least able to accept effluents and freshwater demands are at their peak. In California's climate, the demand for reclaimed water for the average golf course amounts to 0.4 mgd over a 6-month season. At present, water reclaimed from sewage is used to irrigate golf courses, parks, and stretches of freeway right-of-way, mostly in the southern portion of the state. The use of reclaimed water to maintain desired vegetation in fire hazard areas and erosion areas is also being studied.

The use of reclaimed water for landscape irrigation has been instigated in almost every case by an economic advantage favoring its production and use over the combined cost of sewage treatment and freshwater use. It usually has been a case of using reclaimed water or nothing, since the price of fresh water for this use may be prohibitive. The recent growth in this use has been impressive. In 1955, there were only five such operations and today there are 42.

During and after World War II, military installations in the southwest introduced the practice of using sewage effluent for golf course irrigation. The practice was extended to civilian golf courses in areas where freshwater costs were high or water availability was limited. In several cases, the use of reclaimed water for golf course irrigation is practiced with local groundwater users for public relation purposes, as well as to minimize water costs.

A fairly high degree of sewage treatment is automatically required to prevent odors or clogging of spray nozzles. Effective organic removal is also required to ensure adequate disinfection. Even with proper control of the irrigation sprays, use of nighttime irrigation, pop-up spray nozzles, covered drinking fountains, and warning signs, there is some inadvertent and incidental public contact. However, two decades of experience have demonstrated no detectable adverse health effects due to the practice at over 40 public areas.

GOLDEN GATE PARK

John McLaren, the pioneer superintendent and developer of Golden Gate Park in San Francisco, also pioneered in using sewage as a source of water. His first attempts were with raw sewage; however, in 1912, because of numerous and continual complaints, he was forced to build a septic tank to remove the more objectionable solids. He then used the septic tank effluent to fill and maintain a series of ornamental lakes and to irrigate about 250 acres.

The stage was set for the construction of the existing reclamation plant which was placed in operation in 1932. The plant is of historical significance since it was constructed for the sole purpose of water reclamation. Treatment consists of a bar screen and grit removal, preaeration, primary sedimentation, activated sludge, final sedimentation, and chlorination. The Golden Gate Park plant was also the first treatment facility in California that could be operated as an "on line" reclamation facility; that is, it diverts from a sewer as much wastewater as is required and, in periods of plant upset or when the reclaimed water is not needed, the wastewater either is not taken or is returned to the sewer. No sludge treatment or disposal facilities are needed because all sludge is discharged back to the city sewer. At first, the effluent was used only for the ornamental lakes, but a few years later, it was used also for irrigation of the polo field and other areas. This limited use of the water for irrigation without creating a nuisance led to a fuller utilization of the reclaimed water in 1947. The plant now supplies about 1 mgd from the middle of January through November for use in the park's ornamental lakes and irrigation system. This source supplies about 25% of the park's water needs for horticultural purposes.

The reclaimed water is very low in solids and oxygen-consuming substances. It contains about 40–60 mg/liter of total nitrogen and an appreciable amount of phosphate, which makes it desirable for general irrigation when diluted with fresh water.

During the period that the plant has been supplying water to the park irrigation system, there has been no adverse publicity or complaints. Unlike the Santee reclamation operation which has been accompanied by an extensive public relations campaign, few San Franciscans are aware of the source of the park irrigation water. Although the use of reclaimed water in Golden Gate

Park is on a relatively modest scale, this operation has been extremely important in advancing the use of reclaimed water to new uses.

C. Industrial Use

The industrial use of reclaimed municipal wastewater has not been an extensive or well-established practice in California, although the internal reuse of water within an industry is common practice. The extraordinary and often cited example of effective reuse and multiple recycling of water is given by the Kaiser Steel Corporation, Fontana Plant. This inland plant was constructed far from ample sources of process water and was designed to recycle water through successively less-demanding uses—cooling, process, dust arrest, and slag cooling. The plant uses 0.5 mgd of treated sewage together with 5.2 mgd of industrial waste for make-up water. Process water is recycled and reused up to 40 times before being lost through steam or evaporation. Treatment of the recycling water includes chlorination and copper sulfate treatment for slime control, and addition of carbon dioxide to control pH and minimize scale. Oil and suspended solids are removed in clarifiers. Basic quality requirements for total solids are 750 mg/liter for open hearth and cooling towers, 1500 mg/liter for steel-discaling and gas-washing towers, and none for slag pits.

BURBANK

The City of Burbank is one major user of reclaimed water for an industrial purpose. Burbank had contracted with the City of Los Angeles for sewage disposal since the mid-1920s. As increased sewage flows due to development gradually approached the contract limits, Burbank investigated the possibility of wastewater reclamation and reuse to serve the dual function of providing a useful commodity and delaying the need to contract for additional capacity in the Los Angeles sewerage system. A cost comparison demonstrated that a 6-mgd reclamation facility could be constructed with necessary connecting and diversion sewers at a lesser cost than the purchase of additional capacity from Los Angeles. The plant was completed in 1966 and the effluent output is utilized by the Public Service Department of the city as make-up water in the cooling towers of the electric steam-generation plant. After several cooling passes, the water is wasted to a flood control channel.

Several smaller industrial users of reclaimed water include the North American Rockwell operation at Canoga Park which utilizes 0.2 mgd of reclaimed water for cooling and fire control at a static rocket test site. A small cement plant in central California uses reclaimed water in processing cement.

The Contra Costa County Water District supplies fresh water to industry, agriculture, and municipalities in a section of the San Francisco Bay area. A number of heavy industries (steel, oil, and wood processing) purchase large quantities of water from the district for cooling purposes. The district receives its water under contract from the Central Valley Project of the U.S. Bureau of Reclamation. Because of rising water demands and the desire to augment its supply for the future, the district investigated the potential sources and market for reclaimed water in and near its service area in 1967.

The Contra Costa County Water District and the Central Contra Costa County Sanitary District entered into a joint program to demonstrate the feasibility of renovating wastewater for use as industrial cooling water. Pilot plant studies were carried out to determine the treatment processes that would produce water of the required quality and two industries are now testing the reclaimed water in their cooling systems.

Construction is underway on a tertiary treatment with a capacity of 30 mgd. The total treatment process includes lime coagulation followed by primary sedimentation, activated sludge aeration and nitrification, secondary sedimentation, dentrification, sedimentation, and chlorination.

The reclaimed water will be sold to the water district which will distribute and sell it to the industries. It is expected that the yearly average flow of reclaimed water will be 19 mgd. The initial customers will be two oil refineries, two chemical companies, and two power-generating plants.

The present cost to industry for fresh water is $28.00/acre-ft. The cost of producing reclaimed water will be several times this amount. The industries that will use the reclaimed water will not bear the full cost of production—rather, all water users of the district will assist in subsidizing the added cost.

D. Groundwater Recharge by Surface Spreading

The planned artificial recharge of groundwaters by surface spreading of water, particularly polluted surface waters, has been practiced for more than a century in European countries probably as much for the water quality improvement associated with the practice as the groundwater replenishment benefits. In California, major recharge operations have been carried out for several decades along pervious streambeds in the coastal plains of the southern and central portions of the state. For the most part, impounded storm waters or imported waters have been recharged. There have been several reasons for the recharge operations. As water demands in the coastal areas increased over the years, water withdrawals gradually lowered the groundwater levels. There were, of course, attendant expenses associated with increased power demands to obtain groundwater as the groundwater level dropped and, as a result, the construction of deeper wells. The lowering groundwater level led to a reversal of the oceanward flow of water along the coast and the intrusion of seawater into portions of the groundwater basin. Replenishment through recharge by surface spreading in the natural recharge areas did much to counteract these adverse effects. In the central coastal area of Santa Clara County, an added effect of groundwater overdraft was land subsidence. In a 6-year period, ground elevations dropped as much as 3.9 ft near San Jose. Artificial recharge along 90 miles of streambed and 350 acres of ponds has controlled subsidence.

Major off-channel spreading of imported Colorado River water and storm water began in 1953 in the Whittier Narrows area of Los Angeles County. In 1962, the Whittier Narrows Water Reclamation Plant was put into service to provide reclaimed wastewater as an additional source of supply for the groundwater recharge operation. This has become perhaps the best-known facility of its type.

There are, at present, seven planned groundwater recharge operations in California utilizing reclaimed wastewater. The four major operations recharge an estimated 39,000 acre-ft/year (see tabulation below).

	Wastewater in acre-ft (Amount reclaimed)
Whittier Narrows	13,500
San Jose Creek[a]	10,000
Eastern MWD	1000
Camp Pendleton (Military base)	4120
	38,620

[a]San Jose Creek plant is new; main disposal is to the San Gabriel River. Groundwater recharge is intermittent.

In addition to these operations, a percentage of the wastewater flow at Palm Springs and El Mirador is recharged. The City of Oceanside has a reclamation operation involving irrigation and groundwater recharge that was the first in the state of its type; however, no extensive use is made of the recharged groundwater.

WHITTIER NARROWS

The Whittier Narrows area is approximately 10 mi east of downtown Los Angeles. At this point, the entire San Gabriel Valley drainage and subsurface flow passes through a 2-mile gap called Whittier Narrows. A broad floodplain exist below the narrows known as the Montebello Forebay. This is a major recharge zone for two large groundwater basins—the Central Basin and the West Basin. In this area, major planned recharge has taken place either in or adjacent to the river channels of the Rio Hondo and San Gabriel rivers. The Whittier Narrows Water Reclamation Plant was constructed and first delivered water to the spreading areas in 1962. The plant is located in the Whittier Narrows Dams reservoir area and has been constructed to withstand flooding. The operation has taken advantage of its location, which is upstream from major industrial discharges, although isolation of certain industrial wastes was required. Wastewater is diverted from a trunk sewer which is the principal artery for wastewater disposal for the western San Gabriel Valley.

Treatment consists of conventional activated sludge treatment, chlorination, and when nonbiodegradable detergents were a problem in the mid-1960s, foam separation for detergent removal. Sludge is discharged back to the sewer for eventual downstream treatment, a practice that provides one of the operational and economic advantages of upstream reclamation plants. A positive concept in upstream water reclamation is the cost savings which can be credited to the operation through downstream savings in capacity of trunk sewers, treatment facilities, and ocean outfall. At Whittier Narrows, the capital value of this savings was estimated to be $5.75/acre-ft in 1963.

A second treatment plant, the San Jose Creek Plant, began operations in 1971, now treats 24 mgd, and contributes a portion of this water to the spreading operations. In addition, imported Colorado River water and local runoff in the amount of approximately 100,000 acre-ft is spread at Whittier Narrows.

E. Injection of Reclaimed Water

Three groundwater injection projects utilizing reclaimed water have been considered in California. All are directed at controlling saline water intrusion through development of a hydraulic barrier. One project is in Orange County and will be operated by the Orange County Water District. This is the only operation that has passed the construction stage; its history and development are described below. A second proposed injection operation involving reclaimed water is located along the coast in Los Angeles County. A barrier project completed in 1968 presently exists in the area and consists of 93 injection wells and 267 observation wells, various supply lines, and chlorination facilities. Imported Colorado River water is presently used in this West Coast Basin Barrier Project and during the 1970–1971 water year, 30,000 acre-ft of filtered, chlorinated river water was injected. A study investigated the feasibility of using a nitrified secondary effluent, after carbon filtration and chlorination, as a replacement for the imported water for injection. Based on cost and other factors, replacement of the imported river water with reclaimed water for the injection supply was not carried out.

A third injection project involving reclaimed water is proposed for Santa

Clara County in the San Francisco Bay area. Under this proposal, a tertiary effluent would be injected into a saline aquifer and a string of extraction wells located inland from the injection wells would establish a gradient to reduce the extent of present seawater infiltration.

ORANGE COUNTY WATER DISTRICT

Overdraft of the groundwaters in the coastal areas of southern California led to saline water intrusion. By 1935, the entire coastal perimeter of the West Coast basin was infiltrated by seawater which extended more than 5000 ft inland. In the coastal basin of the Santa Ana River in Orange County, overdraft of groundwater resulted in an inland intrustion as much as $3\frac{1}{2}$ miles in the ancient channel of the Santa Ana River.

In 1949, the Orange County Water District began recharge through surface spreading of imported Colorado River water to relieve the overdraft conditions and provide an adequate supply. To prevent further intrusion, the water district investigated the possibility of a hydraulic barrier.

In 1965, the water district presented to the State Department of Health a proposal to inject treated sewage effluent into a water-bearing aquifer known as the Talbert Aquifer. Trickling filter effluent from Treatment Plant No. 1, County Sanitation Districts of Orange County, was to be given treatment by coagulation, sedimentation, rapid sand filtration, and chlorination prior to injection. This initial proposal also included the construction of four observation or monitoring wells. The purpose of the proposal was to study the feasibility of using effluent in a hydraulic barrier system. The State Board of Public Health authorized the project which was limited to the injection of 1000 acre-ft.

Because of technical problems encountered in the injection of both fresh and reclaimed water into the Talbert Aquifer, the water district requested and received an extension of time and modification in the pilot injection study to inject reclaimed water into deeper underlying water-bearing aquifers, designated as Alpha and Beta.

While the studies were successful in describing the effectiveness and cost of the tertiary treatment processes, they were not designed for, nor did they yield information relative to, the long-term effect on reclaimed water as it moves greater distances from the point of injection.

In 1968, an expanded injection demonstration program was proposed and accepted whose objectives were (1) to determine the long-term fate of reclaimed water following injection, including chemical, bacteriological, and hydraulic processes in its transmission through the underground strata; and (2) to demonstrate the acceptability or lack of acceptability, from the standpoint of potability (taste, odor, etc.), public health, and safety, of a treated reclaimed water injected under conditions as proposed for the full-scale barrier.

The project was to use 3 multiple-cased injection wells and 13 multiple-point observation wells. The wells were perforated in the Talbert, Alpha, Beta, and Lambda aquifers (listed in order of depth, the Lambda aquifer being deepest). This made it possible to inject reclaimed water, deep well water or blends of the two into each aquifer simultaneously and to extract from each aquifer as desired.

The resulting study indicated that human intestinal viruses were never found in any of the injection water samples or in the 13 multiple-casing observation wells located at distances up to 1000 ft from the injection wells. The residual organics were the major problems encountered. The threshold odor of the reclaimed water was relatively high and the odor was not significantly reduced or altered by travel through the aquifer. The study report concluded that the odor-causing dissolved organic material and dissolved inorganic substances present in the tertiary treated injection water made it undesirable for a barrier supply. To overcome these undesirable characteristics it was concluded that additional treatment prior to injection was needed.

The Orange County Water District constructed and operated a tertiary treatment pilot plant, employing additional treatment processes, to explore means for removing the undesirable characteristics. The pilot process consisted of clarification with lime and various coagulant aids, ammonia stripping, recarbonation, filtration, carbon adsorption, and chlorination. The pilot plant produced a reclaimed wastewater low in organic material and free of odor.

Utilizing the data obtained from the pilot plant studies, the district proposed a full-scale injection project; construction began in 1972.

The proposed barrier will consist of a line of 22 injection wells located 4 miles inland from the coast and a line of extraction wells (already constructed and in operation) located between the injection wells and the ocean. The Orange County Water District reports that 30,000 acre-ft/year or 26.7 mgd will be required for the full-scale saltwater barrier. The wastewater reclamation project will supply 15 mgd. Initially the remaining water needs, for dilution and blending, were proposed to be obtained from desalted seawater or from well water produced from a deep aquifer which is not subject to saltwater intrusion. Costs and energy requirements may make operation of the desalinization system unfeasible.

The proposed water reclamation facilities are based on the operation of the water district's latest pilot plant studies. Lime coagulation will be used for clarification and phosphate removal, followed by ammonia stripping, recarbonation, mixed-media filtration, carbon adsorption, and breakpoint chlorination. The water district is presently considering the use of reverse osmosis on a portion of the reclaimed water stream to reduce the mineral content of the injected water. Strict quality requirements have been established on the reclaimed water and additional safeguards for health protection have been required.

F. Recreational Impoundments

Development of recreational impoundments which utilize reclaimed water represents an innovative concept in reclaimed water use. Such facilities require an extremely well-treated reclaimed water which is essentially free from biological pathogens. The treatment facility must be effective and reliable. Two operations in the state that have produced water for recreational impoundments have done much to advance knowledge in the field of waste treatment and reclaimed water use. The Santee operation achieved full public acceptance and was the original operation where many new uses were tested. The South Lake Tahoe operation solidified knowledge of advanced waste treatment and established the treatment process chain against which others can be compared.

In recognition of the significance of the systems, expanded coverage is given to the history and development of the Tahoe and Santee systems.

1. THE SANTEE PROJECT

The unincorporated community of Santee lies along the San Diego River Valley about 20 miles from downtown San Diego. The community is largely residential with only a few commercial establishments essential to community needs. There is no waste-producing industry.

The Santee County Water District is a public agency that provides water for the district's 13,000 people, and, in addition, operates the sewerage facility for the same area. During the initial development and study of the reclamation use facility from 1961 to 1965, the sewage was treated in an activated-sludge-type sewage treatment plant of 2 mgd design capacity. After the activated sludge treatment, the sewage effluent was held in a 30-million-gal oxidation pond and was chlorinated prior to pumping to the upstream spreading area. All flow not pumped to the spreading area has been chlorinated prior to disposal from the project area. The large portion of flow leaving the project has been used for irrigation of a nearby golf course.

Sycamore Creek, a tributary of the San Diego River, extends generally northward from the river channel and forms a shallow valley containing both the community treatment plant and the recreational lakes. There is subsurface flow in Sycamore Valley, but only occasional surface flow.

Here five lakes were created by the Santee County Water District to utilize the renovated wastewater from the sewage treatment plant in the Sycamore Valley area after water-bearing materials had been mined for sand and gravel. The lakes were formed by regrading the remaining spoil area into dikes and land-water configurations.

Recovery of reclaimed wastewater from the aquifer constitutes the primary water supply for the lakes, although infrequent heavy rains develop natural groundwater or runoff to slightly supplement the supply.

In 1959, the district was faced with the alternative of abandoning its existing sewage treatment and joining the San Diego Metropolitan System with ocean disposal of treated effluent, or providing additional treatment at the year-old treatment plant and justifying the added cost by putting the wastewater to beneficial use. The district first planned to make recreational use of the reclaimed waters after passage through the oxidation ponds. When public health authorities rejected this plan, the district proposed to include percolation and chlorination of the oxidation pond effluent using the natural aquifer in Sycamore Canyon. On this basis, San Diego health department approval was granted in 1961 for noncontact recreational use of effluent.

Since 1961, the recreational program at Santee was gradually increased by careful short steps from passive aesthetic enjoyment of the lakes and picnicking to boating, through a season of "fishing for fun," to fishing which permitted fishermen to keep and eat their catches, and finally to swimming in a separate area near the uppermost lake, which began in June 1965.

Shortly after the 1961 startup, a several-agency committee, including federal, state, and local authorities, was formed to outline, plan, and conduct a study of the entire public health problem. In response to a request by the district for approval for swimming, an *ad hoc* Advisory Committee on Epidemiology was appointed by the California State Department of Public Health, in May 1964, to review the past program of noncontact water use and to advise the department on the risks, if any, in body contact use of the reclaimed water. The committee, in particular, was to evaluate the feasibility of establishing a swimming area. After reviewing the information from the previous two years of study, the committee sanctioned the swimming program if, among other conditions, the swimming area was developed and maintained to meet the State of California's established water quality standards for artificial pools. The committee emphasized that its recommendation was conditioned upon maintenance of strict safety and sanitary regulations and that filtration through the soil must continue to be a part of the water reclamation process.

Later in 1964, the Santee County Water District advised the San Diego Department of Public Health that it would be necessary to reduce the natural filtration zone to a minimum length of 400 ft. A November 1964 study by the Bureau of Sanitary Engineering and the San Diego County Department of Public Health determined the effects that such a reduction in the filter zone would have on reclaimed water quality. As part of the study, the filtration zone was subjected to a massive dose of attenuated Type 3 polio virus. No virus were detected after 200 ft of travel through the sands and gravel. The major reduction in bacterial concentration occurred within the first 200 ft of filtration. Chemical analyses indicated that high percentages of phosphate, detergent, and certain nitrogen forms in the percolating liquid traveled more than 200 ft.

In February 1965, the Santee County Water District formally proposed to construct a flow-through swimming basin blending in with the other recreational facilities and using reclaimed water. With the understanding that the swimming facility was to be operated on an experimental basis with constant monitoring of water quality and pool operation, variances in standard structural requirements for a public pool were allowed.

The Santee experimental swimming basin, as first constructed was a shallow pool with sand sides and bottom and with depths of $1\frac{1}{2}$ ft at the inlet end and a maximum of $3\frac{1}{2}$ ft near the opposite end. Pool capacity was approximately 80,000 gal, and the dimensions at the waterline were 65 ft × 87.5 ft.

The requirement for operation during the trial swimming program established by the State of California and San Diego County Health Department follow:

Requirements for Santee Experimental Swimming Area Operation

1. Pool water quality must meet the State Swimming Pool Standards for the quality of water in artificial pools.
   a. Pool turnover—8 hours maximum
   b. Not more than 15% of samples shall have:
      (1) Plate count more than 200 per ml
      (2) Coliform bacteria in any of five 10 ml portions
   c. Clarity—a 6″ black disc on a white background shall be visible at the deepest point from 10 yards
   d. pH—7.2 to 8.4
   e. Minimum chlorine residual—0.4 free or 0.7 combined.

   The swimming basin will be immediately closed to use at any time that water quality requirements are not met.
2. Ten days of satisfactory operation, supported by laboratory data, must be obtained prior to opening of the swimming basin to the public.
3. The bathing load shall be controlled. During the first 10 days of use, no more than 25 bathers may use the basin at one time. If suitable sanitary conditions are met with 25 bathers, the load may be increased to 50 persons during the next ten days. If no water quality problems then arise, the basin may be opened to capacity.
4. Until it has been determined that a lesser sampling program is indicated, bacteriological samples will be taken twice daily from several points in the basin.
5. Virus samples are to be obtained from inflowing and outflowing water at the swimming area.
6. All persons using the pool must be registered by name, age, address, date, and time of day of entrance to the pool. The registration shall be maintained as a permanent record.
7. Swimming will be prohibited if apparent illness results from the operation of the swimming program.
8. A qualified lifeguard is to be in attendance at all times that the swimming basin is in use.
9. A sign will be placed at the swimming site which states that the swimming area contains reclaimed water.
10. Natural filtration must be included as part of the reclamation process.

To determine that the water quality consistently met the State Swimming Pool Standards for water quality in artificial swimming pools, a temporary laboratory was set up and operated at the Santee reclamation plant by the California State Department of Public Health. During the entire trial swimming program, conducted with some interruptions from May through September, water samples were collected twice daily from five locations in the pool and were analyzed for coliform bacteria and total plate count. Chlorine residuals were measured at the pool twice daily along with pH determinations. Spot checks were made at other times. The pool alkalinity, temperature, and clarity, as well as water flow, chlorine dosage, and weather conditions were also recorded. Frequent observations during each day of operation were made to ensure that operational requirements were being met. Continuous virologic sampling and analyses were also carried out by the San Diego County Health Department. The results of these analyses showed no isolations of virus in the swim area's waters.

The trial program was interrupted a number of times, due primarily to elements of the swimming basin construction and to the quality of the reclaimed water which resulted in excessive color and/or turbidity in the pool.

The color was chiefly attributed to the presence of manganese and iron in the reclaimed water, which was picked up during the natural filtration step of reclamation. The turbidity was largely due to the sand that was placed in the pool to stimulate natural conditions and was also caused by activity along the collection trench which carried reclaimed water to the chlorine contact chamber and pool. Eventual removal of the sand from the pool and additional treatment facilities corrected the color and turbidity problem.

It was difficult to maintain consistent chlorine residuals throughout the pool, particularly after the filtration equipment was installed and pre- and postchlorination was practiced. The residual in the pool after filtration was primarily a free chlorine residual; there was as much as a 4 mg/liter loss from the inlet to the outlet of the pool. In spite of chlorine residuals of 5–6 mg/liter at the inlet, there were no complaints of eye irritation.

Water samples, particularly from the leeward side of the pool, occasionally had high plate counts and were positive for coliform bacteria. Further investigation revealed that scum and debris such as bird feathers were the cause. The two skimmers could not effectively remove surface scum from all portions of the pool.

Public acceptance of the swimming facility was very good. Even during periods when the district was experiencing difficulty with pool water quality and the water was either turbid or had an unfortunate yellowish-brown color, children brought by their parents eagerly waited their turn to enter the swimming area. The cold weather in June was the only real factor in holding down the number of swimmers. The records showed that more than 3200 individuals registered for swimming. Over 1800 were from the immediate Santee area, 1300 from other San Diego County communities, over 60 from other California communities, and 23 from out-of-state. Approximately 600 local registrants repeatedly used the facility during the swimming program.

In one sense, the experimental swimming program at Santee was not successful. It was the district's expressed objective to create benefits through the practice of wastewater reclamation at lower costs than would otherwise be possible. Unfortunately, more extensive treatment was eventually required for the iron and manganese removal and positive turbidity control than would be required for a conventional pool using domestic water and having recirculation. Also, it was not possible to successfully maintain water clarity with the sand sides and bottom. Consequently, after several years, the use of reclaimed water for swimming was abandoned and a conventional pool using fresh water was constructed.

The program did show that water reclaimed from sewage can be used for a swimming area and can meet the strict water quality requirements for such use. The program also showed that a community that has "grown up" with sewage reclamation readily accepts the use of reclaimed water in a swimming pool.

2. SOUTH TAHOE PUBLIC UTILITY DISTRICT PROJECT

The South Tahoe Public Utility District in California is the largest and most active of the six agencies providing sewerage service in the Lake Tahoe basin. The district, organized in 1951, covers over 21,000 acres and is assessed at more than $80 million.

In 1960, the district started operating a $2\frac{1}{2}$-mgd activated-sludge treatment plant, which replaced two redwood septic tanks that had served as treatment facilities since 1956. The new plant, although capable of producing a high-quality secondary effluent, was not permitted to discharge to the lake. Effluent disposal was accomplished by spray irrigation on land.

Although no official water quality policy had been adopted for Lake Tahoe in 1960, it appeared that land disposal of effluent within the lake basin would be acceptable for many years. This conclusion was based on three factors: California state regulatory agencies would not permit direct discharge of conventionally treated effluent to Lake Tahoe or its tributaries; removal of the treated effluent from the lake basin would be impossible due to legal restrictions on water rights and legal complications involved in disposing of sewage effluent in another state or watershed; and secondary treated effluent would not meet quality standards necessary to protect the high quality water in the receiving streams.

In April 1961, the South Tahoe District Board of Directors investigated alternatives and asked its consulting engineers to recommend a plan for permanent effluent disposal. After a detailed investigation, the engineers recommended the following:

> Continue with land disposal for the next several years. Although the method did not afford

complete protection of the lake, it was the only alternative available for interim use.

Develop an advanced method of waste treatment that might ultimately permit disposal within the lake basin, or permit removal of the effluent from the basin, if necessary.

Study available routes for effluent removal from the basin and seek an agreement that would permit disposal outside the basin.

The plan was approved and its implementation was started.

Existing land disposal areas were enlarged and a new area constructed. This was sufficient for the next several years. With the interim element completed, efforts were devoted to the final two elements of the plan.

For sewage effluent to be acceptable for export, it would have to be essentially of drinking water quality; further, if in-basin disposal were ever to be permanently allowable, maximum removal of phosphate and nitrogen would be required. The district decided to initiate a research program and develop a tertiary treatment process that would meet the objectives.

Research and pilot plant studies were undertaken in 1961 to reveal possible new processes for wastewater reclamation. It was found that a tertiary sequence of treatment, including conventional activated sludge followed by chemical treatment utilizing alum for phosphate removal, mixed-media filtration, and granular carbon adsorption, produced improvements in water quality including virtually complete removal of suspended solids, BOD, bacteria, and other substances only partially removed by secondary treatment. In addition, good removals were obtained of COD, color, odor, viruses, phosphates, MBAS, and other substances relatively unaffected by secondary treatment.

The consulting engineers undertook a program to develop a process utilizing the proposed treatment system. By the end of 1963, pilot plant tests had developed to a point where the district authorized the design of a full scale 2.5 mgd plant incorporating these processes, plus facilities for thermal regeneration of granular activated carbon for flows up to 10 mgd. The plant was constructed in 1964 and placed into service in July 1965. Effluent quality was equal to or exceeded the expected quality and operating costs were within the predicted range.

The completion of the tertiary plant coincided with the culmination of an unprecedented growth within the district. The existing activated sludge plant was overloaded during the summer of 1965. The 2.5 mgd plant was receiving flows up to 4 mgd. Plant performance suffered and the tertiary facility could not continuously perform adequately with poor quality secondary effluent. By the summer of 1966, it was mandatory that the treatment plant be expanded. That fall, plans and specifications were authorized for expansion of plant capacity to 7.5 mgd, including construction of an experimental ammonia-stripping system for nitrogen removal, which was built for one-half plant capacity.

Operation of the original tertiary plant posed a problem in disposing of the alum sludge produced. At first, it was planned to reclaim and reuse the alum, but this proved not to be feasible. To overcome the problem, lime is used as a coagulant in the expanded plant. In addition to the chemical sludge problem, the original plant had trouble disposing of biological sludge because its digesters and drying bed were inadequate, particularly in the winter. For these reasons, a new solids disposal system was incorporated into the expanded plant. Biological sludge is incinerated in a six-hearth furnace. Waste activated and primary sludge is dewatered to about 19% solids by a concurrent flow solid bowl centrifuge. The lime sludge is reclaimed by recalcifying the lime in a second multiple hearth furnace. It is thickened in a gravity flow thickener, then dewatered to about 50% solids by a centrifuge.

The lime centrate contains most of the phosphates, which are removed by feeding the centrate to the primary clarifier and collecting this portion of the lime sludge with the biological sludge. The phosphate sludge is then incinerated along with the biological sludge, and the phosphate is disposed of as an insoluble rock phosphate in the ash.

Treating the secondary effluent with the high lime dose (about 300 mg/liter) necessary for maximum phosphate removal and clarification raises the pH to about 11.0. At this high pH, calcium carbonate could be deposited on the piping and filter beds, so the effluent is recarbonated until the pH reaches 7.5. This is accomplished by a two-stage system utilizing CO_2, which is obtained from the furnace stack gas. Prior to lowering the pH, the lime-treated effluent is pumped through a cooling tower where ammonia nitrogen is stripped from the water and released to the atmosphere as a gas. The stripping tower will remove up to 95% of the ammonia nitrogen during summer temperatures.

The plant expansion, with the exception of the ammonia-stripping tower, has been operating since March 31, 1968. The tower was placed in service in November 1968.

In November of 1963, the governors of California and Nevada met to consider solutions to the potential pollution problem of Lake Tahoe. As a result of the meeting, each state pledged its resources to carrying out a program of effluent export from the Lake Tahoe basin, since this would positively eliminate any danger to the lake from nutrient enrichment by sewage. The South Tahoe Public Utility District was selected as the first target, and California issued an edict telling the district to export by 1965.

The district had available at least three alternative routes for export. All routes were feasible, but each presented problems beyond the district's control.

The route chosen extends southerly and easterly from the water reclamation plant over Luther Pass into Alpine County, California. This route is approximately 75,000 ft long to Luther Pass, and involves an elevation change of 1440 ft.

The district, in 1965, applied to the State of California for permission to dispose of an extremely high-quality effluent into the Hope Valley area adjacent to the West Fork of the Carson River, just over Luther Pass. The Regional Water

Quality Control Board approved the plan. However, the County of Alpine did not agree with the board. After several months of negotiations between Alpine County and the South Tahoe District, final agreement was reached on an effluent disposal project. The agreement contained the following major points:

1. The point of discharge could not be in Hope Valley, but would be in Diamond Valley approximately 12 miles below Luther Pass
2. The effluent must be from a tertiary treatment plant and be essentially of drinking water quality
3. The effluent must be stored in a suitable location, so that discharge will occur only during the irrigation season
4. All effluent must be made available for irrigation use
5. Effluent disposal would be accomplished in Alpine County and done in a manner which would allow recreational use of the impounded water, if approved by the appropriate health agencies
6. The project, as finally designed, must be approved by all state and federal agencies involved in water quality control of Lake Tahoe

Even though the conditions imposed involved appreciably greater expenditures than originally anticipated, the district agreed to accept the project as approved, subject to obtaining financial aid from the state and federal governments.

The design of the export system was authorized by the district in late 1965. The time schedule contemplated placing all elements of the system under construction during 1966, with completion of construction scheduled by the end of 1967.

The purposes of the South Tahoe Water Reclamation Project may be summarized as follows:

1. The preservation of Lake Tahoe against any possibility whatever of pollution or accelerated eutrophication from wastewater discharge
2. Compliance with the export edict of the Nevada, California, and federal governments
3. Compliance with the effluent disposal standards of Alpine County, which provides for water of such quality that unrestricted recreational use of stored water is permitted.
4. The development of a treatment process that will make it possible in the future to beneficially reuse the reclaimed wastewater within the Tahoe Basin

The plant is designated by the Environmental Protection Agency as a National Demonstration Plant under the Clean Waters Restoration Act. Federal grants were made to aid in construction and to assist in the collection of data on the operation of the plant and the associated costs which will be useful in planning other similar works.

Results of a typical monthly report from the Utility District is given in Table V.

3. THE APOLLO PARK PROJECT

Apollo County Park is an aquatic recreational facility built and maintained

Table V
Monthly Report from the South Lake Tahoe Water Reclamation Project

Description	Requirements				Plant performance		
		Lahontan R.W.Q.C.B.					
		percent of time			percent of time		
	Alpine Co.	50	80	100	50	80	100
MBAS, mg/liter, less than	0.5	0.3	0.5	1.0	0.6	0.6	0.1
BOD, mg/liter, less than	5	3	5	10	1.4	2.0	3.1
COD, mg/liter, less than	30	20	25	50	12.2	14.2	15.2
Susp. S. mg/liter, less than	2	1	2	4	0	0	0
Turbidity JU	5	3	5	10	0.4	0.4	0.5
Phosphorus, mg/liter, less than	—	—	—	—	0.26	0.36	0.4
pH, units	6.5 to 8.5	6.5 to 9.0			6.8 to 7.1		
Coliform, MPN/100 ml	Adequately disinfected	Median less than 2; Max. no. consecutive samples greater than 23, 2			Median less than 2; no. of consecutive samples greater than 23, 0		

by the Los Angeles County Department of County Engineer in conjunction with the County Parks and Recreation Department.

The park covers about 56 acres of land of which 26.5 acres consists of a chain of three lakes filled with 80 million gal of polished renovated wastewater taken from the County Sanitation District No. 14 Water Renovation Plant. Here the water receives secondary treatment by eight oxidation ponds with a design capacity of 13.6 mgd and a detention time of about 60 days. From there, 0.5 mgd is treated by a tertiary treatment process consisting of coagulation by the addition of 300 mg/liter of alum followed by sedimentation and filtration through a dual media gravity anthracite sand filter. The water is then chlorinated before being pumped to the park. This process effectively reduces the amount of phosphates in the water to less than 0.5 mg/liter, removes pathogenic agents, and reduces the algae counts from the oxidation pond water to nil.

The lakes make up a closed system. Water entering the lakes stays there except for that lost through evaporation. There are no outlets from the lakes besides three high-level drains and three low-level drains. The bottoms of the lakes are sealed with a plastic liner placed 12–15 in. below the surface of the sediment which effectively stops any losses by percolation.

The lakes have been stocked with fish for recreational and environmental purposes. In 1971, adult bass were planted in the lakes with the idea that they would soon spawn and thereby propogate the species. Later in 1971, trout were planted. Trout unfortunately need a moving stream of water to spawn in and, therefore, have not propagated. Twenty catfish, Gambusia mosquito fish, and sunfish were also planted in 1971. Since then, about 5200 catfish (750 lb) were planted on March 1, 1973.

Prior to opening the park to the public in the spring, the California State Department of Health analyzed some fish samples for various heavy metals. A trout that had been in the lakes for about 16 mo was found to contain 2.0 mg/kg of mercury. The maximum allowable concentration as set by the federal government is 0.5 mg/kg. Since then, more fish samples have been analyzed and the high concentrations have been confirmed.

To determine the source of mercury in the fish, the County Sanitation Districts has conducted a survey of mercury in the environment of Apollo Park. The survey included a study of mercury moving up the food chain of the lake biota and possible sources of where the mercury may have originated from.

The results of the survey indicated that the apparent sources of mercury were the lake sediments and the natural soil outside of the park. Mercury appeared to be entering the lake water via biological methylation of the mercury trapped in the sediments and through the settling into the lakes of windblown soil high in mercury concentration. The investigation of the lake biota showed that the zoo plankton contained considerable amounts of mercury which they had taken up from the water and the food chain. The fish, in turn, had concentrated large quantities of mercury by the consumption of these organisms and by extracting mercury directly from the lake water. A trout "put and take" program has been established. Mature trout are placed in the lakes and within a few days are completely fished out. Tests have shown that the trout do not accumulate any significant levels of mercury in this short period.

The early encouragement of water reclamation has made California a leader in this area for over half a century. It is now the expressed policy of the state to encourage the development of wastewater reclamation so that these waters will be available to meet the state's needs.

It is significant to note that the waste disposal operation at South Lake Tahoe advanced from a septic tank unit to the most advanced treatment system in less than 10 years. In the same time interval, Santee went from waste disposal to a swimming operation. The same logical approach, aggresive enthusiasm, and scientific support can answer many of the remaining problems in wastewater reclamation so that these waters can be put to the broadest uses with assured health protection.

APPENDIX: WASTEWATER RECLAMATION CRITERIA*

Article 1. Definitions

60301. *Definitions.* (a) Reclaimed Water. Reclaimed water means water which, as a result of treatment of domestic wastewater, is suitable for a direct beneficial use or a controlled use that would not otherwise occur.

(b) Reclamation Plant. Reclamation plant means an arrangement of devices, structures, equipment, processes and controls which produce a reclaimed water suitable for the intended reuse.

(c) Regulatory Agency. Regulatory agency means the California Regional Water Quality Control Board in whose jurisdiction the reclamation plant is located.

(d) Direct Beneficial Use. Direct beneficial use means the use of reclaimed water which has been transported from the point of production to the point of use without an intervening discharge to waters of the State.

(e) Food Crops. Food crops mean any crops intended for human consumption.

(f) Spray Irrigation. Spray irrigation means application of reclaimed water to crops by spraying it from orifices in piping.

(g) Surface Irrigation. Surface irrigation means application of reclaimed water by means other than spraying such that contact between the edible portion of any food crop and reclaimed water is prevented.

(h) Restricted Recreational Impoundment. A restricted recreational impoundment is a body of reclaimed water in which recreation is limited to fishing, boating, and other non-body-contact water recreation activities.

(i) Nonrestricted Recreational Impoundment. A nonrestricted recreational impoundment is an impoundment of reclaimed water in which no limitations are imposed on body-contact water sport activities.

(j) Landscape Impoundment. A landscape impoundment is a body of reclaimed water which is used for aesthetic enjoyment or which otherwise serves a function not intended to include public contact.

(k) Approved Laboratory Methods. Approved laboratory methods are those specified in the latest edition of "Standard Methods for the Examination of Water and Wastewater", prepared and published jointly by the American Public Health Association, the American Water Works Association, and the Water Pollution Control Federation and which are conducted in laboratories approved by the State Department of Health.

(l) Unit Process. Unit process means an individual stage in the wastewater treatment sequence which performs a major single treatment operation.

(m) Primary Effluent. Primary effluent is the effluent from a wastewater treatment process which provides removal of sewage solids so that it contains not more than 0.5 milliliter per liter per hour of settleable solids as determined by an approved laboratory method.

(n) Oxidized Wastewater. Oxidized wastewater means wastewater in which the organic matter has been stabilized, is nonputrescible, and contains dissolved oxygen.

(o) Biological Treatment. Biological treatment means methods of wastewater treatment in which bacterial or biochemical action is intensified as a means of producing an oxidized wastewater.

(p) Secondary Sedimentation. Secondary sedimentation means the removal by gravity of settleable solids remaining in the effluent after the biological treatment process.

*Wastewater Reclamation Criteria are exerpted from Title 22, Division 4, Sections 60301–60357 of the California Administrative Code.

(q) Coagulated Wastewater. Coagulated wastewater means oxidized wastewater in which colloidal and finely divided suspended matter have been destabilized and agglomerated by the addition of suitable floc-forming chemicals or by an equally effective method.

(r) Filtered Wastewater. Filtered wastewater means an oxidized, coagulated, clarified wastewater which has been passed through natural undisturbed soils or filter media, such as sand or diatomaceous earth, so that the turbidity as determined by an approved laboratory method does not exceed an average operating turbidity of 2 turbidity units and does not exceed 5 turbidity units more than 5 percent of the time during any 24-hour period.

(s) Disinfected Wastewater. Disinfected wastewater means wastewater in which the pathogenic organisms have been destroyed by chemical, physical or biological means.

(t) Multiple Units. Multiple units means two or more units of a treatment process which operate in parallel and serve the same function.

(u) Standby Unit Process. A standby unit process is an alternate unit process or an equivalent alternative process which is maintained in operable condition and which is capable of providing comparable treatment for the entire design flow of the unit for which it is a substitute.

(v) Power Source. Power source means a source of supplying energy to operate unit processes.

(w) Standby Power Source. Standby power source means an automatically actuated self-starting alternate energy source maintained in immediately operable condition and of sufficient capacity to provide necessary service during failure of the normal power supply.

(x) Standby Replacement Equipment. Standby replacement equipment means reserve parts and equipment to replace broken-down or worn-out units which can be placed in operation within a 24-hour period.

(y) Standby Chlorinator. A standby chlorinator means a duplicate chlorinator for reclamation plants having one chlorinator and a duplicate of the largest unit for plants having multiple chlorinator units.

(z) Multiple Point Chlorination. Multiple point chlorination means that chlorine will be applied simultaneously at the reclamation plant and at subsequent chlorination stations located at the use area and/or some intermediate point. It does not include chlorine application for odor control purposes.

(aa) Alarm. Alarm means an instrument or device which continuously monitors a specific function of a treatment process and automatically gives warning of an unsafe or undesirable condition by means of visual and audible signals.

(bb) Person. Person also includes any private entity, city, county, district, the State or any department or agency thereof.

Article 2. Irrigation of Food Crops

60303. *Spray Irrigation.* Reclaimed water used for the spray irrigation of food crops shall be at all times an adequately disinfected, oxidized, coagulated, clarified, filtered wastewater. The wastewater shall be considered adequately disinfected if at some location in the treatment process the median number of coliform organisms does not exceed 2.2 per 100 milliliters and the number of coliform organisms does not exceed 23 per 100 milliliters in more than one sample within any 30-day period. The median value shall be determined from the bacteriological results of the last 7 days for which analyses have been completed.

60305. *Surface Irrigation.* (a) Reclaimed water used for surface irrigation of food crops shall be at all times an adequately disinfected, oxidized wastewater. The wastewater shall be considered adequately disinfected if at some location in the treatment process the median number of coliform organisms does not exceed 2.2 per 100 milliliters, as determined from the bacteriological results of the last 7 days for which analyses have been completed.

(b) Orchards and vineyards may be surface irrigated with reclaimed water that has the quality at least equivalent to that of primary effluent provided that no fruit is harvested that has come in contact with the irrigating water or the ground.

60307. *Exceptions.* Exceptions to the quality requirements for reclaimed water used for irrigation of food crops may be considered by the State Department of Health on an individual case basis where the reclaimed water is to be used to irrigate a food crop which must undergo extensive commercial, physical or chemical processing sufficient to destroy pathogenic agents before it is suitable for human consumption.

Article 3. Irrigation of Fodder, Fiber, and Seed Crops

60309. *Fodder, Fiber, and Seed Crops.* Reclaimed water used for the surface or spray irrigation of fodder, fiber, and seed crops shall have a level of quality no less than that of primary effluent.

60311. *Pasture for Milking Animals.* Reclaimed water used for the irrigation of pasture to which milking cows or goats have access shall be at all times an adequately disinfected, oxidized wastewater. The wastewater shall be considered adequately disinfected if at some location in the treatment process the median number of coliform organisms does not exceed 23 per 100 milliliters, as determined from the bacteriological results of the last 7 days for which analyses have been completed.

Article 4. Landscape Irrigation

60313. *Landscape Irrigation.* Reclaimed water used for the irrigation of golf courses, cemeteries, lawns, parks, playgrounds, freeway landscapes, and landscapes in other areas where the public has access shall be at all times an adequately disinfected, oxidized wastewater. The wastewater shall be considered adequately disinfected if at some location in the treatment process the median number of coliform organisms does not exceed 23 per 100 milliliters, as determined from the bacteriological results of the last 7 days for which analyses have been completed.

Article 5. Recreational Impoundments

60315. *Nonrestricted Recreational Impoundment.* Reclaimed water used as a source of supply in a nonrestricted recreational impoundment shall be at all times an adequately disinfected, oxidized, coagulated, clarified, filtered wastewater. The wastewater shall be considered adequately disinfected if at some location in the treatment process the median number of coliform organisms does not exceed 2.2 per 100 milliliters and the number of coliform organisms does not exceed 23 per 100 milliliters in more than one sample within any 30-day period. The median value shall be determined from the bacteriological results of the last 7 days for which analyses have been completed.

60317. *Restricted Recreational Impoundment.* Reclaimed water used as a source of supply in a restricted recreational impoundment shall be at all times an adequately disinfected, oxidized wastewater. The wastewater shall be considered adequately disinfected if at some location in the treatment process the median number of coliform organisms does not exceed 2.2 per 100 milliliters, as determined from the bacteriological results of the last 7 days for which analyses have been completed.

60319. *Landscape Impoundment.* Reclaimed water used as a source of supply in a landscape impoundment shall be at all times an adequately disinfected, oxidized wastewater. The wastewater shall be considered adequately disinfected if at some location in the treatment process the median number of coliform organisms does not exceed 23 per 100 milliliters, as determined from the bacteriological results of the last 7 days for which analyses have been complicated.

Article 6. Sampling and Analysis

60321. *Sampling and Analysis.* (a) Samples for settleable solids and coliform bacteria, where required, shall be collected at least daily and at a time when wastewater characteristics are most demanding on the treatment facilities and disinfection procedures. Tubidity analysis, where required, shall be performed by a continuous recording turbidimeter.

(b) For uses requiring a level of quality no greater than that of primary effluent, samples shall be analyzed by an approved laboratory method of settleable solids.

(c) For uses requiring an adequately disinfected, oxidized wastewater, samples shall be analyzed by an approved laboratory method for coliform bacteria content.

(d) For uses requiring an adequately disinfected, oxidized, coagulated, clarified, filtered wastewater, samples shall be analyzed by approved laboratory methods for turbidity and coliform bacteria content.

Article 7. Engineering Report and Operational Requirements

60323. *Engineering Report.* (a) No person shall produce or supply reclaimed water for direct reuse from a proposed water reclamation plant unless he files an engineering report.

(b) The report shall be prepared by a properly qualified engineer registered in California and experienced in the field of wastewater treatment, and shall contain a description of the design of the proposed reclamation system. The report shall clearly indicate the means for compliance with these regulations and any other features specified by the regulatory agency.

(c) The report shall contain a contingency plan which will assure that no untreated or inadequately-treated wastewater will be delivered to the use area.

60325. *Personnel.* (a) Each reclamation plant shall be provided with a sufficient number of qualified personnel to operate the facility effectively so as to achieve the required level of treatment at all times.

(b) Qualified personnel shall be those meeting requirements established pursuant to Chapter 9, (commencing with Section 13625) of the Water Code.

60327. *Maintenance.* A preventive maintenance program shall be provided at each reclamation plant to ensure that all equipment is kept in a reliable operating condition.

60329. *Operating Records and Reports.* (a) Operating records shall be maintained at the reclamation plant or a central depository within the operating agency. These shall include: all analyses specified in the reclamation criteria; records of operational problems, plant and equipment breakdowns, and diversions to emergency storage or disposal; all corrective or preventive action taken.

(b) Process or equipment failures triggering an alarm shall be recorded and maintained as a separate record file. The recorded information shall include the time and cause of failure and corrective action taken.

(c) A monthly summary of operating records as specified under (a) of this section shall be filed monthly with the regulatory agency.

(d) Any discharge of untreated or partially treated wastewater to the use area, and the cessation of same, shall be reported immediately by telephone to the regulatory agency, the State Department of Health, and the local health officer.

60331. *Bypass.* There shall be no bypassing of untreated or partially treated wastewater from the reclamation plant or any intermediate unit processes to the point of use.

Article 8. General Requirements of Design

60333. *Flexibility of Design.* The design of process piping, equipment arrangement, and unit structures in the reclamation plant must allow for efficiency and convenience in operation and maintenance and provide flexibility of operation to permit the highest possible degree of treatment to be obtained under varying circumstances.

60335. *Alarms.* (a) Alarm devices required for various unit processes as specified in other sections of these regulations shall be installed to provide warning of:

(1) Loss of power from the normal power supply.
(2) Failure of a biological treatment process.
(3) Failure of a disinfection process.
(4) Failure of a coagulation process.
(5) Failure of a filtration process.
(6) Any other specific process failure for which warning is required by the regulatory agency.

(b) All required alarm devices shall be independent of the normal power supply of the reclamation plant.

(c) The person to be warned shall be the plant operator, superintendent, or any other responsible person designated by the management of the reclamation plant and capable of taking prompt corrective action.

(d) Individual alarm devices may be connected to a master alarm to sound at a location where it can be conveniently observed by the attendant. In case the reclamation plant is not attended full time, the alarm(s) shall be connected to sound at a police station, fire station or other full-time service unit with which arrangements have been made to alert the person in charge at times that the reclamation plant is unattended.

60337. *Power Supply.* The power supply shall be provided with one of the following reliability features:

(a) Alarm and standby power source.

(d) Alarm and automatically actuated short-term retention or disposal provisions as specified in Section 60341.

(c) Automatically actuated long-term storage or disposal provisions as specified in Section 60341.

Article 9. Alternative Reliability Requirements for Uses Permitting Primary Effluent

60339. *Primary Treatment.* Reclamation plants producing reclaimed water exclusively for uses for which primary effluent is permitted shall be provided with one of the following reliability features:

(a) Multiple primary treatment units capable of producing primary effluent with one unit not in operation.

(b) Long-term storage or disposal provisions as specified in Section 60341.

Article 10. Alternative Reliability Requirements for Uses Requiring Oxidized, Disinfected Wastewater or Oxidized, Coagulated, Clarified, Filtered, Disinfected Wastewater

60341. *Emergency Storage or Disposal.* (a) Where short-term retention or disposal provisions are used as a reliability feature, these shall consist of facilities reserved for the purpose of storing or disposing of untreated or partially treated wastewater for at least a 24-hour period. The facilities shall include all the necessary diversion devices, provisions for odor control, conduits, and pumping and pump back equipment. All of the quipment other than the pump back equipment shall be either independent of the normal power supply or provided with a standby power source.

(b) Where long-term storage or disposal provisions are used as a reliability feature, these shall consist of ponds, reservoirs, percolation areas, downstream sewers leading to other treatment or disposal facilities or any other facilities reserved for the purpose of emergency storage or disposal of untreated or partially treated wastewater. These facilities shall be of sufficient capacity to provide disposal or storage of wastewater for at least 20 days, and shall include all the necessary diversion works, provisions for odor and nuisance control, conduits, and pumping and pump back equipment. All of the equipment other than the pump back equipment shall be either independent of the normal power supply or provided with a standby power source.

(c) Diversion to a less demanding reuse is an acceptable alternative to emergency disposal of partially treated wastewater provided that the quality of the partially treated wastewater is suitable for the less demanding reuse.

(d) Subject to prior approval by the regulatory agency, diversion to a discharge point which requires lesser quality of wastewater is an acceptable alternative to emergency disposal of partially treated wastewater.

(e) Automatically actuated short-term retention or disposal provisions and automatically actuated long-term storage or disposal provisions shall include, in addition to provisions of (a), (b), (c), or (d) of this section, all the necessary sensors, instruments, valves and other devices to enable fully automatic diversion of untreated or partially treated wastewater to approved emergency storage or disposal in the event of failure of a treatment process, and a manual reset to prevent automatic restart until the failure is corrected.

60343. *Primary Treatment.* All primary treatment unit processes shall be provided with one of the following reliability features:

(a) Multiple primary treatment units capable of producing primary effluent with one unit not in operation.

(b) Standby primary treatment unit process.

(c) Long-term storage or disposal provisions.

60345. *Biological Treatment.* All biological treatment unit processes shall be provided with one of the following reliability features:

(a) Alarm and multiple biological treatment units capable of producing oxidized wastewater with one unit not in operation.

(b) Alarm, short-term retention or disposal provisions, and standby replacement equipment.

(c) Alarm and long-term storage or disposal provisions.

(d) Automatically actuated long-term storage or disposal provisions.

60347. *Secondary Sedimentation.* All secondary sedimentation unit processes shall be provided with one of the following reliability features:

(a) Multiple sedimentation units capable of treating the entire flow with one unit not in operation.

(b) Standby sedimentation unit process.

(c) Long-term storage or disposal provisions.

60349. *Coagulation.*

(a) All coagulation unit processes shall be provided with the following mandatory features for uninterrupted coagulant feed:

(1) Standby feeders,
(2) Adequate chemical storage and conveyance facilities,
(3) Adequate reserve chemical supply, and
(4) Automatic dosage control.

(b) All coagulation unit processes shall be provided with one of the following reliability features:

(1) Alarm and multiple coagulation units capable of treating the entire flow with one unit not in operation;
(2) Alarm, short-term retention or disposal provisions, and standby replacement equipment;
(3) Alarm and long-term storage or disposal provisions;
(4) Automatically actuated long-term storage or disposal provisions, or
(5) Alarm and standby coagulation process.

60351. *Filtration.* All filtration unit processes shall be provided with one of the following reliability features:

(a) Alarm and multiple filter units capable of treating the entire flow with one unit not in operation.

(b) Alarm, short-term retention or disposal provisions and standby replacement equipment.

(c) Alarm and long-term storage or disposal provisions.

(d) Automatically actuated long-term storage or disposal provisions.

(e) Alarm and standby filtration unit process.

60353. *Disinfection.* (a) All disinfection unit processes where chlorine is used as the disinfectant shall be provided with the following features for uninterrupted chlorine feed:

(1) Standby chlorine supply,
(2) Manifold systems to connect chlorine cylinders
(3) Chlorine scales, and
(4) Automatic devices for switching to full chlorine cylinders.

Automatic residual control of chlorine dosage, automatic measuring and recording of chlorine residual, and hydraulic performance studies may also be required.

(b) All disinfection unit processes where chlorine is used as the disinfectant shall be provided with one of the following reliability features:

(1) Alarm and standby chlorinator;
(2) Alarm, short-term retention or disposal provisions, and standby replacement equipment;
(3) Alarm and long-term storage or disposal provisions;
(4) Automatically actuated long-term storage or disposal provisions; or
(5) Alarm and multiple point chlorination, each with independent power source, separate chlorinator, and separate chlorine supply.

60355. *Other Alternatives to Reliability Requirements.* Other alternatives to reliability requirements set forth in Articles 8 to 10 may be accepted if the applicant demonstrates to the satisfaction of the State Department of Health that the proposed alternative will assure an equal degree of reliability.

Article 11. Other Methods of Treatment

60357. *Other Methods of Treatment.* Methods of treatment other than those included in this chapter and their reliability features may be accepted if the applicant demonstrates to the satisfaction of the State Department of Health that the methods of treatment and reliability features will assure an equal degree of treatment and reliability.

9

Water Reuse in the Federal Republic of Germany

W. J. Müller

I. WATER RESOURCES AND REUSE OF WASTEWATER

A. Available Water Resources

The Federal Republic of Germany with a population of 61.2 million inhabitants covers an area of 248,000 km^2. The average figure of annual rainfall recorded for the period 1891 to 1930 was found to be 803 mm or about 200,000 million m^3/year for the whole area of the federal territory. On this basis, it was estimated for the year 1965 (Clodius, 1970) that 415 mm or about 103,000 million m^3/year were evaporated by plants, the soil, and other surface areas. About 112 mm of precipitation or 28,000 million m^3/year were accounted for the replenishment of groundwater storage by percolation and 276 mm or 68,000 million m^3/year for direct runoff flowing into rivers and streams. During dry weather periods about 83 mm or 21,000 million m^3/year were estimated to run from groundwater storage into open watercourses which maintains low water flows in rivers and streams and makes the total runoff figure rise to, say, 90,000 million m^3/year.

The catchment areas of the German sections of the Rhine, Elbe, and Danube rivers extend to other countries and, therefore, these rivers receive additional rates of runoff from other countries, such as from France, Luxembourg, Switzerland, Austria, Czechoslovakia, and the German Democratic Republic, which amount totally to about 80,000 million m^3/year. These quantities substantially increase the flows in the Rhine, Elbe, and Danube rivers and in some of their tributaries. On the other side, some of the rivers of Germany are flowing into other countries—the Rhine into the Netherlands and the Danube into Austria; others are discharging into the North Sea—the Weser and Elbe rivers—or into the Baltic Sea.

These natural conditions have lead to international cooperation and agreements between the Federal Republic of Germany and its neighboring countries with regard to problems arising in rivers which are flowing through or bordering several countries. This cooperation has become desirable or even necessary especially in problems of water resources management, water utilization for supply purposes, water pollution control, navigation, etc. Examples are the "International Commission for the Protection of the Rhine River against Pollution" and the "International Commission of the Danube."

B. Use of Natural Waters

The natural waters in Germany are used for many purposes, e.g., for water supply, recreation, sport, fishing, navigation, irrigation, etc. The most important use from the standpoint of water quality is that for public water supply. All measures of water resources management are governed by the principle that high-quality water should primarily be reserved for the supply of drinking water; preferably, this supply will come from ground-water resources since these waters usually are relatively clean and protected from pollution, and therefore hygienically safe and aesthetically satisfactory. Surface waters such as rivers and streams were mainly regarded as the natural receiver of storm-water and wastewater effluents apart from other uses, such as fishing, recreation, navigation, irrigation. In the last decades, the position has been changed. With the increasing demand for domestic (potable) water and industrial supplies in the Federal Republic of Germany (Table I) and the limited quantities of ground-water available in the federal territory, more and more water for public supply has to come from surface water resources such as rivers and lakes.

On the other side, the quantity of wastewater effluents resulting from the use of supply water also is increasing. Most of these effluents have to be discharged into inland watercourses as the natural receiver. Only small amounts of waste-water effluents from towns and industry of coastal areas—in 1957 about 1.2% of the total volume of wastewater effluents—can be discharged to the North Sea or to the Baltic Sea. Wastewater disposal onto land also is limited (see Section IV,B). Therefore, more than 95% of wastewater effluents produced in the Federal Republic of Germany have to be discharged into inland water-courses, i.e., into the same natural waters that partly have to be used for water supply purposes and, for this reason, must be kept reasonably clean and free from polluting substances.

C. Inevitability of Wastewater Reuse

In the densely populated and industrialized area of the Federal Republic of Germany, the natural watercourses have to be used for both water supply and wastewater disposal in almost all parts of the country. It is practically impos-sible—apart from special cases such as in headwaters of river systems—to

Table I
Quantity of Water Supplied in the Federal Republic of Germany[a]

Year	1957	1963	1969
Public water supplies (Siegmund, 1970)	3138	3725	3990
Those for industry	952	1040	1035
Industrial water supplies (Statistisches Bundesamt, 1972)	6830	9595	11,346

[a] In million m^3/year.

develop water supply schemes based on surface water without any connection with wastewater disposal systems. Water obtained for supply from a river usually contains some part of wastewater discharged into the river system upstream of the intake. Thus indirect reuse of wastewater is inevitable. This situation not only requires proper water pollution control but also involves problems of water reclamation from wastewater and reuse of wastewater effluents with regard to hygienic and aesthetic points of view.

II. RECLAMATION OF WATER FROM WASTEWATER FOR PUBLIC SUPPLY

A. The Water/Wastewater/Water Cycle

The reuse of wastewater implies recycling. Water after being used for supply purposes becomes wastewater. This wastewater may be reused for other supplies, for irrigation, or for other purposes after it undergoes specific treatment to remove contaminating substances so it will be suitable for the second use. The wastewater of the second use may be recycled again, etc.

In practice, water quality deteriorates as the same water passes repeatedly through the water/wastewater/water cycle and is reused no matter how elaborate the process of treatment may be. This is because the concentration of certain compounds and substances produced during the breakdown of organic compounds increases as the cycle is repeated. The salt content, for example, goes up. The carbon dioxide content may rise, with a consequent increase in corrosion and chemical action. Nutrients accumulate and encourage the growth of algae. The cycle also leads to enrichment in substances such as sulfates and phenols found in industrial effluents, trace elements, and other substances that may reach such a concentration that the water becomes unfit for use. This sets limits to the water/wastewater/water cycle.

Therefore, wastewater cannot be reused directly for drinking and other domestic purposes such as bathing. Even with the best treatment methods available, there are also objections of a subjective or psychological nature, and the possibility of danger to health cannot be excluded. For the reclamation of potable water from wastewater, a substantial proportion of pure natural water should be added to the water/wastewater/water cycle and, accordingly, a similar quantity of wastewater effluent removed from the cycle and disposed of elsewhere (Müller, 1973). This will limit the enrichment of certain substances, as just mentioned, and maintain certain concentration of substances contained in the water/wastewater/water cycle.

If the cycle includes dilution by natural water and self-purification processes in, say, a river, the water produced may be suitable for direct supply to the public, if, during the cycle, the rate of dilution by natural water is satisfactory and the water receives careful and safe treatment. A higher degree of river pollution, which means less dilution of the wastewater by natural water from this source, may be regarded as acceptable if the water undergoes further natural self-purification during the cycle. One way of achieving this is to make use of the self-purification capacity of natural soil by abstracting the water from the river or other open watercourses not directly, but indirectly, after filtration by seepage through the riverbank, and using riverside wells located sufficiently faraway from the river itself. Suitable geological formations are an essential part of any such system. The period of time that the water takes to seep through from the river to the wells from which it is abstracted should be sufficiently long, e.g., 40 to 100 days, or even longer, because natural self-purification underground is a slow process. Similarly, "artificial" groundwater can be produced by introducing river water into the ground with the aid of infiltration basins, trenches, or infiltrations wells. Here, too, the path of the water through the ground to the supply wells has to be long enough to provide sufficient detention periods for adequate self-purification.

B. Water for Public and Industrial Supply

In the Federal Republic of Germany *water for public supply* is taken from various water resources (Table II). Spring water and true groundwater are the

Table II
Water Obtained for Public Supply in 1968[a]

Origin of water	Amount[b]	Percent of total
Spring water	359	11%
True groundwater	1817	54%
Surface water	1170	35%
From bank infiltration	474	(14%)
Artificial groundwater	388	(12%)
River water	45	
Lake water	75	(9%)
From reservoirs of dams	188	
	3346	100%

[a] From Verband der Deutschen Gas- und Wasserwerke (1970).
[b] Measured in million m^3/year.

main sources for drinking water and account for about 65% of the supply. In addition, surface water is also used, directly or indirectly, at a rate of about 35% of the total demand. Considerable quantities of water are being obtained by "bank filtration" and by the production of "artificial groundwater" by the recharge of groundwater with river water. Direct abstractions of surface water for public supply purposes are applied only to a minor extent, a little less than 10% of the total supply. Water from lakes and reservoirs of dams accounts for the greater part of it (about 85% in the year 1968). It can be assumed that in these instances there is virtually no reuse of wastewater. Only about 15% of the directly abstracted surface water comes from sources that are polluted to some degree. Such raw water must be properly purified before using as drinking water.

Water for industrial supply is obtained mainly directly from surface waters at a rate of about 63.5% of the total supply of industry (Table III). The remaining supply uses water from springs and groundwater storage. In addition, industry obtains water from public supply and other resources.

The total figure of water quantities obtained from surface waters, directly or indirectly, for public and industrial supply in the Federal Republic of Germany is estimated for the year 1969 in the following tabulation:

Public supply, 35% of 3,990	1,400 million m^3/year
Industrial supply	7,205 million m^3/year
Total surface water used	8,605 million m^3/year

Table III
Water Obtained for Industrial Supply in 1969[a]

Supply and use	Amount[b]	(%)
Own supply		
Spring water and groundwater	4141	(36.5%)
Surface water from rivers and lakes	7205	(63.5%)
Supply from other sources		
Public supply and others	1370	
Total output	12,716	
Waste of water (without use)	1494	
Supply to other consumers	497	
Consumed by industry	10,725	

[a] From Statistisches Bundesamt, (1972).
[b] In million m^3/year.

As the surface runoff from precipitation within the federal territory amounts, on average, to about 90,000 million m³/year (see Section I,A) in 1969, about 10% of the runoff was used to provide water to public and industrial supplies.

C. Bank Infiltration Water

In the Federal Republic of Germany, water from bank inflitration of surface water accounts for some 14% of the total consumption for public supply. For this method to be effective, the part of the riverbed that is near the bank and the underlaying stratum must consist of suitable geological formations, such as diluvial and alluvial strata of sand and gravel. Water for use is drawn from the river or other surface waters only indirectly from wells or infiltration galleries separated from the watercourse. Since the water first flows through a sufficient volume of sand or gravel before entering the intake wells to make it acceptable, the increasing pollutional load on German rivers has made the intake of water from bank infiltration more difficult. Clogging of the riverbed reduces the infiltration capacity and the quality of the water is affected by higher odor content, varying contents of iron, manganese, ammonia, and other substances (Herrig, 1970).

Along the Rhine, most of the water used for supply is abstracted from the river by bank infiltration. As the river is heavily polluted with municipal and industrial effluents, the waterworks on the lower Rhine that obtain their water by bank infiltration are faced with the problem of applying suitable methods of processing the water that will eliminate the substances responsible for taste and odor. Large-scale experiments using activated carbon for this purpose have been carried out in Düsseldorf. A combined ozone and activated-carbon technique has proved successful and is in use in Düsseldorf and Duisburg (Hopf, 1970). This combination also successfully removes the content of manganese.

D. Recharge of Groundwater by Infiltration

The production of artificial groundwater by infiltration of river water is applied at many waterworks in Germany. In the Federal Republic as a whole, production of water by this method amounts to about 12% of the total output from public waterworks (Table II). In the Ruhr River area the production of artificial groundwater by infiltration of water from the Ruhr River is the most important source for water supplies, since it accounts for more than 60% of the total.

The principal methods employed in Germany for recharging groundwater are infiltration basins, filter trenches, and recharge wells. Among these, infiltration basins are by far the most widely used; they account for about 75% of the artificial production of groundwater.

1. INFILTRATION BASINS

Infiltration basins in the form of sand filters have been constructed at many places in the federal territory for recharge of groundwater. The best-known infiltration basins of this type are those operated by the waterworks in the Ruhr valley, which supply water to about 5 million people and a great number of industrial undertakings of the industrial region of North Rhine/Westphalia. Among these are the waterworks at Dortmund, Hamm, Essen, Gelsenkirchen, and other places. Similar installations are to be found also in other parts of Germany. When groundwater is produced artificially in this way, the river water to be filtered must not be too heavily polluted. According to the German experience (Brix *et al.*, 1963), the upper limits are concentrations of about 5 mg/liter of settleable solids, 10 to 20 mg/liter of plankton, and a bacteria content of less than 100,000/ml. If the river is too heavily polluted, the infiltration water may have to undergo preliminary treatment, such as settling, or even biological treatment, so as to increase the ratio of oxygen to carbon dioxide, i.e., to bring the oxygen content above 6 mg/liter and the carbon dioxide content below the calcium carbonate/carbon dioxide balance (Frank, 1965).

2. FILTER TRENCHES

Filter trenches are in use, for example, in Hamburg (Brix *et al.*, 1963). At the Hamburg waterworks "Curslack Plant," water from the Elbe and Bille rivers is distributed through a large branched network of trenches over some 2200 ha of the Elbe marshes in the Vierländer district. The total length of the trenches is about 1000 km and the water infiltrates into alluvial strata. The raw river water does not undergo any primary treatment. The waterworks intake system is 85 m away from the trench system, and the water takes about 60 days to travel through the underground formations from the trenches to the 277 supply wells.

3. INFILTRATION WELLS

Infiltration wells were installed at the Wiesbaden-Schierstein waterworks for the infiltration of pretreated water from the Rhine (Herzberg, 1965). The Rhine water is taken from midstream and it is treated by prechlorination, chemical precipitation using ferric chloride, settling in sedimentation tanks, and filtration through rapid gravel filters. For final filtration, dechlorination, and deodorization, the water passes through closed activated-carbon filters before being finally conducted to the distribution system in the infiltration well area. These wells are located about 180 m away from the water-supply intake wells. The water remains underground for about 4 weeks, and this is long enough for the temperature of the infiltrated water, which in summertime may be as high as 20°C, to come down to that of the artificially augmented groundwater, i.e., to between 12° and 13°C.

E. Wastewater Disposal and Reuse

Pollution of natural waters such as rivers and lakes is mainly produced by sewage effluents from cities and other residential areas and by wastewater discharges of industry and power stations. In the area of the Federal Republic of Germany, these effluents amounted to figures in 1969 as shown in Table IV according to the statistics published recently (Bundesministerium des Innern, 1972).

The municipal sewage effluents and polluted industrial wastewater discharges add up to, say, 24 million m^3/day or about 8,800 million m^3/year and cooling water discharges from industry and power stations came to a figure of 52.57

Table IV
Wastewater Effluents in 1969[a]

Effluent	Amount[b]
Municipal sewage	
Including effluents from local trade activities collected by public sewerage systems	8,706,000
Surface water entering these systems	2,330,000
Sewage effluents from isolated buildings, institutions, and fringe areas of larger communities (estimated)	1,300,000
	12,336,000
Industrial wastewaters	
Discharged into public sewerage systems	5,762,000
discharged directly into natural waters	5,900,000
Cooling water discharges of industry	18,540,000
Other effluents	4,405,000
	34,607,000
Cooling water from power stations	
Intake of 34,200,000 m^3/day from surface waters and groundwater and after use discharged into surface waters	34,030,000

[a] From Bundesministerium des Innern (1972).
[b] Measure in m^3/day.

million m³/day or about 19,000 million m³/year. These discharges represent a substantial portion of the runoff from the federal territory, which is estimated at 90,000 million m³/year (see Section I,A). In the Rhine, Danube, and Elbe rivers and in some of their tributaries, additional dilution of the wastewater discharges is obtained from tributary flows that are running from other countries into these rivers and accounting for about 80,000 million m³/year as stated above (see Section I,A).

Generally, the direct and indirect use of water from German rivers and lakes for supply purposes implies the reuse of wastewater discharges to a large extent which, on average, may be estimated to amount to about 10% in proportion to the natural runoff regarding polluted wastewater discharges and to about 30% in proportion to the natural runoff if the cooling water discharges are included. The ratios of wastewater discharges to river flows will vary widely in the different river regions of the federal territory depending on the flow of the river in question and the location and size of the wastewater outfalls. The river flow varies throughout the year from a minimum (dry weather flow) to a maximum (flood flow), with variations also from year to year, but wastewater effluents are discharged at relatively constant rates. Sometimes and at some places the proportion of wastewater in supply water obtained from surface resources will be practically nil, while in other extreme cases it may be well above the average. Under these circumstances, it will be necessary that before discharge into surface water systems wastewater has to be treated accordingly to remove all polluting substances that could be harmful to drinking water supplies which natural processes of self-purification and known methods of water purification will not remove. Proper wastewater treatment is prerequisite to the reuse of wastewater.

For example, the reuse of wastewater discharges is of great importance to the water resources management in the Ruhr river area where river water is used as the main source of public and industrial supply to the heavily industrialized area of North Rhine-Westphalia and where the Ruhr River is used as the receiving water of large quantities of wastewater discharged from this area. Under the dry summer conditions experienced in 1929, it has been observed (Imhoff and Hyde, 1931) that a large part of the water of the Ruhr River that was used for supply purposes had for a short period been passing through the water/wastewater/water cycle three times over. More recently, it is reported (Koenig *et al.*, 1971) that in the interest of water supply from this source the ratio of quantities of polluted wastewater discharges to that of the Ruhr River flow is controlled by appropriate water resources management with the aim of keeping the ratio below a certain level. During periods of minimum dry weather river flow, this ratio is maintained at a maximum of 22% which has proved satisfactory. This control is achieved by an artificial increase of the minimum dry weather flow of the Ruhr River as required by the addition of more water taken from the same river system by using floodwaters previously stored in reservoirs of dams constructed in the upper regions of the catchment area of the Ruhr River. In 1959, when such additional dilution water at minimum river flow was not available during a long, dry spell, the ratio of wastewater discharges to river flow went up considerably and reached 86% for a short period. The quality of supply water obtained from artificial groundwater production from river water deteriorated substantially. In spite of the normal application of water purification processes using settling basins, slow sand filters as infiltration basins, soil passage, and disinfection before use, drinking water delivered from these waterworks had high contents of organic matter, ammonia, detergents, chlorides, and other harmful substances.

III. REUSE OF WASTEWATER FOR INDUSTRIAL PURPOSES

A. Reuse by Recycling of Wastewater

A distinction should be made between the reuse of wastewater originating outside the factory where it is to be used and the reuse of wastewater within the factory where it is produced. In the first case, wastewaters from elsewhere may be reused indirectly, as is practiced for public supply by using river or other surface water mixed with wastewater effluents discharged from public sewerage systems or from industries upstream of the intake of supply water for the factory concerned; or wastewater may be reused directly by making use of treated effluents from another industry or public sewerage system. In the Federal Republic of Germany, there are no installations for direct reuse of wastewater from outside the factory, but indirect reuse of wastewater effluents discharged into surface waters applies in the same way as in the case of public supply from surface waters. About two-thirds of the water used for industrial supply comes from surface water resources (see Table III), which carry substantial amounts of wastewater discharges. The rules applying to the reuse of wastewater for public supply are generally followed. However, the requirements regarding the quality of water for use in the factory may differ considerably from those of public supply and vary according to the type of industry. The requirements of food industries, for example, are just as strict as those for potable water supplies, but other industries need only lower-grade water, e.g., for cooling purposes. On the other hand, boiler feed water must meet special requirements that are different from those for drinking water.

The reuse of wastewater arising within the factory is a different matter. It may be an effective means of reducing the volume of liquid wastes discharged into a public sewerage system or into natural watercourses, and therefore, has been applied in many industries. Recycling of wastewater within an industrial plant may be especially feasible where water is used for cooling, material transport, washing of raw materials or of manufactured products, etc., and where the treatment of the wastewater for reuse is not too costly. In particular, industries that require large volumes of water have resorted to treatment and reuse of wastewater.

In Germany, reuse of wastewater by recycling is common practice in many industrial undertakings including coal washeries, oil refineries, paper industry, steel mills, gas works, and electroplating process plants. Figures of water used and reused in major water-consuming industries in 1969 (Statistisches Bundesamt, 1972) are listed in Table V. In industry as a whole, over 60% of the water consumed has been recycled.

B. Recovery of Waste Materials from Waterborne Wastes

In addition to the reclamation of water, the recovery of specific constituents from industrial wastewater for reuse may be important and is applied in many cases. The process of recovery of substances may be advantageous to industry since a reduction in the amount of waste materials contained in wastewaters may increase the efficiency and economy of manufacturing processes and reduce the required wastewater treatment and, consequently, the cost of treatment and disposal. The recovery of valuable substances from wastewater will be more worthwhile the higher the concentration in wastewater and the less the wastewater contains of other matter. Also, it will become more economical the higher the legal demand on the quality of treated effluents in the interest of pollution control in natural waters. In any case, the process of recovery must be relatively cheap and the recovered material must have some market value.

There are many examples of useful recovery of materials from wastewaters already in use in the Federal Republic of Germany (Müller, 1973). It is common practice to remove coal slurry from washing wastes and to make use of this coal. In the steel and metal industries, pickling wastes are neutralized in processes that will permit recovery of acids and iron and metal salts. Products such as ferrous sulfate may be recovered and marketed. Phenols contained in wastewaters from coke ovens, gasworks, and similar carbonization processes are recovered by special methods including solvent extraction, adsorption by activated carbon and distillation. In the Ruhr and Emscher industrial districts of

Table V
Water Used and Recycled for Reuse within the Factory of Major Industries in the Federal Republic of Germany in 1969[a,b]

Branches of industry	Water supplied	Total quantity used	Recycled within the works	
			Amount	Percentage
Coal mining	1223	8746	7523	86.0%
Chemicals	3838	6309	2471	39.2%
Iron and steel	1751	5306	3555	67.0%
Oil refining	485	2485	2000	80.5%
Wood pulp, cellulose, paper and board	923	1892	969	51.2%
Stone and earth	354	756	402	53.2%
Motor vehicles	262	554	292	52.7%
Metal industry	174	329	155	47.1%
Engineering works	135	281	146	52.0%
Sugar industry	47	262	215	81.9%
Textiles	238	244	6	2.4%
Dairy industry	101	220	119	54.0%
Breweries	122	217	95	43.7%
Electrical industry	103	204	101	49.6%
Other industries	969	2199	1230	55.9%
Industry as a whole	10,725	30,004	19,279	64.3%

[a] From Statistisches Bundesamt (1972).
[b] In million m^3/year.

Germany, phenol recovery has been the subject of much study. In 1969, there were 13 plants in the Emscher area mainly using the benzole sodium phenolate process and operated successfully with a phenol output of 3825 tons/year. The effluents from paper mills and cellulose factories carry substantial amounts of fibers which usually are recovered with the aid of screens and savealls. Wastewaters from fruit and vegetable canning normally are screened at the factory. At many canning factories, screenings are used for feeding stock. Also screenings recovered from brewery wastewaters are utilized as cattle feed. Fats, greases, and oils entrained in industrial wastewaters may be recovered by means of separating tanks or by special processes. For example, the partial recovery of wool greases is common practice in wool-washing plants. Similarly, grease recovery from wastewater of abattoirs amounts to about 0.4% by weight of animals killed, while blood and other refuse yield valuable fertilizers.

IV. REUSE OF WASTEWATER FOR AGRICULTURAL PURPOSES

A. The Natural Cycle of Elements

Agricultural utilization of wastewater is based on the principle of recycling waste materials to the soil that originally came from the land in the form of food and other useful products. It was Justus Liebig who more than a hundred years ago first emphasized the idea that the maintenance of the natural cycle of elements is required for the conservation of soil fertility. Many substances contained in wastewater effluents represent one of the available resources of fertilizing constituents apart from solid waste materials such as garbage and other refuse and in addition to artificial fertilizers. The wastewater nutrients may be returned to the soil in admixture with the wastewater by irrigation or as sludge separated from the water by wastewater treatment.

B. Wastewater Irrigation

Agricultural utilization of wastewater by irrigation is based on the use of both the water content of wastewater and its constituents that may be of value as fertilizer. Wastewater irrigation will be of particular interest in areas where the land is already watered or where this is desirable. The need for irrigation will depend on many local factors such as the properties of the soil, the use made of the soil, on rainfall, evaporation, and other climatic conditions, on groundwater, and other factors.

For most satisfactory operation, the wastewater should be treated before irrigation to remove substances that could produce odors and be harmful to the soil, to the crops to be grown, to grazing animals, and to public health. It is essential to ensure that pathogenic organisms are prevented from contact with plants used for food supply. Even after biological treatment, domestic wastewater still contains considerable quantities of fertilizing constituents such as nitrogen and phosphates which are very important for the plant life on the land to be irrigated. The organic matter contained in domestic wastewater consists of humus and other compounds which also give humus in the end. The principle fertilizing constituents of domestic wastewater (Sierp, 1950) are given in Table VI. In addition to these substances, many metallic salts are also present in the wastewater in very small quantities (trace elements) and these, too, are important as fertilizers or as essential elements for pastures and other agricultural land. High contents of salts, toxic metallic compounds, and boron may be harmful. Irrigation water should not contain more than 0.3 mg/liter arsenic, 0.3 mg/liter

Table VI
Fertilizing Constituents of Domestic Wastewater[a,b]

Fertilizing constituent	Raw sewage	Biologically treated sewage
Nitrogen	12.8	10.9
Phosphate (P_2O_5)	3.5	2.8
Potash (K_2O)	7.0	6.7
Organic matter (loss on ignition)	55	19

[a] From Sierp (1950).
[b] Measured in gm/person/day.

boron, 2 mg/liter zinc, 5 mg/liter copper, 0.5 to 1.0 mg/liter cobalt, and 0.5 to 1.0 mg/liter nickel (Meinck *et al.*, 1968). Therefore wastewater carrying high concentrations of these and other harmful substances are not suitable for agricultural utilization. This usually applies to wastewater discharges of the chemical and metal industries.

The most appropriate wastewater for agricultural use by irrigation is domestic sewage, but also many industrial wastewaters, preferably those containing organic substances, have proved suitable for irrigation purposes including those from dairies, breweries, starch factories, textile mills, sugar factories, dyeworks, and tanneries.

Since domestic wastewater such as sewage does not contain the fertilizing constituents in optimum proportions, it may be necessary to supplement them with artificial fertilizers to ensure that they are fully utilized for agricultural purposes. In any event, the fertilizer substances should not be lost by drainage into the ground or into surface waters. To maintain this balance as far as possible, the loading of the soil with these substances should be kept fairly low. In practice, it has been found in Germany (Imhoff and Imhoff 1972) that under German climatic conditions the load allowance of domestic sewage should, on average, not exceed a level corresponding to 30 persons or equivalent per hectare of irrigated land. This is the only way to ensure the complete disposal of domestic sewage by land irrigation. Such "low-rate irrigation" is different from high-rate irrigation usually applied as "broad irrigation" in sewage farming (Imhoff *et al.*, 1971).

Low-rate irrigation corresponding to the domestic sewage from 30 persons/ha yields, on the basis of 200 liters/persons/day, about 220 mm/year of irrigation water. Since from the standpoint of agriculture, irrigation cannot continue all the year around, alternative means of sewage disposal facilities must be available to be used during the intervals between irrigation, using, for example, intermittent sand filters draining into groundwaters or surface waters or some other suitable treatment and disposal systems; or providing storage or other ponding systems.

Special regulations on sewage irrigation have been introduced in Germany to protect the population, and also animals, against possible nuisances and health hazards. These take the form of general public health rules to be observed when domestic sewage irrigation is practiced. The most important provisions may be summarized as follows (Müller, 1973):

(1) Sewage and other suitable wastewater to be used for irrigation must be pretreated by passing through settling tanks with sufficiently long detention periods and additional biological treatment at least partially is desirable. When necessary, especially in the case of spray irrigation, this should be followed by chlorination of the effluent before irrigation for odour control.
(2) Housing, transport facilities, recreation facilities and other adjacent public areas, together with grazing land, orchards, vineyards and market gardens must not be affected when sewage spray irrigation is practised on nearby land. There must be no objectionable odour. These conditions are to be met by sprinkling only at a sufficient distance from the area in question, and by planting trees or hedges, laying out protective strips of land, and other measures.
(3) Sewage irrigation of land used for water supply purposes or of land adjacent to such area, should not be permitted.
(4) The time-schedule for irrigation of an area must be such as to stop the irrigation of land at least 14 days before grazing or harvesting.
(5) Beets for use as feeding stuffs or in sugar manufacture, potatoes for industry, oil-seed and fibre producing plants may be irrigated with sewage only up to four weeks before harvesting. With potatoes for human consumption and cereals sewage may be used only up to the flowering stage. Growing vegetables may be watered only with pure water.
(6) Highly infectious sewage from certain establishments should not be used for agricultural irrigation. This applies in particular to hospitals for infectious diseases, sanatoriums, factories processing animal carcasses, special abattoirs dealing with epidemics and certain diseases, mortuaries and quarantine establishments.

The planning of a sewage and wastewater irrigation scheme of the low-rate irrigation type as developed in Germany is principally a matter for agricultural experts. Various irrigation methods are employed to spread the wastewater, such as flood irrigation, surface irrigation, land filtration using shallow trenches, and the more elaborate and costly system of spray irrigation which can be used on any natural configuration of ground and applied in small doses of water. However, spray irrigation may give rise to objectionable odors due to gases given off as the wastewater is split up into small droplets if the sprayed wastewater is not pretreated accordingly. Also, the health hazards are greater using spray irrigation, because the sprinkling of sewage can spread bacteria over great distances. This is why biological purification of sewage followed by chlorination or the admixture of pure water is necessary before sprinkling (Reploh and Handloser, 1957).

In the Federal Republic of Germany, the natural supply of water from precipitation to areas used for agriculture and horticulture amounts to about 55,000 to 60,000 million m^3/year and that to forest areas to about 25,000 million m^3/year. Only a small proportion of the agricultural area—about 8%—needs irrigation. At the end of 1958 (Bundesminister für Ernährung, Landwirtschaft, und Forsten, 1961), about 1,227 million m^3/year were used for irrigation to an area of about 250,000 ha and were taken from the sources enumerated in the following tabulation.

Surface water	1,069 million m^3/yr
Groundwater	55 million m^3/yr
Wastewater	103 million m^3/yr
	1,227 million m^3/yr

It is estimated that in the future additional 800,000 ha may need water irrigation with about 1,320 million m^3/year from the sources enumerated in the following tabulation.

Surface water	850 million m^3/yr
Groundwater	370 million m^3/yr
Wastewater	100 million m^3/yr
	1,320 million m^3/yr

Spray irrigation is by far the mostly applied method of irrigation.

At present about 3% of the total quantities of wastewater collected in sewerage systems is disposed of by irrigation (Müller, 1969). It must be assumed that in the future there will be no substantial increase in the disposal and use of wastewater discharges by irrigation in the federal territory.

C. Utilization of Wastewater Nutrients Contained in Sludge

Wastewater sludge produced at wastewater treatment works by sedimentation, biological, and other processes may be disposed of by utilization as fertilizer in agriculture and horticulture if it is suitable for this purpose (Bundesgesundheitsamt, 1972).

Usually, the sludge is applied to land in a dewatered state, but in some recent installations, a liquid digested sludge was used (Triebel, 1969). Before use, the sludge must be tested to make sure that it does not contain toxic and other harmful substances at concentrations above the maximum permissible limits. The sludge may be used raw or after digestion or after composting in admixture with waste solids, such as garbage or other refuse materials. The solids of raw sludge include bacteria, viruses, parasitic ova, and other organisms and constituents, some of which are pathogenic or otherwise harmful. Generally, digested sludge carries less dangers to health than raw sludge, while suitably composted sludge is relatively safe. More recently pasteurization of sludge by heat treatment at 65° C for 30 min is being used before utilization in agriculture and horticulture (Kugel, 1972).

In the Federal Republic of Germany, quite a substantial portion of wastewater sludge produced at municipal wastewater treatment works is used in agriculture and horticulture.

REFERENCES

Brix, J., Heyd, H., and Gerlach, E. (1963). "Die Wasserversorgung," 6th ed. Oldenbourg, Munich.

Bundesgesundheitsamt. (1972). Mitteilungen aus dem Bundesgesundheitsamt. Merkblatt Nr. 7. "Die Behandlung und Beseitigung von Klärschlämmen unter besonderer Berücksichtigung ihrer seuchenhygienisch unbedenklichen Verwertung im Landbau. *Bundesgesundheitsblatt* **15**, 234–237.

Bundesminister für Ernährung, Landwirtschaft und Forsten. (1961). "Der zusätzliche Wasserverbrauch in der Landwirtschaft im Bundesgebiet durch Bewässerung." Bonn.

Bundesministerium des Innern. (1972). "Abwasser, Anfall, Behandlung und Beseitigung in Gemeinden und Industriebetrieben in der Bundesrepublik Deutschland." Umweltschutz, Bonn.

Clodius, S. (1970). Wasserversorgung in Jahre 2000. *Z. Kommunalwirtschaft*, pp. 216–221.

Frank, W. H. (1965). Vorfiltertechnik bei der Wasserversorgung der Stadt Dortmund. *Gas- Wasserfach* **106**, 268–270.

Herrig, H. (1970). Flusswasser-Inhaltsstoffe und Trinkwasserbeschaffung. *Gas- Wasserfach, Wasser/Abwasser* **111**, 32–35.

Herzberg, H. (1965). Die Grundwasseranreicherungsanlagen im Wasserwerk Wiesbaden-Schierstein. *Gas- Wasserfach* **106**, 617.

Hopf, W. (1970). Zur Wasseraufbereitung mit Ozon und Aktivkohle (Düsseldorfer Verfahren). *Gas- Wasserfach, Wasser/Abwasser* **111**, 83–92 and 156–164.

Imhoff, K., and Hyde, C. G. Possibilities and Limits of the Water-Sewage-Water Cycle. (1931). *Eng. News-Rec.* **106**, 833.

Imhoff, K., and Imhoff, K. R. (1972). "Taschenbuch der Stadtentwässerung," 23rd ed. Oldenbourg, Munich.

Imhoff, K., Müller, W. J., and Thistlethwayte, D. K. B. (1971). "Disposal of Sewage and other Water-borne Wastes," 2nd ed. Butterworth, London.

Koenig, H. W., Rincke, G., and Imhoff, K. R. (1971). "Water re-use in the Ruhr Valley with particular reference to 1959 drought period." *Proc. Int. Conf. Water Pollut. Res.*, 1970, Vol. 1, p. I-4/1.

Kugel, G. (1972a). Liquid sludge disposal. *Water Res.* **6**, 555–560.

Kugel, G. (1972b). Pasteurization of raw and digested sludge. *Water Res.* **6**, 561–563.

Meinck, F., Stooff, H., and Kohlschütter, H. (1968). "Industrie-Abwässer," 4th ed. Fischer, Stuttgart.

Müller, W. J. (1969). "Re-use of Waste Water in Germany." Organization for Economic Cooperation and Development, Paris.

Müller, W. J. (1973). "Report on Problems of Water Pollution Control," Wasser und Abwasser in Forschung und Praxis, Vol. 6. Erich Schmidt Verlag, Bielefeld.

Reploh, H., and Handloser, M. (1957). Untersuchungen über die Keimverschleppung bei der Abwasserverregnung. *Arch. Hyg. Bakteriol.* **141**, 632.

Siegmund, H. (1970). Die öffentliche Wasserversorgung 1969—Ergebnis der Leistungen während eines Jahrzehnts. *Gas- Wasserfach, Wasser/Abwasser* **111**, 677–682.

Sierp, F. (1950). Der derzeitige Stand der Abwasserforschung. *Zentralbl. Bakteriol., Parasitenk., Infektionskr. Hyg. Erste Abteilung* **155**, 318.

Statistisches Bundesamt. (1972). "Wasserversorgung und Abwasserbeseitigung der Industrie 1969." Industrie und Handwerk, Vol. 5. Kohlhammer, Stuttgart.

Triebel, W. (1969). Möglichkeiten der landwirtschaftlichen Verwertung von Faulschlamm. *Ber. Abwassertechnische Vereinigung* **23**, 277–289.

Verband der Deutschen Gas- und Wasserwerke. (1970). "80. Wasserstatistik, Berichtsjahr 1968." ZfGW-Verlag, (Verlag für Gas- und Wasserverwendung) Frankfurt am Main.

10

Water Reuse in India

Sorab J. Arceivala

I. INTRODUCTION

India is a land of contrasts. It has the primitive and the modern. While in the villages of India sanitation may appear to be primitive or totally absent, some of the larger cities have modern sewerage and sewage-treatment systems with all the equipment manufactured within the country.

Out of 246 towns with a population larger than 50,000, nearly 190 towns are presently sewered. However, in terms of population served, this amounts to nearly 40 million people (about 7% of the total population). Vast numbers of people live in very small towns and villages and using night soil for agricultural purposes makes good sense for them.

With the rapid population growth (12 million/yr) and an urbanization rate at which some cities are doubling in population every 15 to 20 yrs, the water resources development programs and city water distribution systems have not been able to keep pace with growth. Some of the larger cities like Bombay and Calcutta supply only half of their actual needs, and even less in a year of poor rainfall. Industries in water-hungry cities have therefore welcomed every new idea for water conservation and reuse. It is estimated that at least a hundred examples of intentional reuse of water on a substantial scale, some using quite sophisticated methods, can be found today in industries.

Some typical examples of reuse of wastewater in India are given below.

II. MUNICIPAL WASTEWATER TREATMENT FOR INDUSTRIAL USE

Reuse of treated municipal wastewater for industrial and other purposes is practiced in some cases in India, notably in Bombay where supplementation from groundwater sources is not feasible as they are generally insufficient or brackish. One typical installation will be described to illustrate the methods used and results obtained.

Figure 1 shows a tertiary treatment plant installed in 1970 by an industry in Bombay to augment its supplies (Arceivala, 1967; Bannerjee, 1972). The plant is capable of treating 5000 m^3/day of raw sewage which is obtained from a municipal sewer located 2.75 km away. The sewer tapping point was chosen with a view to avoiding industrial wastes from other industries and seawater infiltration which is quite substantial as the sewer deepens. The sewer carried nearly two or three times the flow desired to be tapped and, hence, was considered as a dependable source of "water." An agreement was entered into with the municipal authorities for a 25-yr period upon payment of a nominal annual charge.

The raw sewage characteristics are given in Table I. Design criteria were developed on the basis of a pilot plant study (Arceivala, 1967). The raw sewage is pumped in a cast-iron main over 2.75 km to the factory premises for treatment. After screening and grit removal, the sewage enters an aeration tank as shown in Fig. 1. It was decided to use the extended aeration process to ensure odor-free aerobic treatment throughout, an important consideration in a well-developed

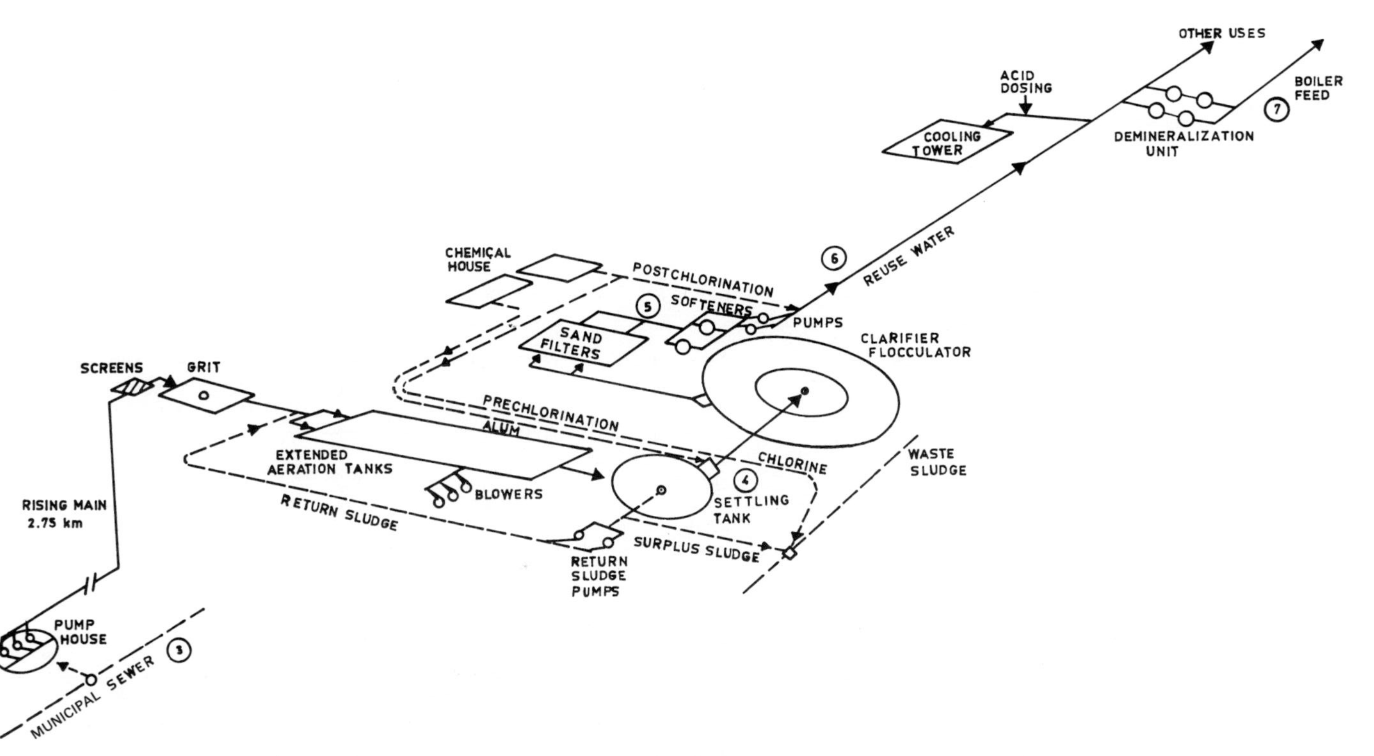

Fig. 1. Flowsheet for treatment of municipal wastewater for reuse in cooling, boiler feed, and other uses at an industry in Bombay, India. (Arceivala, 1967). (Circled numbers refer to corresponding analysis results in Table I.)

Table I
The Range of Change in Water Quality as Fresh Water Becomes Wastewater and Is Gradually Renovated for Reuse in Industry and Tall Buildings in Bombay (Arceivala, 1967, 1969)

			Water quality at different treatment steps			
Item (1)	Fresh municipal water (2)	Raw domestic sewage from the area (3)	After extended aeration and settling (4)	After coagulation and filtration (5)	After softening and chlorination (6)	After demineralization (7)
pH	7.6–7.8	7.15–7.65	7.2–7.8	7.1–7.3	7.1–7.2	8.75
Total hardness (mg/liter as $CaCO_3$)	35–40	120–160	120–160	120–170	4.0 [a]	Nil
M.O. Alkalinity (mg/liter as $CaCO_3$)	40–45	125–200	125–200	110–180	110–180 [b]	5.0
Chlorides, mg/liter as Cl,	15–20	60–130	60–130	60–130	60–130	Nil
Sulfates, mg/liter as SO_4	1.5–2.5	10–20	10–15	15–25	15–25	Nil
Phosphates, mg/liter as PO_4	Traces–0.1	6–16	3–5	0.2–0.5	0.2–0.5	Nil
Nitrates, mg/liter as NO_3	1.0–2.0	1.0–3.0	13.–19	13–19	13–19	Nil

Silica, mg/liter as SiO_2	8–24	10–24	10–24	10–20	10–20	Nil
Total solids, mg/liter	80–90	500–800	300–500	300–450	320–480	5.0
Suspended solids, mg/liter	5–10	150–250	15–30	Nil	Nil	Nil
Turbidity, SiO_2 units	5–10	Turbid	10–20	2.0–3.0	2.0–3.0	0.2
BOD_5 20°C, mg/liter	0.5–1.5	200–250	6–10	1.0–2.0	1.0–1.5	Nil
COD, mg/liter	1.0–2.0	250–350	16–40	4–6	3.5–5.0	Nil
Bacteriological quality (as per coliform standards)	Safe	Unsafe	Unsafe	Safe	Safe	—
Specific conductance	—	—	—	—	—	10 micromhos

[a] Softened water is blended with unsoftened water to give a final hardness of 40 mg/liter as in fresh municipal water.
[b] Alkalinity is reduced by acid treatment just prior to use in cooling towers. This increases sulfate content somewhat, since H_2SO_4 is used.

commercial and industrial area in a warm climate. Another advantage would be the high BOD removal efficiency of the plant giving a well-nitrified effluent and a relative ease of operation and maintenance. These considerations have been justified in practice. With a raw sewage BOD of about 200 mg/liter, it has been consistently possible to obtain low BOD values of about 5 to 10 mg/liter in the settled effluent, which gives about 95 to 97% efficiency of removal at a loading of about 0.15 kg of BOD/kg MLSS* or about 0.25 kg of BOD/kg of MLVSS†. Winter ambient temperatures average about 23°C. The sludge volume index (SVI) approximates 100 and return sludge flows have to be kept between 80 and 100% to maintain a mixed liquor suspended solids (MLSS) concentration of about 4000 mg/liter.

The aeration tank is followed by a mechanically scraped settling tank with a surface loading of about 30 m^3/m^2/day. The characteristics of the settled effluent have been shown in Table I. Alum is now dosed, and the water treated as in a typical water treatment plant: coagulation, sedimentation, and rapid gravity filtration. The filtrate is pumped partly through softeners (sodium chloride regeneration cycle) and partly bypassed to give a hardness of about 40 mg/liter after blending as in fresh municipal water. The softened water characteristics are also shown in Table I and are sufficient to meet the needs of cooling and other uses except boiler feed. If coagulant aids were available, and if settling tank operation could be well controlled, it might have been possible to dispense with filtration. Filtration was included in the flowsheet as an additional safeguard, particularly since resins in the softeners could get fouled. Lime treatment was not preferred as it was felt that zeolite softeners would be easier to operate and would give a satisfactory water for the uses contemplated. Further treatment is given to about 15% of the flow in a demineralization plant for use as boiler feed. This would have been necessary even with fresh municipal water. Demineralized water characteristics are also given in Table I.

About 90 to 93% of the raw sewage inflow is converted into reusable water. The capital cost of the plant has been nearly $700,000 (U.S.)‡ (1970 prices), including pumping main, land, and all the units described above. The plant is so designed as to be readily expanded to treat 10,000 m^3/day in the future and, hence, a few structures are larger than necessary for present requirements. Operating costs, including repayment of capital and interest charges (which is the major item), are reported to amount to about $0.09 (U.S.)/1000 liters of softened water prior to demineralization (Bannerjee, 1972). Similar plants installed in tall buildings—described in the next section—reported lower unit costs.

*MLSS, mixed liquor suspended solids.

†MLVSS, mixed liquor volatile suspended solids.

‡$1.0 (U.S.) = Rupees 7.50.

Other Attempts to Treat Municipal Wastewater for Reuse

Two other attempts to treat municipal wastewater for reuse can be cited. The conclusions arrived at were based on pilot studies. One was an attempt at Bombay where the Bombay Municipal Corporation itself set up a pilot plant to give tertiary treatment to the effluent from one of its existing activated sludge plants.

The raw sewage at this plant was, however, so heavily mixed with industrial wastes that the final result was discouraging. BOD and suspended solids were certainly well removed, but the final effluent had a persistant yellowish color that could not be removed by the chemical treatment adopted. Activated carbon treatment was not possible to consider as carbon regeneration equipment was not available in the country and importation was not desired. If successful, the Corporation contemplated distributing it to adjoining industries through a special pipeline with due safeguards against cross-connection (L. P. Borkar, personal communication, 1968).

Another study was undertaken in Madras city with its water in short supply. Comparative studies were made between activated sludge and trickling filters followed by a polishing treatment. The latter consisted of three alternative methods: lagoon of 5 days detention, alum coagulation followed by sedimentation, and excess lime treatment followed by recarbonation. The latter in combination with activated sludge gave the best results. Split treatment with lime gave nearly as good results as excess treatment and was, in fact, cheaper to adopt. Problems of lime recovery are being explored (Raman *et al.*, 1973).

III. REUSE OF WATER IN TALL BUILDINGS

A rather unusual application of reuse techniques is to be seen in nearly seven tall commercial buildings ranging from 20 to 25 stories that were constructed in recent years in Bombay. The need for such reuse stems from the need to fully air-condition the buildings. Cooling water is recirculated through a mechanical draught tower located at the top of the building, and make-up water requirements amounting to 150 to 250 m^3/day depending on the size of the building have to be obtained from sources other than municipal water supply because of acute shortage.

A way was soon found to meet this urgent need. Based on feasibility studies, it was shown that the most economical and dependable way of meeting these requirements was to take the sewage of the building itself to the basement for treatment in a compact plant to make the water fit for reuse as cooling water (Arceivala, 1969). The capital cost for construction of a 200 m^3/day capacity tertiary plant is about $50,000 (U.S.) (1970 prices) and the running cost (including repayment of loan and interest charges) approximates $0.04 (U.S.)/1000 liters, but excluding rental loss due to area occupied by the plant in the basement. Fresh municipal water cost at that time $0.058 (U.S.)/1000 liters in Bombay.

Figure 2 gives the flowsheet developed for reuse of water in tall buildings (Arceivala, 1969). The building drain is carried to the basement with an arrangement for bypassing the sewage to municipal sewer if desired. Likewise, an arrangement is provided for tapping municipal sewage if necessary to augment the flow through the treatment plant. As the latter arrangement is not routinely used, the plant design has to take into account the fact that practically all the building wastewater is received within the usual office hours of working, since these are mostly office buildings. Treated water thus needs to be stored.

Raw sewage enters the aeration tank after screening. No grit removal is found necessary in these plants. Once again, the process adopted is the extended aeration process owing to the reasons enumerated earlier in Section II. Odor-free operation is most important at all times since the plant is located within the

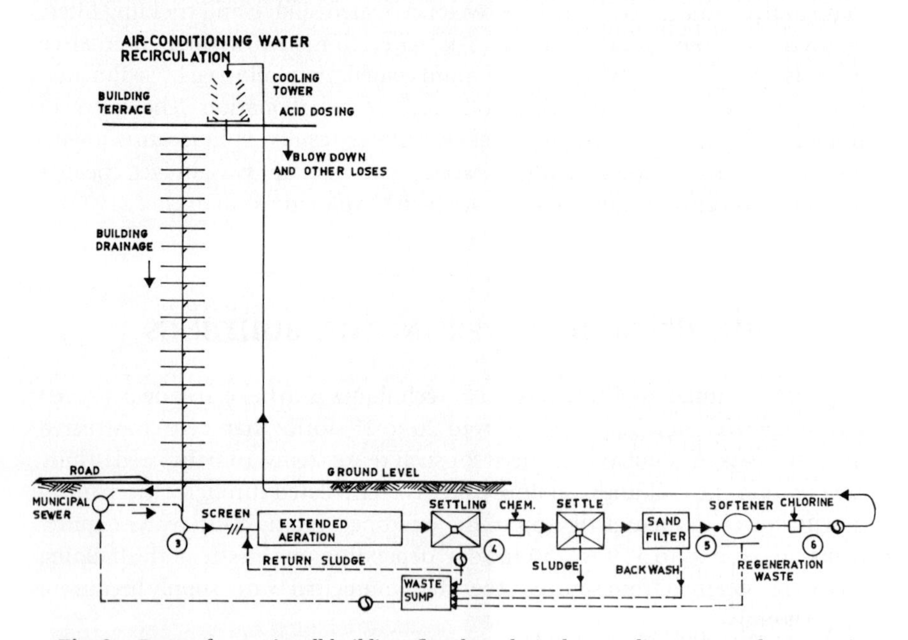

Fig. 2. Reuse of water in tall buildings flowsheet shows how cooling water make-up requirements are met in some tall buildings in Bombay, India, by treating wastewater from the building itself. Arrangement for augmenting the flow by tapping adjoining municipal sewer is provided. Treatment plant sludge and other wastes are pumped back to municipal sewer downstream of tapping point. (Arceivala, 1969.)

building. For the same reason, an anaerobic sludge digester with gas production would not be acceptable. In sizing the aeration tank and the units that follow, the flow pattern of the incoming wastewater is kept in view.

The remaining sequences of treatment are similar to the ones described in Section II, except that demineralization is not provided. The whole range of water quality as it changes from fresh municipal water to wastewater and is then gradually renovated for reuse is shown in Table I.

To determine the quality and quantity of water required for reuse in the tall buildings the following approach is followed: An open recirculating system is generally adopted for air-conditioning cooling water and the amount of water to be kept recirculating in the system is approximately 11 liters/min for every ton of refrigeration capacity when the temperature drop is 5°C in the cooling tower. For such a situation, the water lost in evaporation (E) is about 1% of the recirculating water.

Windage loss (W) is of the order of 0.1 to 0.3% of the recirculating water when mechanical draft towers are used, but increases to 0.3 to 1.0% for atmospheric towers. Blowdown requirement (B) is estimated from the following equation if the maximum permissible cycles of concentration (C) are known.

$$B = \frac{E + W(1 - C)}{C - 1}$$

where B, E, and W are in liters/min.

For trouble-free operation and minimum use of water quality-control chemicals in the recirculating water, the cycles of concentration are generally kept at 2.0 to 3.0 and, in no case, more than 4.0 in cooling towers such as these in Bombay. Hence, for a 100-ton air-conditioning plant recirculating 1100 liters/min of water with a temperature drop of, say, 10°C through a mechanical draft tower where cycles of concentration are to be restricted to 2.0,

$$E = 2\% \times 1100 = 22 \text{ liters/min}$$
$$W = 0.2\% \times 1100 = 2.2 \text{ liters/min}$$
$$B = \frac{22 + 2.2\,(1 - 2)}{(2 - 1)} = 20 \text{ liters/min (approx)}$$

The total make-up water requirement thus equals 44.2 liters/min (= 22 + 2.2 + 20) or 63.4 m^3/day for 24-hr working of this 100-ton plant.

Similarly, if 3.0 cycles of concentration are permissible, the total requirement of make-up water reduces to 47.7 m^3/day for a 100-ton plant.

If cycles of concentration equal 3.0, the various stable constituents in the make-up water are theoretically increased by a factor of 3.0 in the recirculating water. If the concentrations of various constituents in the make-up water lie within the range of values given in column 6 of Table I, the corresponding concentrations in the recirculating water can be readily estimated. However, the pH of the recirculating water cannot be estimated in this manner. The assumption is frequently made that in the absence of phenolphthalein alkalinity, the pH of the water leaving the cooling tower will be between 8.0 and 8.3 due to elimination of free carbon dioxide in the tower. Sometimes, for other reasons, a lower or higher pH may be observed. Thus, knowing the pH, the concentrations of calcium, alkalinity, and total dissolved solids in the recirculating water, and the temperature in the hottest part of the system, one can determine the Langelier Index or Ryzner Stability Index and note the tendency of the water to scale or corrode. Assuming that the recirculating water shows a tendency for deposition of scale, reduction in hardness and alkalinity is the usual means of control since nothing can be done to reduce temperature, and reduction in total solids would not have much effect on the Index.

For this reason, partial zeolite softening (by blending the softened water with bypassed hard water), plus acid feeding if required for reduction of alkalinity, provides a relatively simple and flexible means of preventing excessive scaling in this type of installation. The blending ensures a certain amount of hardness in the water which is useful to protect against corrosion of ferrous heat-exchange surfaces. The acid treatment (using H_2SO_4) depends for its functioning on the fact that calcium and magnesium sulfates are much more soluble than the carbonates. With the usually adopted dosages and the cycles of concentration obtaining in the system, calcium sulfate concentrations are well below the solubility limit. Similarly, calcium phosphate is also kept within the solubility limit.

Automatic dosing and control equipment is normally not provided in the plants just described. The clear water storage tanks help to maintain uniformity of quality of water pumped to the cooling towers. Storage ensures that pH, total dissolved solids, etc., do not vary much from hour to hour, and the wide variations in inflow quantities are balanced out.

Prechlorination is done as the water enters the coagulation tanks, while postchlorination is mainly in the form of periodic "shock" doses to control slime and algal growths. The latter are likely owing to the presence of nitrates and phosphates in the treated water and the warm and sunny climate of Bombay. However, this has never been a difficult problem to handle.

This method of reuse has proved successful wherever installed and more tall buildings in Bombay are going in for similar plants. Some industries with high refrigeration requirements are also going in for similar plants to produce the required cooling waters. One deterent, in some cases, has been the unavailability of municipal wastewater free from industrial wastes and brackish-or seawater infiltration.

IV. REUSE OF WATER IN THE COTTON TEXTILE INDUSTRY

The cotton textile industry is one of the major consumers of water and provides at the same time much scope for its conservation and reuse. The following paragraphs summarize the experiences gained in developing methods of reuse in 25 textile units in Bombay (Arceivala *et al.*, 1969). Some surveys were sponsored by individual mills, some by the industry's own association, and some by municipal authorities, and practically all of Bombay's 70 mills were covered gradually. Such interest on the part of all concerned speaks for the severity of the problem and the success of reuse techniques.

A. Water Consumption

The consumption of water per unit production in different manufacturing steps gives *prima facie* evidence of the extent of water conservation practices followed in a mill, and this section will describe how this is done.

The overall consumption of water for cotton textile mills in India where spinning, weaving, processing, and finishing operations are all performed under one roof is in the range of 120 to 280 liters/kg of goods produced. This is smaller than the textbook value of 350 to 750 liters/kg given for mills in the United States and some other western countries. Yet, there is good scope for water conservation and reuse as will be seen later. The breakup of consumption is given in Table II. The major water consumer is the wet processing and finishing department.

There is a wide variation in the consumption pattern from mill to mill. This is due partly to use of old and new technologies and, partly, to differences in the processing steps followed. Some machines require larger volumes of solution

Table II

Water Consumption Pattern in Textile Mills, India[a,b]

Department	Percentage of total water consumed	liter/kg of goods
Humidification in spinning	3–6	5–10
Humidification in weaving	4–7	7–12
Boiler house	10–16	10–25
Sanitary and miscellaneous (about 50 liter/worker)	4–7	—
Wet processing including finishing	60–80	100–220
		120–280

[a]Source: Arceivala *et al.*, 1969.
[b]Typical values without reuse.

and wash waters than others for treating the same weight of cloth. For example, a winch requires around 15 to 40 liters of solution for 1 kg of cloth treated while a jigger requires 3 to 5 liters only. Yarn-treating machines generally require more volume than cloth machines. Manual operations and batch-type production require larger volumes than continuous type processing machines with countercurrent flow of washwaters. High-pressure spray rinses consume less water than solid jets or dip tanks.

The sequence of processing operations followed depends on the type of finish required, the quality of material, and other factors. It is generally not possible to suggest modifications in the sequence from the point of view of water conservation. Reuse possibilities have, therefore, to be worked out without disturbing the present setup in a mill. It is also evident from this that the product-mix has an important bearing on the water-consumption pattern.

Where new mills or extensions to existing mills are contemplated, the washing efficiency of a machine or water requirements of different types of processes can be kept in view from the planning stage (see Table III).

The major consumption of water is in the processing department and it will be appropriate to discuss it in greater detail. The principal operations involved are singeing, desizing, scouring, bleaching, mercerizing, dyeing, printing, and finishing. Singeing is a dry operation. In some singeing machines, however, water is used for cooling the gas burners. This is usually a small quantity of water–5 to 10 liters/min—and can be easily collected for reuse because of its purity.

The quantity of water used in desizing is 1 to 4 liters/kg of cloth and is irrecoverable. The washwaters after desizing are too full of impurities to be considered for reuse.

After desizing, the two steps, scouring and bleaching, together with their related washes account for the largest fraction of water consumed in the processing department. Scouring and bleaching can be done in three possible ways, each requiring different quantities of water as shown in Table III.

Using the conventional method of bleaching, a sequence like the following would need a minimum water supply of 87 liters/kg of cloth based on the water-consumption figures for individual steps given in Table III:

> desize—external wash—external wash—kier boil—internal wash—internal wash—external wash—external wash—chemick—external wash—peroxide kier—internal wash—internal wash—external wash—blue

This sequence may be varied in many ways. For example, additional gray souring and two washes may be given before kier boiling. In some cases, one wash after desizing, kiering, and chemicking may be dropped. With such variations, it is not surprising that water consumption in bleaching can vary from 90 to 130 liters/kg without reuse.

Table III
Water Requirements of Different Operations in Bleaching in India[a]

Method of bleaching	liter/kg of goods (without reuse)	Percent saving possible with conservation and reuse
Conventional method with cloth in rope form in batches of 2 to 5 tons		
Caustic kier	4	
Peroxide kier	4	
Internal washes in kier	4 each	
Chemicking, souring, antichlorine, and bluing	1 each [b]	
All external washes	10 each	
Overall for the sequence	90–130	25–40
Continuous bleaching of cloth in rope form by J-box method		
Washing units	15–20 each	
Overall	60–80	25–50
Bleaching in open-width form, of cloth		
Overall	35–40	20–30

[a]Source: Arceivala *et al.*, 1969.
[b]Four liter/kg for each item if liquor circulating machine is used.

It will be observed from Table III that continuous bleaching systems of the J-box and open-width type are more economical with regard to water consumption. Whichever method is adopted, there is considerable scope for reuse of water as shown in the table.

B. Direct In-Plant Reuse in Cotton Textile Mills

The scope for in-plant reuse without treatment is given below for each of the major operations in a cotton textile mill.

1. REUSE OF WATER IN SCOURING AND BLEACHING

The scheme for reuse is simple and direct. No treatment is involved. It is based on the consideration that the relatively clearer wastewaters from the later washes in a bleaching sequence can be reused in some of the earlier washes where water quality is not so important. Table IV gives the typical chemical analysis of the wastewaters from different washing operations.

The washwaters that can be reused are shown in Table IV. Reuse arrangements are also shown diagrammatically in Figs. 3, 4, and 5 for the three common types of bleaching methods. If these arrangements are followed the percent savings in water consumption shown in Table III can be achieved.

2. REUSE IN MERCERIZING

On the mercerizing range, the scope for water conservation and reuse lies in adopting a countercurrent flow pattern. The water requirement can then be reduced to 7 to 10 liters/kg of mercerized cloth. Even most of this water can be recovered as steam condensate from the multiple effect evaporators of the caustic recovery plant provided suction in the evaporators is carefully controlled to avoid "boiling over", otherwise the condensate will be alkaline and yellowish in color (Fig. 6).

3. REUSE IN DYEING

The scope for reuse is least in this section owing to the colors and chemicals involved. The overall consumption of water for cloth dyeing on jiggers comes

Table IV
Some Typical Chemical Analyses of Washwaters from Cotton Textile Mills, India[a]

		Chemical analysis				
Source of wastewater	Turbidity units	Suspended solids (mg/liter)	Dissolved solids (mg/liter)	pH	Alkalinity as $CaCO_3$ (mg/liter)	Reusable or not
Wash after desizing	410	2150	1560	6.0	—	No
First external wash after caustic kier	170	290	1980	11.9	890	No
First wash after chemicking	340	225	300	9.7	450	No
Second wash after chemicking	140	92	142	7.8	95	Yes
First wash after white souring	130	110	195	5.9	—	Yes
Second wash after white souring	48	52	78	7.0	—	Yes
First external wash after peroxide bleach	22	43	300	7.7	400	Yes
Second external wash after peroxide bleach	10	20	270	7.5	300	Yes

[a]Source: Arceivala *et al.*, 1969.

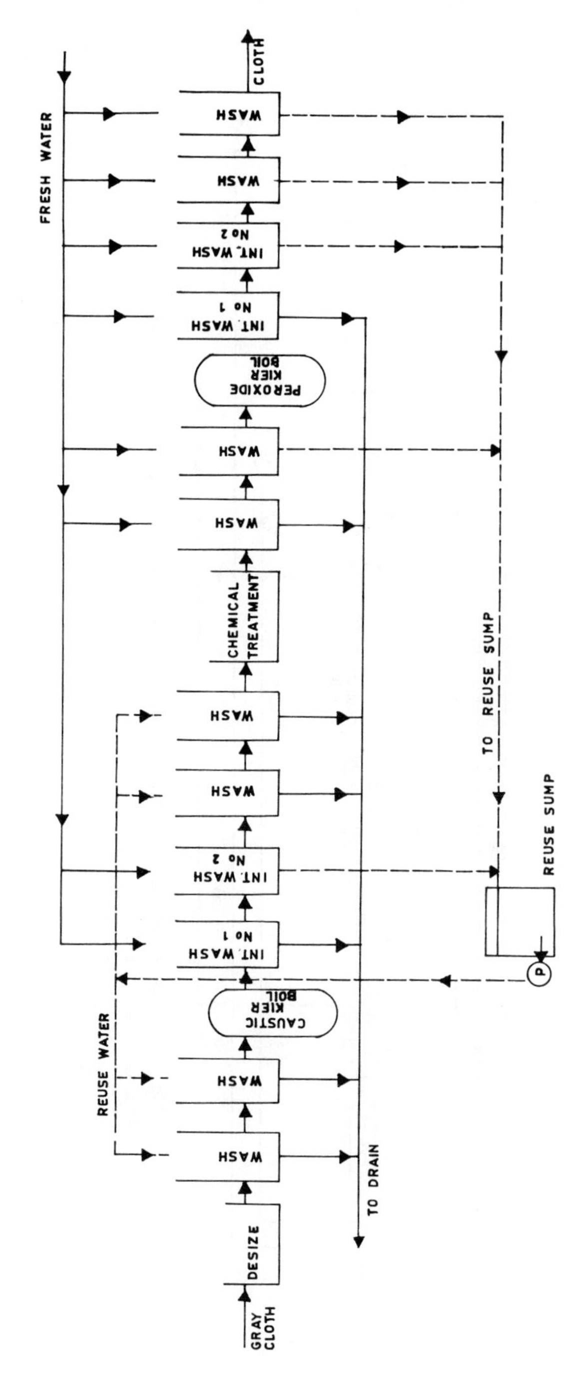

Fig. 3. A scheme for reuse of water in batch type conventional bleaching for cotton cloth (Arceivala *et al.*, 1969).

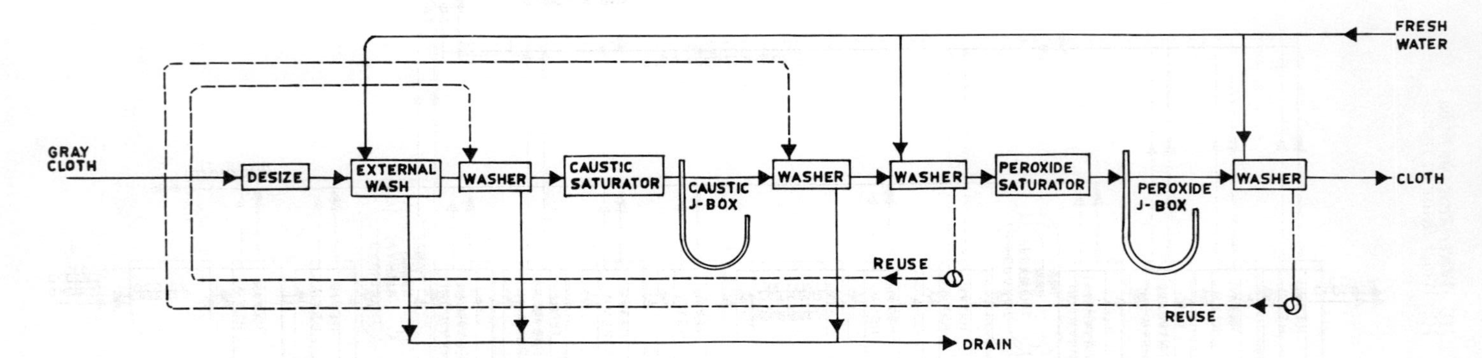

Fig. 4. A scheme for reuse of water in continuous bleaching by J-box method for cotton cloth (Arceivala *et al.*, 1969).

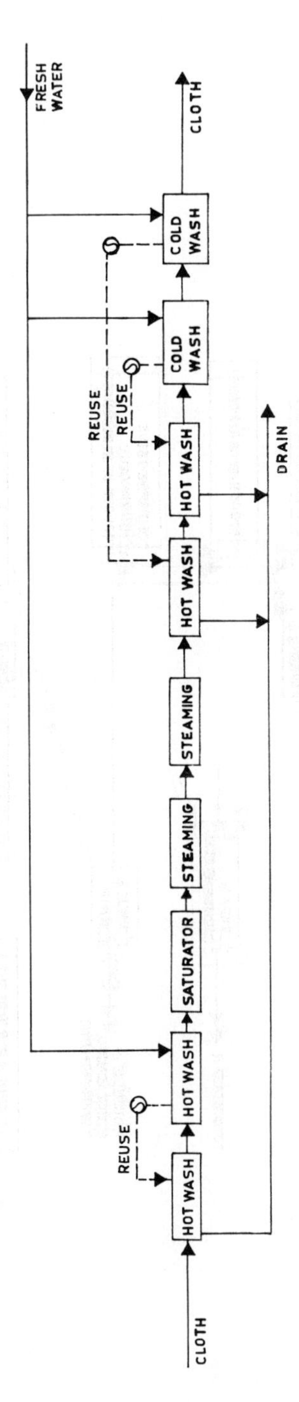

Fig. 5. A scheme for reuse of water in open-width bleaching of cotton cloth (Arceivala *et al.*, 1969).

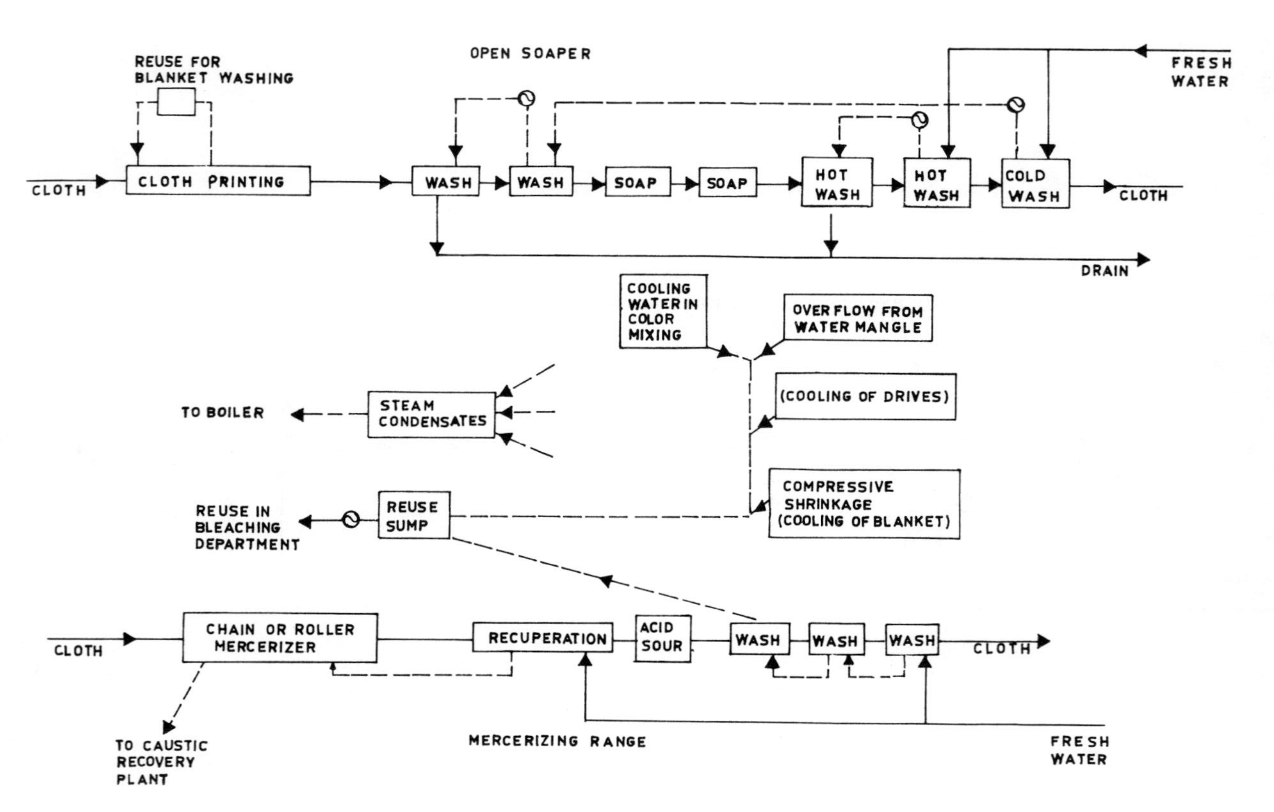

Fig. 6. A scheme for water conservation and reuse in mercerizing, printing, soaping, and finishing of cotton cloth (Arceivala *et al.*, 1969).

to about 40 liters/kg of cloth dyed. Small savings can be effected if running washes are replaced by static ones wherever possible. Further, the batching or wetting water need not be drained out. It can be retained for use in the next operation like dyeing or naptholating.

4. REUSE IN PRINTING AND FINISHING

This section uses water for various cooling and washing operations and consumes most of the steam used in a mill. The principal uses and the manner in which water conservation and reuse can be practiced are given in the following tabulation (see Fig. 6):

Item	Reuse
Cloth printing	
(1) Cooling water for color mixing	Reuse in item (2) or lead to general reuse sump
(2) Blanket washing in printing machine	Reuse for same purpose, drain out at end of day
(3) Washing of printed cloth on open soapers	Adopt countercurrent which can yield saving up to 66% in this item
Cloth finishing	
(4) Cooling water for drives of stentering machine (10 liters/min/drive)	Lead to general reuse sump
(5) Cooling of rubber blanket on compressive shrinking machine (about 45 liters/min)	Lead to general reuse sump
(6) Overflow from water mangle before drying range (about 15 liters/min)	Lead to general reuse sump
Steam condensates	
(7) Steam condensates from drying ranges, stenters, hot flues, agers, curing machines, and in other sites in all departments where steam is used in indirect form	Reuse steam condensates in boilers

Generally, some amount of steam condensates are recovered in most mills because of their heat value. But the possibilities for stepping up recovery can be explored more fully. Maximum recovery ranges from 50 to 70% of the total steam depending on the proportion used in direct and indirect form.

The total scope for direct reuse of water, i.e., without treatment, as estimated on the basis of work done in 25 mills in Bombay is about 18 to 22% of the normal daily requirement of water (Arceivala *et al.* 1969). It should be borne in mind that most cotton textile mills in India are already consuming less water than their counterparts in some of the western countries. Yet, a saving of this magnitude can be achieved. The reuse schemes suggested here have been installed by a number of mills and found successful. To the extent reuse could be done, it has helped mills to tide over difficulties caused by short supply of water without reducing cloth production and laying off workers.

5. COST OF DIRECT REUSE

The cost of reuse arrangement given above essentially consists of (a) remodeling of sewers to separate reusable wastewaters and lead them to a sump(s), (b) construction of a sump of adequate size to balance inflows and outflows (a sump volume of 10 to 15% of the flow to be reused is generally enough), and (c) pump and distribution piping back to reuse points.

Installation costs in India have ranged from $5000 to $15,000 (U.S.)/mill depending on the size of mill and extent of modifications required. Operating costs, including depreciation, are found to be about $0.01 (U.S.)/1000 liters of water reused which is far cheaper than fresh municipal water.

6. WATER QUALITY STANDARDS

The standards of water quality as existing today for industrial purposes are generally based on the requirement of the most "critical" process in that industry. Consequently, they are adequate for unrestricted use of water in the whole plant.

Most of the industrial water quality standards, therefore, need to be reviewed and stated in such a manner as to encourage usage of lower quality waters in less critical operations. The standards themselves should not become a deterent to conservation and reuse.

C. In-Plant Reuse after Treatment in Cotton Textile Mills

Figure 7 shows a scheme developed for reuse of water after treatment in some cotton textile mills in Bombay. It is naturally more elaborate and meant for saving an additional 20 to 50% of the daily intake over and above that which can be saved by direct reuse.

The scheme shown in Fig. 7 is based on a fire-insurance requirement that mills be provided with large open ponds or covered reservoirs to meet emergency fire-fighting requirements. The open ponds suffer from evaporation and percolation (although at times the latter could be in reverse where groundwater is high) and mills replenish them with *fresh* water. The ponds could as well be replenished by selected process wastewaters that cannot be directly reused

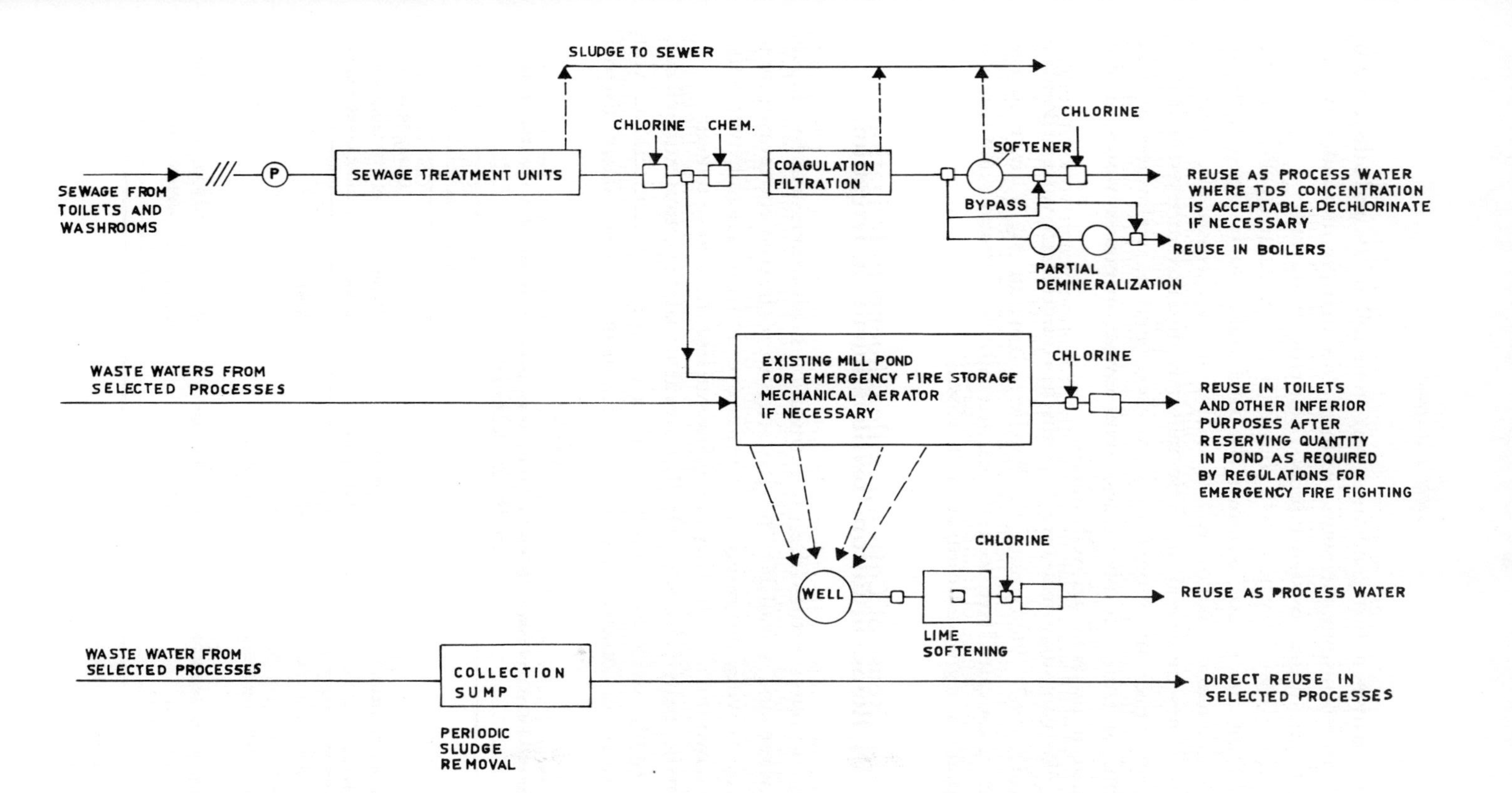

Fig. 7. Scheme for reuse of water partly with and partly without treatment. Existing mill pond can be replenished with wastewater instead of fresh water (Arceivala, 1967–1968).

and have no dyes in them. Sanitary sewage from mill toilets and baths can also be used for replenishing after some treatment. Alternatively, the sanitary wastes can be treated further as shown in Fig. 7 for reuse in the mill.

The existing pond may need to be converted into an aerated pond before the pond effluent can be used as shown. The pond could also be tapped indirectly through an adjacent shallow well in suitable soil and reused after treatment. The possibilities are several and a few mills have installed such systems in part, but not in whole. For example, in a few cases, pond and well waters are used after lime softening. In other cases, some sewage or process wastes are also contained in the pond and well waters.

In a pilot plant study made over 7 months in a cotton bleach works in Bombay (Arceivala, 1967–1968), using the flowsheet shown in Fig. 7, the reuse possibilities investigated have been summarized in Table V. It shows how the costs rise rapidly as degree of treatment is increased.

D. Reuse of Cotton Textile Effluents in Irrigation

Reuse of cotton textile effluents in irrigation is being done at a few mills in India where land is available. The characteristics of the combined wastes from spinning, weaving, processing, and finishing of cloth are generally within the range shown in Table VI. They are highly variable from hour to hour.

From the analysis given in Table VI, it is evident that the cotton textile effluent is not readily usable for irrigation and needs either sufficient dilution with plain water or domestic sewage or pretreatment to meet effluent disposal

Table V

Estimated Cost versus Degree of Treatment before Reuse at a Textile Unit in Bombay, India[a]

Degree of treatment before reuse	Scope (m^3/day)	1970 Costs in U.S. $	
		Capital cost	Operating costs per 1000 litres, including depreciation
Reuse *without* treatment	540	12,000	0.01
Reuse *with* treatment, but without demineralization[b]	600 (additional over item 1)	140,000	0.06
Reuse *with* treatment including demineralization[b]	20% additional over item 2)	50,000 (additional over item 2)	1.30 for the fraction demineralized

[a]Source: Arceivala, 1967–1968.
[b]See Fig. 8.

standards (see Table X). All these methods are used in India, since reuse of wastewater for irrigation is given the highest priority wherever it is possible. Some dilute the waste and irrigate, while some discharge into public sewers for mixing with domestic sewage followed by irrigation with or without pretreatment.

Notably in the city of Ahmedabad, which has a large concentration of cotton textile mills, the municipal authorities have prescribed a pretreatment of textile wastewaters before discharge to public sewers with the intention of minimizing harmful effects on the sewers and on the irrigated land since irrigation is the most favored method of disposal of city wastewaters.

The municipal requirement for pretreatment specifies that the mixed wastes from the bleaching and dyeing departments should be treated with alum or lime, settled, and passed through a contact bed filled with gypsum pieces before discharge to the public sewer. In this manner, the calcium content in the effluent is increased and percent sodium decreased.

Unfortunately, this method creates a sludge disposal problem for each mill and the requirement of gypsum, which may run into several tons per day, makes the treatment expensive. The TDS is not reduced but, as would be expected, somewhat increased which makes dilution with municipal sewage essential before irrigation. For a large textile mill having an average flow of 8150 m^3/day, the capital cost was estimated in 1971 to be $120,000 (U.S.) and the operating cost (power and chemicals only) at $200,000 (U.S.)/yr. Other methods of pretreatment have also been used elsewhere. Where public sewer systems exist, the preference is for discharging untreated or minimally pretreated wastes to the public sewer for treatment and disposal along with the domestic wastewaters—more often than not on land.

Table VI
Characteristics of Combined Wastes from Cotton Textile Mills, India

Characteristic	Range of values[a]	
	Maximum	Minimum
pH	9.6	4.7
Alkalinity as $CaCO_3$	980	50
Total dissolved solids (TDS)	6600	980
Suspended solids	3200	300
BOD_5	600	120
COD	1400	400
Chlorides	1500	20
Percent sodium in total cation content	0.9	0.4

[a]All values are expressed as mg/liter except pH and percent sodium.

V. REUSE OF WATER IN OTHER INDUSTRIES

In-plant reuse possibilities in various industries besides cotton textiles have also been investigated. In an effort to reduce water consumption in a fast-growing industrial city, the Bombay Municipal Corporation is perhaps the only municipal body that has actually ordered systematic surveys to be carried out at its own cost in all major water-consuming industries within its jurisdiction. Water conservation and reuse possibilities were assessed by the corporation before curbing usage.

Some of the types of industries investigated and the scope for reuse found in each case have been summarized here (Arceivala, 1967–1968). Unlike cotton textiles, only broad indications are given because a sufficient number of plants of each type do not exist to be able to generalize water-consumption figures. Suitable treatment may need to be provided so that water quality is carefully controlled in order to protect the health of the workers and product consumers and meet process and other requirements in each case.

1. AERATED WATER MANUFACTURE AND BOTTLING

Adopt countercurrent flow in bottle washing, and, if desired, treat its effluent by neutralization, coagulation, filtration, and chlorination for reuse partly in bottle washing and partly in other housecleaning operation. For refrigeration, use cooling tower and provide make-up water only. Plant effluent can be reused in irrigation if land is available.

2. AUTOMOBILE MANUFACTURE

Reuse all cooling waters required for welding guns, compressors, transformer cooling, tempering, hardening, rotary and gas furnaces, electronic devices, cutting operations, radiators for engine testing, and general building air conditioning.

Take to individual sump and reuse in same operation (discharging once a day or once a week to drain) waters used in (a) alkaline hot wash before spray painting of autos (b) providing "water curtain" in spray painting, (c) quenching of parts, (d) cooling of quenching oil, and (e) cooling of fans in oil furnaces.

Major use of water is in heat treatment, sheet-metal department, forging, paint shop, compressors, and boiler house. Make-up water for cooling towers can be obtained from treated domestic sewage as explained in Section II.

3. ASBESTOS CEMENT PRODUCTS

Reuse water required for curing asbestos products. Also, recycle overflow from sheet-forming machine with or without treatment to asbestos cement-mixing unit.

4. ASBESTOS CLUTCH LININGS

Reuse water from drum-cooling stations in brake- and clutch-lining department. Reuse chilled water from calendering, mixing kettles, and solvent recovery plant. Reuse backwater in asbestos mill board department by providing settling or "save-all" type rotary drum filter.

5. CANNING OF FRUITS, VEGETABLES, ETC.

Adopt countercurrent in bottle washing. Reuse cooling waters from refrigeration. Recover steam condensates from indirect uses. Reuse final effluent for crop irrigation if land is available.

6. MILK DAIRY

Reuse cooling waters from refrigeration, pasteurization, and jacket cooling. Adopt countercurrent in washing of bottles and cans. Treat and reuse in the first two compartments of washing machines. Reuse final effluent for irrigation if land is available.

7. PAPER-BOARD MANUFACTURE

Reuse possibilities that exist are similar to those in a paper mill, for example, reuse water in log flumes, reuse filtrates from various consistency control operations, screens, washers, and white water from paper-board machine. Reuse of final effluent for irrigation is possible.

8. PHARMACEUTICAL MANUFACTURERS

Use cooling tower and provide only make-up water for various cooling operations such as the following: refrigeration, humidity control, cooling of pharmaceutical kettle jackets, fermenter jackets, seed vessels, air-compressors, distillation units, etc. Ensure steam condensate recovery wherever indirect use of steam is made. Adopt countercurrent in bottle- and vial-washing operations; treat and reuse in first two compartments of washing machines.

9. CHEMICAL AND PETROCHEMICAL

Some of the major industries like fertilizers, synthetic fibers, PVC and other plastics, caustic and other chemicals, refineries and petroleum-based industries are included in this large group.

Reuse possibilities are similar to those described under pharmaceutical manufacturers. Additionally, if some treatment is done, slightly polluted wastewaters from the following operations can be collected in a separate network and reused: (a) vacuum pump seals, (b) rotary vacuum driers, (c) barometric condensers (d) degasifiers, (e) vacuum jets, (b) multiple effect evaporators, (g) scrubbers, (h) external cooling of sulfuric acid towers in PVC manufacture, and such other operations.

These relatively polluted waters can be reused directly at times for other uses where quality is not important. Even boiler blowdown may be reused in this manner. At other times, the wastewaters may be treated for reuse. Treatment along with domestic sewage is often quite feasible and treated effluent can be reused as make-up water in cooling towers and in vacuum-creating devices, etc. For water conservation, replace barometric condensers with surface condensers, and generally discourage use of water for creating vacuums. Treated effluents can also be used for flushing of toilets, gardening, etc.

Reuse of wastewaters from some industries, like fertilizers, can be done for irrigation after dilution with plain water or domestic sewage.

10. TEXTILES

Cotton textiles have been discussed at length in Section IV. Synthetic fiber-manufacturing industry has good scope for reuse along the general lines indicated for chemical industries. However, the scope for reuse in the synthetic textile finishing industry is limited to about 10% of normal consumption. The corresponding figure for wool weaving and finishing is about 5% only.

11. RAILWAY YARDS, CITY BUS DEPOTS, DOCKS, AND HARBORS

Large quantities of washwaters consumed in washing of wagons, carriages, and public transport buses can be collected and treated to remove solids, grease, detergents, etc., and reused for the same purpose. Watering stations for steam locomotives can be supplied with reclaimed water obtained by treating domestic sewage.

As a water conservation measure, hydraulically operated cranes and machinery can be gradually replaced by electrically operated units to conserve city supply.

VI. REUSE IN IRRIGATION

Direct reuse of municipal and industrial wastewater for irrigation purposes is extensively practiced in India. The warm climate, the practically negligible

Table VII
Characteristics of Municipal Wastewaters in India

Item	Range of values
BOD_5, gm/cap/day	45–54
Total nitrogen, as N, gm/cap/day	6–8
Phosphorus as P, gm/cap/day	0.9–2.5
Potassium as K_2O, gm/cap/day	2.0–5.5
Detergents (ABS) mg/liter	0.65–3.6
Chlorides, gm/cap/day	4–8
Dissolved solids, mg/liter	800–1500
Suspended solids, mg/liter	200–560
Total bacterial count, per 100 ml	10^9–10^{10}
Coliforms, per 100 ml	10^7–10^9
Fecal streptococci, per 100 ml	10^5–10^6
Salmonella typhosa, per 100 ml	10^2–10^4
Protozoan cysts, per 100 ml	10^2–10^3
Helminth eggs, per 100 ml	10^2–10^3
Virus (plaque-forming units/100 ml)	10^1–10^3
Flow, liters/cap/day	100–250

flow in most of the streams during the summer months, and the need to grow more food for its teeming millions make irrigation the favored mode of disposal of wastewaters. The first sewage farm in India was established as far back as 1895; today there are over 132 farms covering more than 12,000 hectares and utilizing over 1 million m^3 of sewage per day. There are several more farms that receive industrial effluents, particularly from the sugar, distillary, food-processing, fertilizer, and other industries. Some authorities, in fact, encourage the utilization of sewage for farming purposes by extending grants and loans on special terms for sewerage schemes which include sewage irrigation.

A. Sanitary Aspects

Raw municipal sewage characteristics are shown in Table VII. The range of values given are typical of municipal wastewaters from small and large communities in India. Some of the municipal and industrial wastewaters used for irrigation are treated, while others are not. Disinfecting effluents prior to irrigation is not generally practiced. There is concern about the possibility that vegetables and foods which may be eaten raw are also, at times, cultivated on these farms. Repeated examinations in the past have failed to show the presence of pathogens on farm products, presumably due to the fierce tropical sun. Lately, renewed efforts, using better techniques, are being made to detect pathogens and

Table VIII
Disease Condition of Sewage Farm Workers in India (C.P.H.E.R.I., 1971a)

Diseases observed	Test group (%)	Control group[a] (%)
Gastrointestinal	45.6	13.0
Respiratory	19.6	4.3
Anemia	50.3	23.6
Skin condition	22.3	4.0
Worm infections (stool examinations)	79.5	29.0

[a]Farm workers in similar areas, but where sewage is not used as an irrigant.

a few vegetables examined from a farm at Nagpur that used raw sewage were found positive for *Salmonella* and parasites during the summer. Almost all the vegetables were positive for coliforms and fecal streptococci (C.P.H.E.R.I., 1972).

However, greater concern has been expressed over the health status of sewage farm workers and some useful studies have been made in recent years by Patel (1964), Kabir (1968), and C.P.H.E.R.I. (1971a). Table VIII reveals, in no uncertain terms, the poor status of their health.

This is mainly due to the traditional (and relatively insanitary) methods of working still used by the Indian farmer. The farms are not mechanized and the farmers do not wear any protective clothing or footwear.

Settlement of sewage prior to irrigation has been shown to reduce hookworm eggs by 70% (Bhaskaran, 1958), while treatment in waste stabilization ponds has given practically complete removal of *Salmonella* and helminths as shown in Table IX.

To provide some form of treatment before irrigation, the country has nearly 75 stabilization ponds and over 100 sewage treatment plants some with only primary treatment and some with secondary treatment (mostly trickling filters)

Table IX
Removal of Organisms in Stabilization Ponds in India[a]

Organism	Percent removal
Coliforms	90–99.9
E. Coli	90–99.9
Salmonella	99–100
Helminths	99–100
Virus	80–95[b]

[a]Source: Rao, 1973.
[b]From limited studies.

to treat wastewaters from municipalities, industries, and housing developments. (For further details on the health aspects of sewage irrigation see Chapter 2.)

B. Chemical and Other Aspects

Land requirements for irrigation naturally vary widely depending on type of soil, nature of crops, and local climatic conditions. In India, they range between 75 and 250 m^3/day/ha of land and generally average around 100 m^3/day/ha. Additional land may be brought into use during harvesting and other agricultural operations on the cultivated land. Farmlands need time to dry out and reaerate.

Irrigation requirements often require that the crops be watered to a depth of about 20 cm, once in 10 days in the growing season. Each part of raw sewage is generally diluted with one to three parts of fresh water irrigating the field. If the raw sewage BOD_5 is 500 mg/liter and if it is diluted with an equal amount of water and irrigated to a depth of 20 cm once in 10 days, the BOD_5 applied per day per hectare is

$$\frac{500}{10^6 \times 2} (10{,}000 \times 0.2 \times 10^3) \frac{1}{10} = 50 \text{ kg/ha/day}$$

The BOD_5 loading on the soil is reduced proportionately when the sewage is treated or the dilution is increased. BOD_5 loading often ranges between 25 and 150 kg/ha/day. A practice often followed in some parts is to dilute the concentrated raw sewage up to three times and spread over the desired area.

Since industrial effluents are also used extensively, it has become necessary to develop some guidelines, and the Indian Standards Institution has issued a standard for the discharge of wastewater on land (see Table X).

C. Nutrients in Effluent and Sludge

The nutrients of agricultural importance, N, P, and K, contained in sewage have been presented in Table VII. About 50% removals of N and P take place in biological treatment. The commercial value of nutrients contained on an average in biologically treated sewage is estimated to be $10 (U.S.)/1000 m^3 of effluent at current prices in India, whereas the effluent is generally sold at the price of plain irrigation water, which accounts for the popularity of the former.

Dried sludge is likewise in demand as manure and soil conditioner and is a source of some revenue at most sewage treatment plants. Liquid sludge from Pasveer-type oxidation ditches is being used experimentally as an irrigant

Table X
Indian Standards Institute Standard No. IS 3307 of 1965 "Tolerance Limits for Industrial Effluents Discharged on Land for Irrigation"

Item	Maximum permissible value
BOD_5 mg/liter	500
pH	5.5–9.0
Total dissolved solids (inorganic) mg/liter	2100
Oils and grease, mg/liter	30
Chlorides, mg/liter	600
Sulfates, mg/liter	1000
Boron	2.0
Percent sodium [a]	60

[a] Percent sodium, percentage of sodium in total cation content.

to save cost of drying beds. Nutrient values are listed in Table XI. The nutrient value of aerobically stabilized sludge is not significantly affected by the sludge age as observed in experiments with sludge age varying from 1.0 to 30 days (Handa, 1973).

D. Cultivation of Nonedible Crops

The standards given in Table X cover only the chemical aspects of wastewaters. A survey of several existing farms in the country is in progress to evaluate the nature and extent of standards needed for control of bacterial quality and sanitary conditions of working in sewage farms.

Meanwhile, successful attempts have been made to cultivate nonedible crops which, by their special commercial value, would attract the farmers away from growing vegetables and other edible crops. Citronella (*Cymbopogon winterianus*) and Mentha (*Mentha arvensis*) are two such crops of commercial value in India since about $1 million (U.S.) are spent annually in importing their oils for use in

Table XI
Nutrients in Sludge after Aerobic and Anaerobic Treatment

Nature of sludge	Percent dry weight basis		
	Nitrogen	P_2O_5	K_2O
Anaerobically digested and dried	0.8–5.0	0.6–5.0	0.3–1.5
Aerobically stabilized	5.8–7.8	2.9–5.14	0.8–1.9

medicinal and cosmetic preparations. Experimental farms have demonstrated not only that they can be grown on municipal wastewater, but that the net yield of oil is actually 30 to 40% more per hectare than when plain water is used (C.P.H.E.R.I., 1970).

VII. USE OF NIGHT SOIL

Nearly 80% of the Indian population lives in villages where sanitary methods of excreta disposal are necessary. Migration from the villages to the towns is progressing at a rapid rate and practically in every large town or city where a sewerage system exists the coverage is not complete. The suburban areas are generally the "gray" areas where water is supplied mainly through standpipes and excreta disposal facilities are lacking.

In all such areas, a dry-conservancy system of some sort exists. Much effort has been spent on developing rural latrines, such as the bore-hole, pit-privy, and aqua privy latrines, to avoid the necessity of manual handling of excreta. It is estimated that over a million latrines have been constructed in the country over the last 20 yr. A pit or bore of about 1.3 m^3 vol suffices a family of 5 persons for about 4 to 5 yr, after which the superstructure needs to be shifted to another pit or bore and the original one dug up after a few months to use the dried and stabilized night soil as manure.

Many of the older latrines are of the "basket" type where the excreta accumulated daily in a "basket" placed below the toilet seat is cleared manually and carted away to night-soil dumping grounds. There, the night soil along with the community refuse is filled in trenches for composting and subsequent use in agriculture.

A variant of this disposal method is night-soil digestion, which is provided to give an inoffensive and stabilized sludge, undiminished in its fertilizer value, along with gas that may be profitably used in the rural areas for cooking or lighting. Another advantage of night soil digestion is that pathogenic organisms are eliminated in the process.

A pilot digester of 6.5 m^3 capacity operated for about a year has yielded the essential design criteria given in Table XII (Mohanrao, 1970). Digester capacity required depends upon the minimum water temperatures during the year and varies from about 3 m^3/100 persons in South India where winter ambient temperatures average 25°C, to 5 to 6 m^3/100 persons in North India with winter temperatures around 15°C.

Several night-soil digesters have been constructed in the country, the largest unit serving 20,000 people. The night soil arriving at the plant is normally recommended for dilution with a volume of dilution water two to three times the

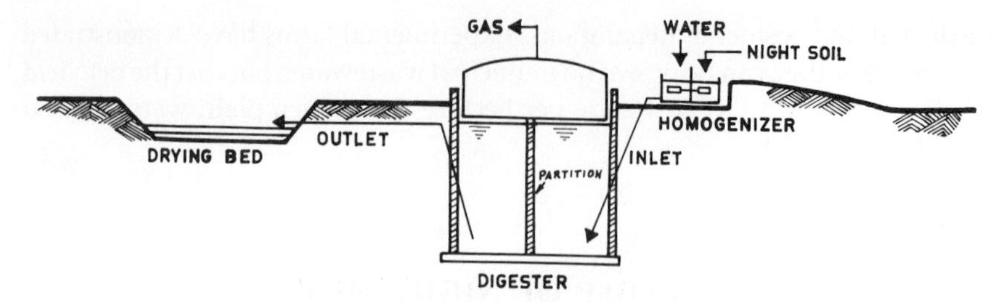

Fig. 8. A night soil digestion arrangement. (Mohanrao, 1970).

night soil volume. This is carried out in a small tank, where it is slowly stirred (either manually or mechanically) and then fed to the digester. Night soil digesters need to be desludged to make space for the fresh incoming flow, with the withdrawn liquid being spread on a typical sludge-drying bed of about 3 to 5 m^2/100 persons, which is sufficient in the tropical climate. Dried sludge is used as manure (Fig. 8); filtrate from the drying bed can be disposed of through soak pits or subsurface disposal fields or treated further in a small waste stabilization pond. Alternatively, the liquid withdrawn from the digester can be lagooned for drying.

Cow dung can also be similarly digested and, in fact, can substantially add to the gas yield. Cow dung can be digested either in separate digesters or in the same digester with night soil. Cow-dung digestion is feasible where a number of cattle are kept at one place facilitating collection of dung. Design criteria for cow-dung digestion are also given in Table XII.

A night-soil digestion unit capable of serving 1000 persons is estimated to cost about $500 to 700 (U.S.) to construct (Mohanrao, 1970).

Table XII
Design Criteria for Night Soil and Cow Dung Digestion in India[a]

Item	Night soil	Cow dung
Digester capacity (m^3/100 persons or cattle)	3–6	100–130
Volume of dilution water required (vol/vol)	2–3	1–1.5
Gas yield (m^3/100 persons or cattle)	3	50
Calorific value of gas (KCal/m^3)	5558	5130
Approximate HP generated per 1000 persons or 1000 cattle	2.0	36
Manurial value (% dry basis)		
Nitrogen, N	3.0–5.0	1.4–1.8
Phosphorous, P_2O_5	2.5–4.4	1.1–2.0
Potash, K_2O	0.7–1.9	0.8–1.2

[a]Source: Mohanrao, 1970.

Where a partial sewerage system exists and the sewered flow is treated in a waste stabilization pond, the night soil collected from the remaining part of the town can be disposed of in the pond itself, which must be designed accordingly. Waste stabilization ponds in India need desludging once in 5 in 10 years, and the humus removed from them is in great demand with farmers as a soil conditioner.

VIII. CONCLUSION

Traditionally, water reuse in India had been limited to agriculture until the late sixties when industrial reuse began.

Reuse of water in Indian industries has come to be accepted sooner and more readily than at first expected. Necessity is often the handmaiden of technological progress and industries located in water-scarce cities have out of necessity welcomed every feasible method for reusing water. The municipal authorities have also participated in this effort. In attempting to find new solutions for new problems, the opportunity to enhance the image of the profession has also been harnessed.

The writer can state from personal experience that one of the beneficial results of working in this relatively new and exciting field of water reclamation and reuse in industries is the unique opportunity to project a better image of the sanitary engineer on the public mind because for once the beneficial results of his work can be readily assessed in terms of clear cost-benefit ratios.

REFERENCES

Arceivala, S. J. (1967). "Pilot Plant Studies, Feasibility and Design for Water Reclamation Plant." Carbide Chemicals Company, Division of Union Carbide, Bombay, India.

Arceivala, S. J. (1969). "Reuse of Water in Tall Buildings." Report on Pilot Studies and Design. A.I.C., Pollution Control Division, Bombay, India.

Arceivala, S. J. (1967–1968). "Water Conservation and Reuse Surveys in 20 Different Industries in Bombay," Reports Submitted to Bombay Municipal Corporation, India, through Associated Industrial Consultants, Bombay.

Arceivala, S. J., Kapadia, J. R., and Wadekar, V. R. (1969). Economy and reuse of water in textile mills. *Proc. All-India Text. Conf., 1969* pp. E-9 to E-17.

Bannerjee, R. N. (1972). Purified wastewater for industrial use. *J. Indian Water Works Ass.* **4**, 250.

Bhaskaran, T. R. (1958). "Sewage Farming". Indian Council Medical Research, New Delhi, India.

C.P.H.E.R.I. (1970). "Sewage Farming," Tech. Dig. No. 3. Cent. Pub. Health Eng. Res. Int., Nagpur, India.

C.P.H.E.R.I. (1971a). "Health Status of Sewage Farm Workers," Tech. Dig. No. 17. Cent. Pub. Health Eng. Res. Inst., Nagpur, India.

C.P.H.E.R.I., (1971b). Annual Report. Cent. Pub. Health Eng. Res. Inst., Nagpur, India.

C.P.H.E.R.I. (1972). Annual Report. Cent. Pub. Health Eng. Res. Inst., Nagpur, India.

Handa, B. K. (1973). Ph.D. Thesis, Nagpur University.

Kabir, S. A. (1968). "Health Status of Sewage Farm Workers". Indian Council Medical Research, New Delhi, India.

Mohanrao, G. J. (1970). "Digestion of Night-soil and Cowdung," Tech. Dig. No. 8. Cent. Pub. Health Eng. Res. Inst., Nagpur, India.

Patel, T. B. (1964). Report to Indian Council Medical Research, New Delhi, India.

Raman, A., Sundaramoorthy, S., and Varadarajan, A. V. (1973). "Treatment of Sewage at Madras for Industrial Use", Symp. Proc. Indian Association for Water Pollution Control, and Cent. Pub. Health Eng. Res. Inst., Nagpur, India.

Rao, N. U. (1973). "Removal of Pathogenic Microorganisms from Sewage," Symp. Proc. Indian Association for Water Pollution Control, and Cent. Pub. Health Eng. Res. Inst., Nagpur, India.

11

Water Reuse in Israel

Gedaliah Shelef

I. INTRODUCTION

Israel has reached a stage where over 90% of its "conventional" water resources (i.e., surface and groundwaters) are being used. Meeting future national goals in industrial development, population growth, and agricultural production is dependent, to a critical extent, on the introduction of new, non-conventional water resources such as reclaimed wastewater and desalinated seawater. Due to economic reasons, which have recently become even more marked by the increased cost of energy, it is assumed that the stage of wastewater reclamation will precede the stage of large-scale seawater desalination by at least a decade. Israel has therefore adopted a rather rigorous program, in recent years, of research, planning, and design of various schemes of water reclamation aimed at renovation of wastewater for various beneficial uses ranging from agricultural reuse to reuse of wastewater for industrial and drinking purposes.

The justification for wastewater reclamation programs stems also from the need to provide more sophisticated wastewater treatment in order to meet sanitary and environmental requirements.

The growing pollution hazards of groundwater and surface receiving bodies of water in Israel create a situation that make mandatoray more rigorous effluent quality requirements. Therefore, the resultant effluent quality should be closer to what is required by various consumers of reclaimed wastewater. Thus, a favorable cycle is created in which rigorous water pollution control programs bring about the production of high-quality effluent, which, in turn, can be more extensively reused, further reducing pollution threats.

II. RENOVATED WASTEWATER AS PART OF OVERALL RESOURCES

Table I summarizes the present annual water demand according to the various uses, as compared to estimated demand in 1985. It shows that from an annual total water yield of 1620 mcm (million cubic meters), serving a popula-

Table I
Water Usage in Israel at Present (1974) and Predicted (1985)

	Annual demand			
	1974		1985 (est.)	
Water use	mcm[a]	Percent	mcm[a]	Percent
Domestic	310	19.1	460	23.2
Industrial	100	6.2	340	17.2
Agricultural	1170	72.2	1120	56.6
System's loses	40	2.5	60	3.0
	1620	100.0	1980	100.0

[a]Million cubic meters.

tion of approximately 3.5 million in 1974, an estimated increase to an annual total water yield of 1980 mcm is expected in 1985. The increase in demand for domestic and industrial purposes versus a standstill, or even a slight decrease, in agricultural use is also evident from Table I. Nevertheless, total agricultural production is still expected to increase significantly through better water-saving methodology such as trickle (drip) irrigation, greenhouse and plastic tunnels crops, and selection of crops with the highest cost return per unit volume of water.

Table II summarizes the various water sources which would meet the water demand indicated in Table I according to information and estimations of the Water Commissioner Office (1975). It is clearly seen that the "conventional" water resources, such as groundwater and surface water, have already approaching their maximum utilization.

The most significant new source of water to meet the increasing demand in the coming two decades is renovated wastewater which, through various wastewater reclamation and reuse schemes, should yield approximately 310 mcm/year by 1985. This includes 50 mcm/year of reclaimed industrial

Table II
Water Sources in Israel at Present (1974) and Predicted (1985)

	Annual supply					
	1974			1985 (est.)		
Water source	mcm[a]		Percent	mcm[a]		Percent
Groundwater						
Sand aquifer	280			240		
Limestone aquifer	630			660		
Total		910	56.2		900	45.5
Surface water						
Jordan watershed	590			570		
Flood & storm water	60			100		
Total		650	40.1		670	33.8
Renovated wastewaters						
Domestic	45			260		
Industrial	5			50		
Total		50	3.1		310	15.7
Desalinated waters[b]		10	0.6		100	5.0
		1620	100.0		1980	100.0

[a]Million cubic meters.
[b]Including seawater and brackish water desalination.

wastewater, but excludes industrial in-plant water recycling and reuse schemes. Due to its importance in solving the pressing short-term water deficit, wastewater reclamation projects are of a high priority and it is estimated that some 1200 million IL (approximately $200 million U.S.) will be invested in wastewater reclamation schemes between 1975 and 1980 (Water Commissioner Office, 1975).

Around 1985, seawater desalination and brackish water demineralization will provide between 80 to 100 mcm/year with steep increases expected toward the turn of the century. It is assumed that the increase in wastewater reclamation schemes will be parallel to further increase in desalination, since any increase in the marginal cost of water supplies will make reuse of wastewater more economically justified. In other words, it is inconceivable that the costly desalinated waters, which are of high quality and low salinity, will have only one single use.

III. REUSE OBJECTIVES AND REGULATIONS

A. Agricultural Reuse

Over 70% of Israel's waters are used for agricultural irrigation. Since the degree of treatment needed to produce adequate reclaimed wastewater for agricultural purposes is significantly less than what is required for more sophisticated types of reuse (such as for domestic and industrial purposes), it is logical that reuse for agricultural irrigation will be the first priority.

Israel's Ministry of Health (1965) issued regulations that allowed the reuse of secondary effluent (including oxidation ponds effluent) in irrigation of crops with the exclusion of vegetables that are consumed without cooking. The regulations specified conditions such as the method of irrigation and harvest or fruit picking. In 1970, following a short outbreak of cholera caused by irrigation of vegetables (specifically lettuce) with raw untreated sewage, the Ministry of Health also excluded vegetables that are eaten cooked from the list of crops that could be irrigated with wastewater effluent to prevent cross-contamination of the food-handling and preparation facilities such as kitchens.

New draft regulations have been proposed recently by the Ministry's of Health Committee (Shelef, 1975) which broaden the scope of crops to be irrigated on one hand and specify required treatment and disinfection practices on the other hand. These draft regulations indicate in essence the following requirements: (a) crops excluded from being irrigated by wastewater effluents are lettuce, cabbage, spinach, watercress, and strawberries; (b) crops that are eaten raw could be irrigated by well-treated effluent (dissolved BOD of not more than 20 mg/liter), following disinfection (at least 1 hr contact) that will bring fecal coliform concentration to less than 25/100 ml at least in 80% of the monthly samples, provided that the time lapse between irrigation and picking is at least 72 hours; (c) fruits should not be picked prior to 2 weeks after irrigation with secondary disinfected effluent; (d) industrial crops (such as cotton, sugar beet, soybeans, and cereals) could be irrigated by secondary effluent provided that the distance between sprinklers and residence area will be at least 400 m to minimize airborne pathogens.

B. Reuse in the General Water Supply System

This type of reuse refers to highly treated effluent, usually after groundwater recharge, which could be mixed with other sources of water and become part of the general water supply, including domestic and industrial use.

The basic requirement is that the reclaimed wastewaters will not affect adversely the general water supply in meeting existing drinking water standards and that the highest degree of quality control will be maintained in the system.

Since the current drinking water standards in Israel do not specify concentration of viruses nor concentrations of persistent organics, special provisions will be required to control these concentrations at the allowable levels.

C. Industrial Reuse

The possibility of reusing municipal wastewater for industrial purposes, particularly for cooling waters, is very promising and advantageous due for the following reasons: (a) proximity of industry to municipal treatment plants; (b) consistency of demand over the various seasons; (c) high control of quality and health aspects that can be achieved by industry, and (d) the higher economical return of such reuse as compared to agricultural reuse.

No official regulations exist today in Israel regarding industrial reuse of municipal wastewater effluents. The quality of effluent should be such that it will reduce the development of slimes and algae (low BOD and phosphates, respectively), reduce the possibility of heat-exchange corrosion (low ammonia nitrogen), and reduced health hazards (low coliform concentrations).

The realm of intraplant industrial reuse is also of high importance in both increasing the efficiency of water utilization and reducing environmental pollution associated with industrial wastes; this subject, however, is beyond the scope of this chapter.

IV. PRACTICES AND PLANS OF WASTEWATER REUSE

A. Agricultural Reuse

1. LOCALIZED REUSE

The direct reuse of treated or partially treated wastewater for irrigation of agricultural crops is the most common in Israel at present. This reuse can be practiced by the same community that produces the wastes or by a neighboring community.

According to the Water Commissioner Office (1971) survey, 20% of the effluent from sewered communities in Israel are used for irrigation of agricultural crops. This amounts to 37.5 mcm/year of which 6.5 mcm/year are effluents from agricultural communities (mostly kibbutzim and other cooperative settlements) that are reused at the same communities, while 31 mcm/year are effluents from urban communities. The lower overall percentage of reused effluents results from the decrease in water demand during the rainy season and because of restricted crop rotation imposed by the quality of the effluent.

Most rural communities as well as small urban communities (up to populations of about 40,000 inhabitants) base their wastewater treatment on a series of ponds. Two alternatives of pond systems are commonly used in the following manner:

a. Anaerobic Ponds followed in Series by Facultative Ponds. This system is composed of a series of ponds as schematically illustrated in Fig. 1 where the first pond is anaerobic and no oxygen is present at the entire cross-section of the pond, while the successive ponds are facultative where the lower levels remain anaerobic and the upper levels are aerobic. The degree of aerobiosis generally increases from pond to pond in the series as the organic loads decrease. The last ponds in the series may reach full aerobic conditions at most of the cross-section and the biodegradative activity is reduced. These ponds are referred to as maturation or polishing ponds.

The anaerobic primary ponds can be regarded as a combination of a primary settling basins, where the sludge and grit accumulate at the bottom of an unheated, opened, and partially mixed digestor. They are built usually in pairs

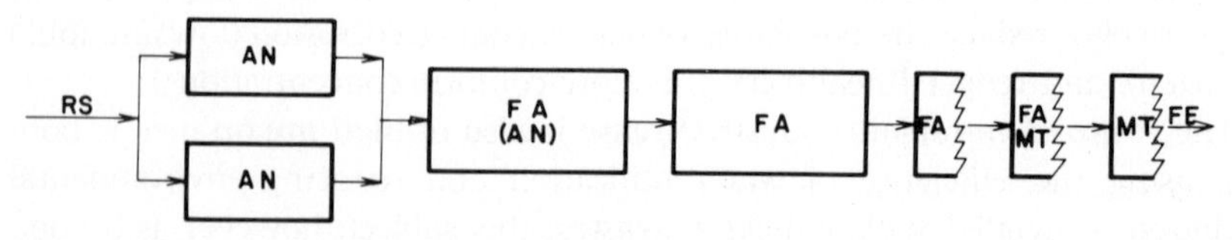

Fig. 1. Flow scheme of anaerobic pond (AN) followed by facultative ponds (FA) and maturation ponds (MT). RS, raw sewage; FE, final effluent.

of parallel ponds to enable the removal of sludge every year or every few years. The depth of the pond ranges between 1.5 to 4.0 m with a trend to increase the depth to increase detention time and sludge storage.

Organic loadings on anaerobic ponds might range between 200 to 2500 kg BOD_5/ha/day with detention time of between 1.0 to 2.5 days where loadings of between 500 to 1400 kg BOD_5/ha/day and detention periods of between 3 to 12 days are most common. Loadings and detention time depend on many factors, among them: (a) the average temperature and seasonal temperature variations, since the rate of anaerobic biodegradation increases very steeply with temperature changes between 14° to 40°C, while it virtually stop at temperatures below 12°C; (b) the depth of the pond and sludge storage volume; (c) the "strength" of the sewage—the higher the strength the higher the permissible loadings within a given pond volume; (d) the proximity of the neighboring community and its sensitivity.

Experience in rural areas in Israel, mostly in Kibbutzim (communal agricultural settlements), showed that organic loadings of between 800 to 1400 kg BOD_5/ha/day gave excellent BOD removal efficiencies and very little, if any, offensive odors.

The initial operation period of a new pond system is of a critical importance, since the proper establishment of the complex anaerobic process is quite slow, particularly at lower temperatures.

The Dan Region Sewerage Authority (Greater Tel-Aviv) commenced, for example, the operation of a series of ponds in the winter of 1969/1970 without taking the necessary precautions of gradual increase of loadings. The anaerobic pond with an area of 8 hectares was abruptly loaded with 800 kg/hectare/day of BOD_5. Although the removal efficiency surpassed 60% the pond produced offensive odors. Had the pond commenced its operation at summertime with careful gradual increase of the loadings, many of the nuisances could have been avoided.

Once the organic load has been reduced by the anaerobic ponds, the successive ponds in the series will become facultative, i.e., an aerobic zone will be formed at the upper part of the pond. The depth of the aerobic zone strongly depends on the oxygenation (which, in turn, depends on solar irradiance levels).

For rural areas in Israel, particularly in Kibbutzim, the following system has been successfully used in the past 15 years: Two parallel anaerobic ponds loaded with between 1000 to 1500 kg BOD_5/ha/day are followed by a single facultative pond loaded with between 150 to 300 kg BOD_5/ha/day. The effluent is then discharged into another pond which serves both as a semi-maturation pond (this is because the pond is still fairly loaded) and as an operative reservoir for the irrigation system which is usually composed of a pump and a sprinkling reticulation network.

The detention period in both anaerobic ponds is between 1.5 to 4 days and the degree of BOD_5 removal is assumed to be between 40 and 55%. The detention period in the facultative ponds is between 5 to 15 days and the operative reservoir pond provides additional detention of between 2 to 4 days. These ponds occasionally produce slight odors that can still be tolerated in the agricultural community. The quality of the final effluent was found to be adequate for wide nozzles sprinkling irrigation of restrictive agricultural crops.

b. Facultative Ponds followed by Maturation Ponds. This pond system is similar to the preceeding one except that it omits the preceding anaerobic ponds and limits the BOD_5 loadings on the first pond in the series to between 50 to 180 kg/ha/day. This system was developed in the United States, South Africa, Canada, New Zealand, Israel, India, and elsewhere where the threat of offensive odors that were too often produced by anaerobic ponds had to be eliminated or reduced.

The final effluent still contains pathogenic bacteria and viruses as well as organic suspended matter (usually algae and bacteria). According to official regulations, the effluent can be used for industrial crops (cotton, sugar beets, etc.), fodder, grains, and fruit trees (out of fruit-picking season).

As reported by Kott (1970), a chlorination dosage of about 20 mg/liter with about 60 min contact time can free the effluent from bacteria and viruses and render its use for unrestricted crop rotation and can facilitate the use of advanced irrigation methodology such as trickle irrigation.

2. CENTRALIZED AGRICULTURAL REUSE

A central conduit which will collect secondary effluent from the densely populated area at the central coast and transport it southward to a centralized irrigation area in the Negev was proposed as an alternative or supplement to localized agricultural reuse. According to a preliminary plan by Tahal (1972) between 100 to 160 mcm/year can be transported by an asbesto-cement conduit (20 in. to 48 in. in diameter) from Netania in the north to Dorot in the south, which covers a distance of about 140 km. The climatic conditions in the Negev are such that the irrigation season is prolonged while the geological conditions facilitate surface and subsurface seasonal storage. Some 6000 ha of cotton, 20,000 ha of wheat, and 7500 ha of sorghum can be irrigated by annual flow of 120 mcm. It should be noted that without irrigation this land is of very little use due to frequent droughts.

The estimated cost of the water (excluding treatment) at the point of use is between 20 to 30 cents (U.S.) per 1000 gallons, which renders the scheme economical. If substituting from this cost the damage that would have been created by the effluent, or else the final disposal costs that would have been needed for environmental and health protection, the cost of the reclaimed water could be significantly reduced.

This scheme will collect only surplus effluents that cannot be used locally or that can be used elsewhere with higher cost return. This is because preference is given to utilization of wastewater by existing consumers within their regular water allotment, rather than creating new large scale farms in the Negev.

The treatment for any centralized reuse scheme should be of higher quality than that provided by oxidation ponds. For communities of between 30,000 to 100,000 inhabitants, the method of aerated lagoons is recommended. These lagoon systems, which are usually aerated by surface mechanical aerators, are divided into two main types: (a) Completely mixed lagoons with a detention period of between two to five days and aerial loadings of between 2000 to 7000 kg of BOD_5/hectare/day. The energy requirement for aeration in these lagoons is approximately one kilowatt-hour/kg of BOD_5 removed or between 8 to 25 W/m^3 of basin volume. (b) "Facultative" or aerobic-anaerobic aerated lagoons where settleable solids and some of the bacterial biomass are allowed to settle at the less turbulent zones at the bottom of the lagoon. Energy input in such a system is between 0.4 to 0.8 kWh/kg of BOD_5 removed and power requirements per m^3 of aeration basin volume range between 2 to 8 W. Detention time in such lagoons range from 5 to 20 days and surface loadings from 700 to 3000 kg BOD_5/hectare/day. Polishing ponds are usually required in succession to the aerated lagoons in order to provide a higher effluent quality.

3. REUSE FOLLOWING SEASONAL IMPOUNDMENTS

To fully reuse wastewater from individual communities or clusters of communities, seasonal impoundments near the site of the agricultural reuse scheme have been planned where treated wastes are collected during the rainy season to provide irrigation waters during the summer season. The most commonly used crop is cotton which requires only between 60 and 90 days of irrigation.

Municipal trickling-filter and activated-sludge treated wastewaters of Haifa are planned for such seasonal impoundment some 20 miles away for the irrigation scheme at the Ezraelon Valley. This scheme will add approximately 18 mcm of irrigation water per year.

Similarly, some 10 mcm/year of activated-sludge treated effluent from the City of Jerusalem will be diverted to the Gazaza area some 30 miles southeast of the city. Following seasonal impoundment, the water will irrigate about 1800 hectares of cotton and 600 hectares of cereals per year by 1985 (Balashe-Yalon 1973). The seasonal reservoir will have storage capacity of about 3 million m^3, depth of about 10 m, and area of about 38 hectares.

Another seasonal wastewater reservoir exists for the partially treated wastewater of Nazareth where the reservoir provides an important part of the treatment prior to irrigation. Similar seasonal reservoirs are planned for the cities of Migdal-Ha'emek and Ramat-Ha'sharon.

B. Industrial Reuse

The reuse of treated municipal effluent as cooling water for industry has received considerable attention in Israel. Studies by Gottesman *et al.* (1973) using secondary municipal effluent from Haifa showed that lime (dosages of 500–600 mg/liter) and sodium aluminute (dosages of 20–40 mg/liter) produced adequate quality cooling water for the oil refinery and fertilizer plants near Haifa. Detergents were removed by defoaming and ammonia by stripping. It was found, however, that nitrification, taking place within the cooling system, can eliminate or reduce the need for ammonia stripping. The quality of the final effluent reached levels of 10 mg/liter BOD, 40–80 mg/liter COD, and less than 1 mg/liter of phosphate.

The existing plant adds about 4 mcm/year of cooling water to the oil refinery in Haifa, while further increase in plant capacity is expected in the future.

C. Reuse by Groundwater Recharge and Subsequent Recovery (The Dan Region Project)

The Dan Region Project is designed to treat and reclaim the effluent from the Greater Tel Aviv area. The Region's population is forecasted to about 1.3 million in 1985 producing effluent at the rate of 150 mcm/year. The treated effluent is designed for recharging the basin-like, sandy aquifer at the Rishon LeZion area, south of Tel Aviv, by intermittent spreading over leveled sand dunes. A series of 29 wells, at a distance of between 500 m to 1500 m from the spreading area will recover the reclaimed water which will then be pumped into the National Water System (NWS) conduits or will be used for various irrigation schemes.

In the year 2000, the population of the region is estimated to reach 1.7 millions producing about 200 mcm/year.

The Dan Region Wastewater Treatment and Reclamation Project is divided into two stages:

1. THE DAN REGION PROJECT—FIRST STAGE

This stage is in operation now with current capacity of 15 mcm/year serving the southern communities of the region. It is expected finally to treat and reclaim between 20 to 30 mcm/year.

The train of processes schematically illustrated in Fig. 2 consists of bar screens and approximately 120 hectares of recirculated facultative oxidation ponds followed by maturation ponds. The recirculation ratio is about 2 to 1 (pond effluent to raw sewage). The ponds unique configuration of "half-orange" as proposed by P. G. J. Meiring (private communication, 1970) following the experience in Capetown in South Africa is illustrated in Figs. 3 and 4.

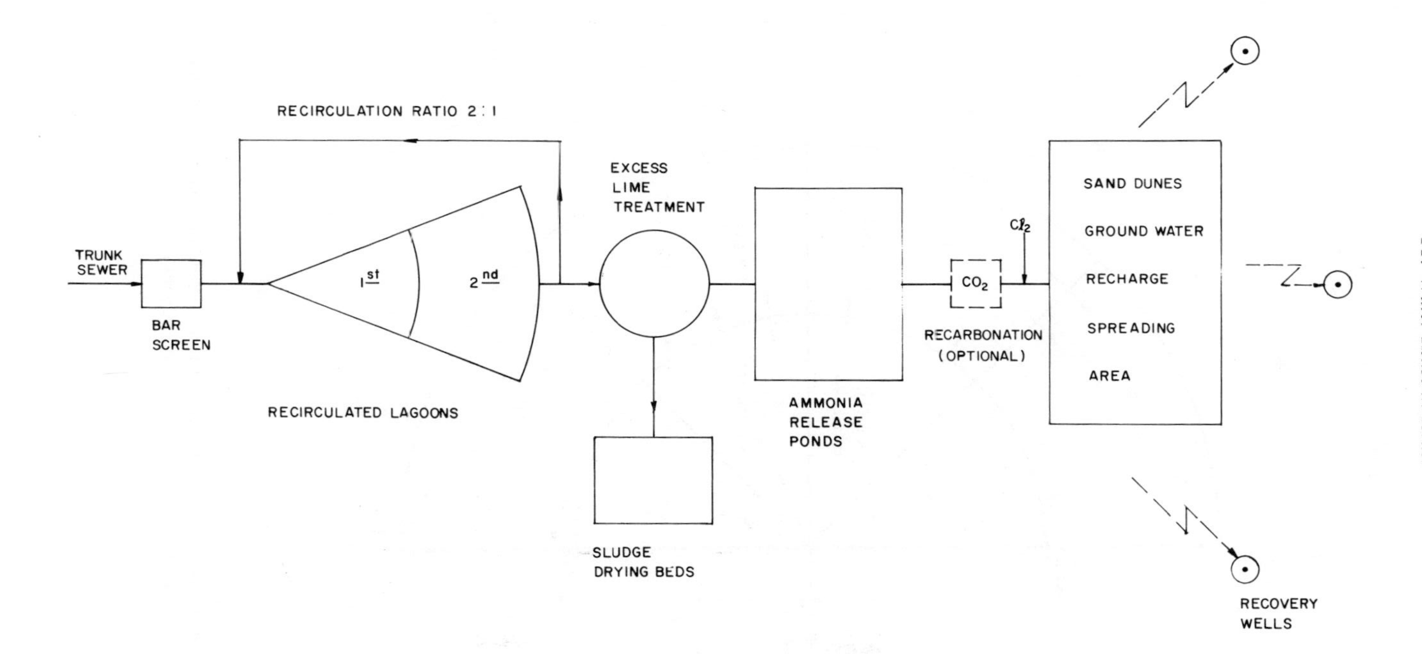

Fig. 2. Schematic flow scheme of The Dan Region Wastewater Treatment and Reclamatioı Project—first stage.

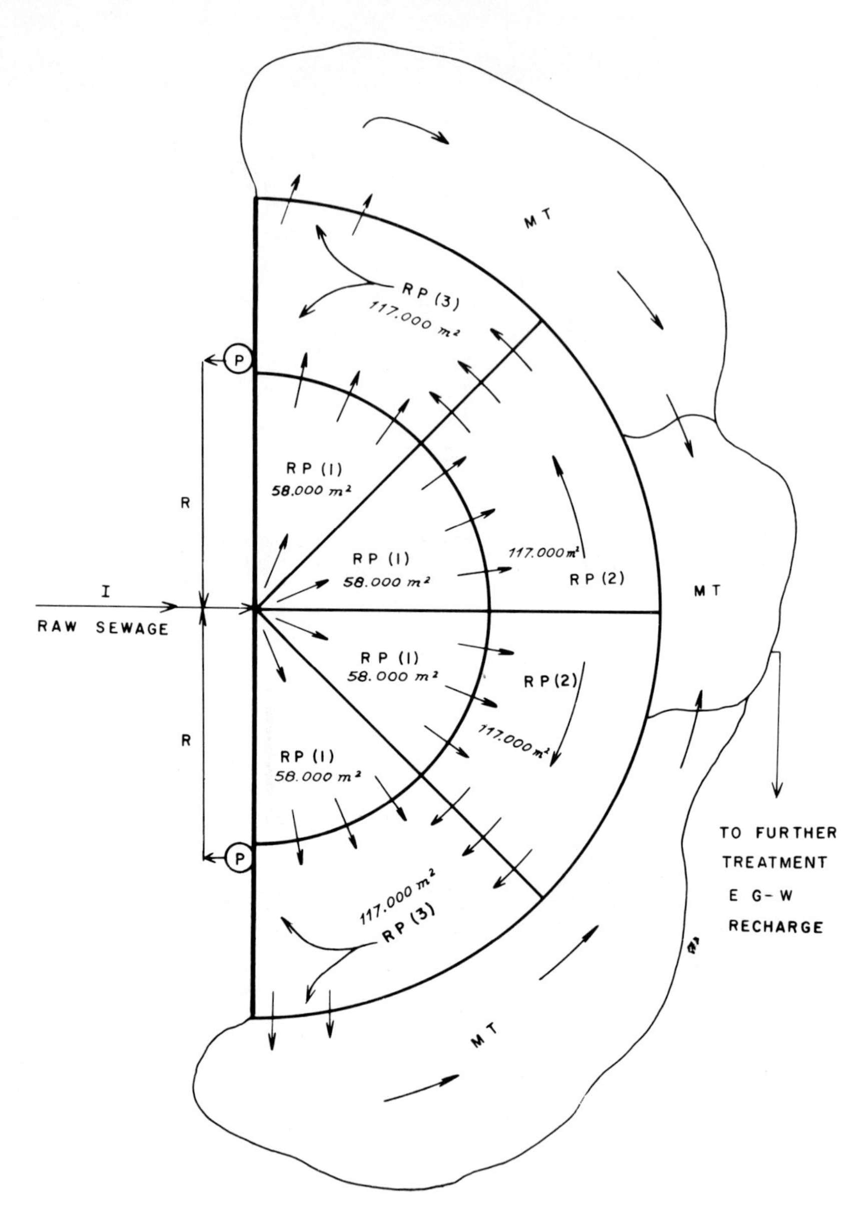

Fig. 3. Schematic plan of The Dan Region recirculated ponds system. I, influent; recirculation; RP, recirculated ponds; MT, maturation ponds; P, pump.

Fig. 4. Aereal photograph of The Dan Region ponds system during construction.

Algae-rich effluent from the ponds system is treated by 35 m in diameter sludge-blanket clarifier where approximately 400 mg/liter of lime is added together with $MgCl_2$ and polyelectrolytes serving as flocculation aids.

The lime-clarified effluent with pH of about 11 is diverted into 60 ha of open ammonia-stripping ponds. The precipitated sludge is dewatered in drying beds and transported off-site.

Following neutralization by CO_2 and disinfection, the final effluent is diverted into 25 hectares of sand dunes spreading and infiltration area.

The quality of the effluent from various stages of the plant is given in Table III (Amir and Idelovitch, 1974).

2. THE DAN REGION PROJECT—SECOND STAGE

The second stage with 1980 capacity of 100 mcm/year will provide screening, grit and flotables removal, and nitrification-denitrification modified low-rate activated sludge with or without simultaneous precipitation with ferric or sulfate. The flow scheme of this plant is illustrated in Fig. 5.

The core of the plant consists of four rectangular aeration basins each 80 m wide and 220 m long and each equipped with six rotor (brush) aerators which guarantee full oxidation-nitrification zones followed in succession by "anoxic" zones where denitrification occur. Once the second stage becomes fully operational, the first-stage recirculated pond system will serve for the treatment of peak flows only.

The aeration tanks are followed by 12 circular clarifiers, each one is 4.7 m in

Table III
Composition of Effluent from Various Treatment Units of The Dan Region Project—First Stage[a]

Composition	Raw wastewater (mg/liter)	Ponds effluent (mg/liter)	Final effluent (mg/liter)
BOD_5	200–450	30–120	10–20
COD	400–1000	150–450	40–80
Suspended solids	150–450	100–300	10–20
Total–N	70–100	20–60	2–5
NH_4–N	40–70	10–30	0.5–2.0
NO_3–N	0.5–1.0	0.5–2.0	0.5–2.0
Phosphorus	10–20	5–15	0.5–1.0
pH	6.5–7.5	7.5–8.0	7.5–8.5
Chlorides	100–300	150–350	150–350
Coliforms (per 100 ml)	10^7–10^9	10^6–10^8	10–10^2

[a]Source: Amir and Idelovitch, 1974.

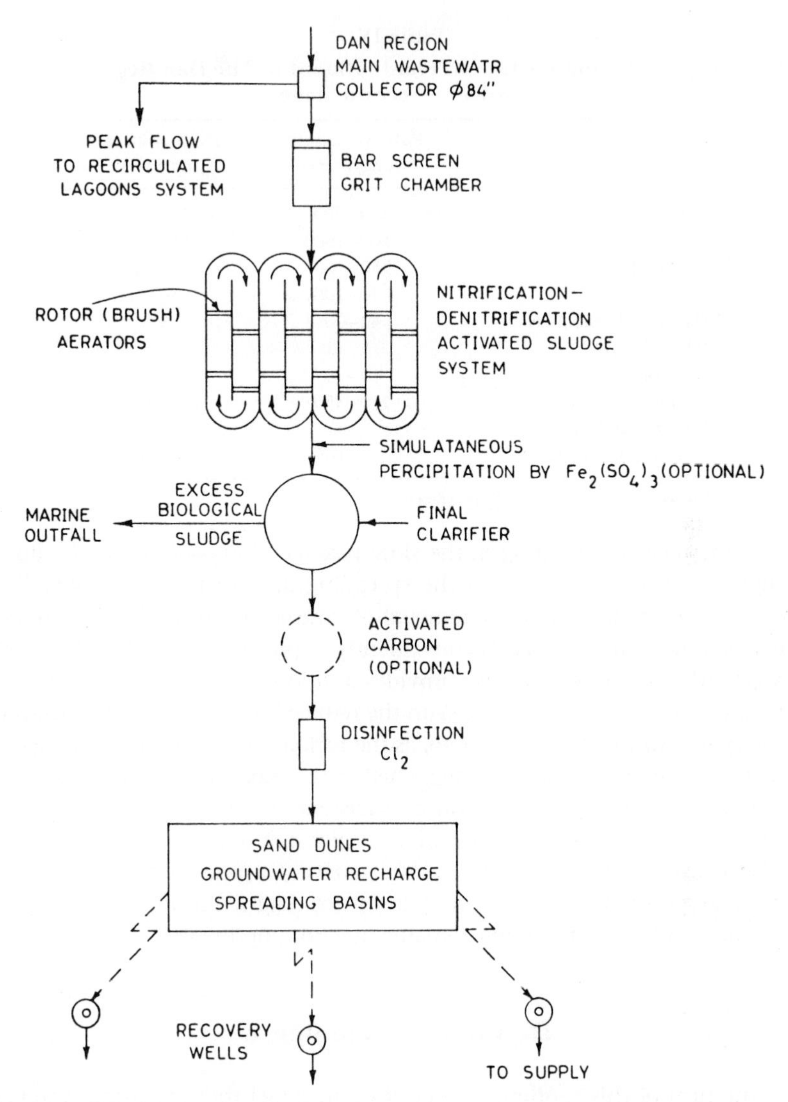

Fig. 5. Flow scheme of The Dan Region Wastewater Treatment and Reclamation Project—second stage.

diameter. Excess biological sludge is to be disposed of into the sea by a long marine outfall.

The effluent will be postaerated, chlorinated, and recharged to the aquifer through sand dune spreading-infiltration basins.

The expected composition of the plant final effluent, based on experimental bench-scale work, is given in Table IV.

Table IV
Expected Composition of Plant Effluent of The Dan Region Project—Second Stage[a]

	Raw wastewater (mg/liter)	Plant effluent (mg/liter)
BOD_5	300–700	10–15
COD	600–1600	60–120
Suspended solids	350–500	10–20
Total N	70–100	10
NH_4–N and organic N	40–70	8
$(NO_3 + NO_2)N$	0.5–1.0	2
Phosphorus	10–20	1.5
Chlorides	150–350	150–350
Dissolved oxygen	–	4
Coliforms per 100 ml	10^7–10^9	10–500

[a]Source: Amir and Idelovitch, 1974.

In both stages of the project, the slow passage (300–400 days) of effluent through the sandy aquifer from the spreading area to the recovery wells is assumed to provide additional removal of viruses and some removal of persistent organic matter which is conveniently expressed by COD or TOC (total organic carbon). The aquifer also provides dilution with natural groundwater, particularly in the first few years. With the reduced dilution with the incoming waters of the National Water System, as the reclamation project reaches its full capacity, it is not certain at this time whether the mixed waters will be suitable for drinking without prior removal of persistent organics by activated carbon. Due to the high cost of activated carbon sorption, it might be decided that the reclaimed wastewater will not be used for general purposes, including drinking, but will serve solely for irrigation of unrestricted agricultural crops. The addition of an activated carbon unit remains therefore optional.

D. Closed-Cycle Reuse

At the turn of this century, communities in Israel that are using waters of high marginal cost, such as desalinated waters or water pumped to a high elevation, will have to consider the use of highly treated reclaimed wastewaters as part of their water supply. This is particularly true in cities such as Eilat which relies on expensive desalination and the city of Jerusalem where most of its water supply is pumped from a far distance to a considerable elevation.

The train of processes will probably include full tertiary treatment including nitrogen removal, filtration, disinfection, and activated carbon sorption. The recycled wastewater will furnish only a certain proportion of the city needs and, as shown by Shelef *et al.* (1972), no desalination is needed whenever recycled wastewaters consist of less than 50% of the total municipal supply.

V. ECONOMIC ASPECTS

The economic justification of wastewater reclamation stems from two basic premises: (a) health and environmental damage created by untreated or partially treated wastewaters; this damage diminishes as a function of the degree of the applied treatment and of the increased quality of the final effluent; (b) the higher the quality of the effluent produced by the higher degree of treatment, the higher the value of the reclaimed wastewater. The total benefit of wastewater treatment and reclamation program is therefore the sum of damage prevented by the treatment and the value of the reclaimed water. This value is affected by the restrictions and limitations put upon the utilization of the reclaimed water which, in turn, is a function of the final effluent quality.

Figure 6 summarizes schematically this concept. Applying primary and secondary sewage treatment will reduce surmountable health and sanitary hazards. Tertiary treatment, with processes such as chemical clarification, chlorination, filtration, nutrient removal, and activated carbon sorption, will reduce residual pathogens, eutrophication factors, and refractory organics. Quaternary treatment, which consists mainly of removal of soluble mineral ions by desalination techniques such as ultrafiltration, reverse osmosis, ion-exchange, or distillation, will remove excess salinity contributed to wastewater through domestic and industrial uses.

As summarized by Channabasappa (1972) and Shelef *et al.* (1972), the incorporation of desalination techniques within the treatment process is possible, substituting some of the processes at the tertiary and quaternary stages of treatment. Incorporation of such techniques within the train of processes can therefore render the production of waters, which as far as dissolved minerals content is concerned, is even superior to the original water supply.

As summarized by Fig. 6, the reuse benefit increases as the degree of the treatment and the quality of effluent are increased. For example, it is possible to reuse effluent of partial secondary treatment for restricted agricultural use such as irrigation of cotton and grains. Irrigation of produce without any restrictions will require a higher degree of treatment in which full oxidation followed by chlorination, chemical clarification, and/or filtration is required in order to reduce residual pathogens. Obviously, the value of the renovated water for restricted agricultural use is much lower since many limitations are imposed; namely, (a) limited crop rotation; (b) limited season of irrigation; (c) limited method of irrigation (clogging of sprinklers and trickel irrigation); (d) health hazards to farmers and neighboring communities.

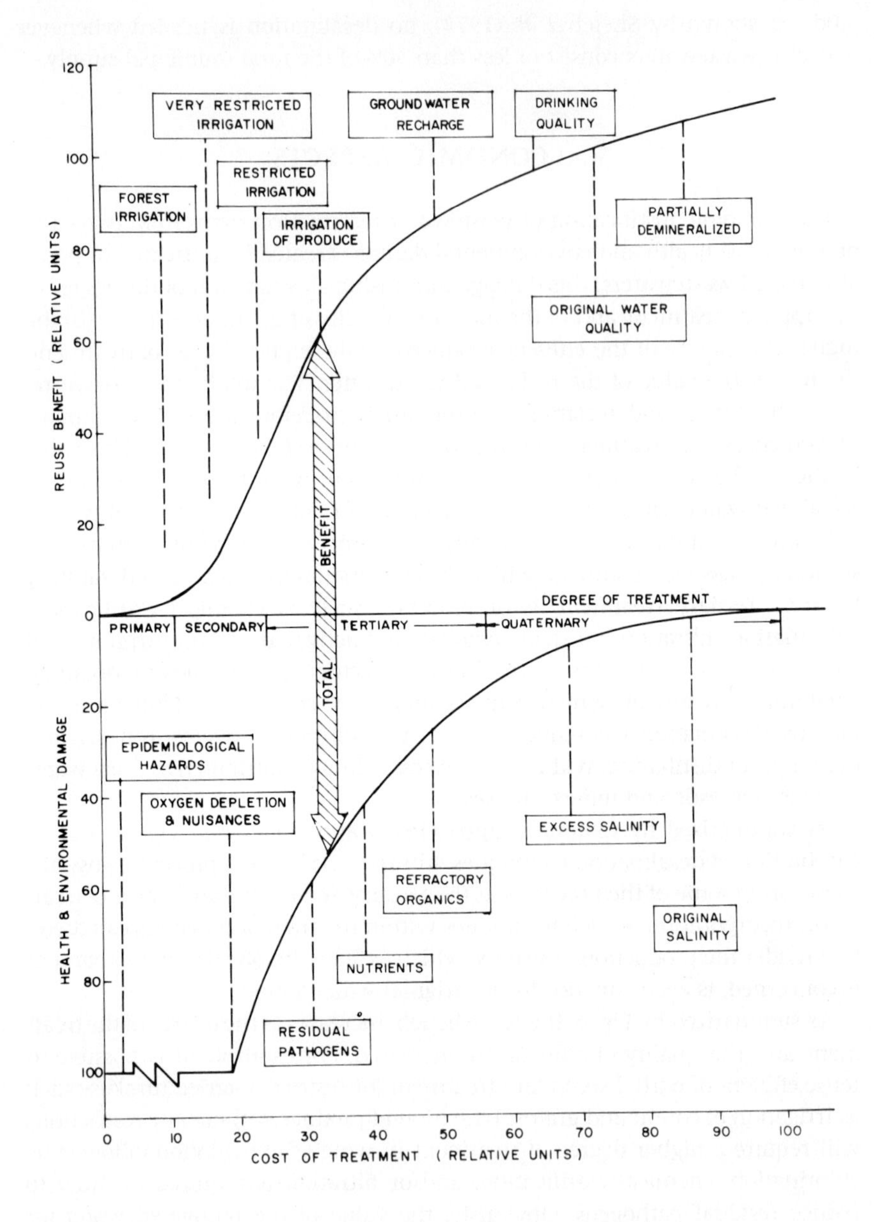

Fig. 6. Total benefit of wastewater treatment and reclamation as a function of the degree of treatment.

Higher level of treatment, for example nutrient removal, not only prevents eutrophication of receiving bodies of water and increase of nitrate concentration in groundwater, but it renders possible the use of such effluent for groundwater recharge which, after dilution and recovery, adds to the overall nonrestricted water resources. Removal of residual refractory organics by activated carbon or removal of both excess salinity and refractory organics by ultrafiltration, for example, will bring the quality of the effluent to meet drinking water standards. The value of such renovated waters increases not only due to the basic high cost of domestic water supplies, but also because the source of the renovated waters is in closer proximity of the community, which then reduces the costs for pumping and conveyance.

The shape and the actual dimensions of the curves in Fig. 6, describing both the damage and the reuse benefits, will obviously vary according to the actual cost and damage functions in a given locality and will reflect treatment costs, value of reclaimed waters, and damage externalities to health and environment.

Cost data of various stages of wastewater treatment and reclamation were given by Clayton and Pybus (1972) for the Windhoek plant in South Africa and by Culp (1968) for Lake Tahoe, in the United States. Similar data were compiled by Frankel (1971) and Sachs (1972) in Israel. Estimated costs of complete reclamation range between 32 to 78 cents (U.S.)/1000 gal, according to the scale of plant and technique used. Unfortunately, the cost analysis has not included the value of damage prevention to assess the total benefit of such reclamation schemes.

The value of a combined wastewater treatment and wastewater reclamation as far as gross environmental quality is concerned is schematically shown in Fig. 7. The approach is similar to the one used by Kneese and Bower (1968) for the economic analysis of environmental protection schemes in general and of water quality management schemes in particular. The degree of contaminant removal will tend to approach the value indicated by X, which corresponds to the intersection A between the incremental damage curve and the curve delineating the marginal cost of treatment. Substituting the marginal added value of the reclaimed wastewater from the marginal treatment costs will shift the intersection to a new point A′, corresponding to a degree of contaminant removal X′, which is much superior to X. The higher the value of the reclaimed water, the higher the level of treatment that is economically justifiable, with the end result being less damage to the environment. It can be shown that complete recycling of wastewater will tend to bring the damage curve to point X″, where complete removal of contaminants is accomplished.

Table V summarizes the estimated cost of some of the wastewater treatment processes practiced or planned in Israel, according to the mode of wastewater reuse. The differences in cost of treatment per m^3 of wastewater (including

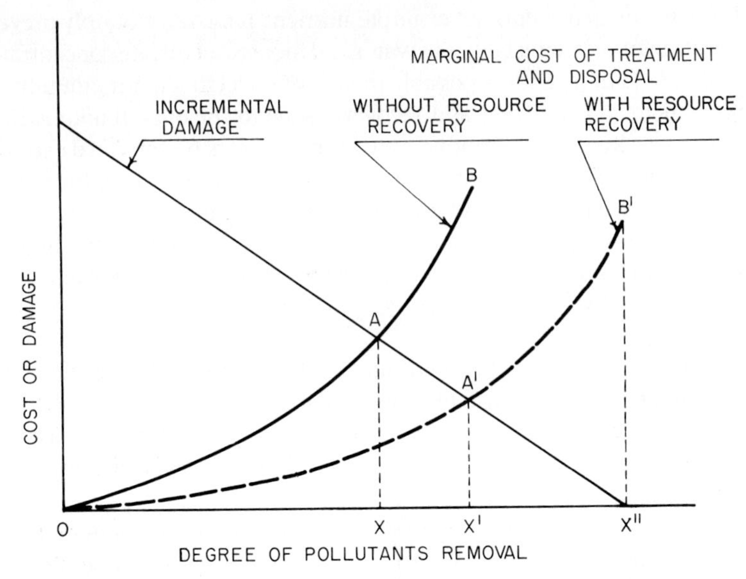

Fig. 7. Effect of wastewater reuse on economically optimal degree of pollutants removal.

Table V
Cost of Various Wastewater Treatment Processes According to Possible Mode of Reuse (1974 Cost Estimates)

Type of treatment	Mode of reuse	Estimated cost of treatment, 1974 (IL/m³)
Anaerobic—facultative ponds	Restricted localized agricultural irrigation	0.20
Recirculated ponds	Restricted agricultural irrigation	0.30
Aerated lagoons	Centralized agricultural irrigation with seasonal impoundment	0.50
Standard rate activated sludge treatment and disinfection	Irrigation of unrestricted crop rotation	0.80
Nitrification—denitrification activated sludge system and disinfection	Groundwater recharge and reuse for unrestricted irrigation or general supply when well-diluted with other supplies	1.10
Nitrification—denitrification activated sludge system followed by filtration, activated carbon sorption and disinfection	Groundwater recharge for recovery as unrestricted water supplies (including drinking), direct recycling for domestic reuse providing proper dilution with fresh water to prevent salinity buildup	1.60

investment and running costs) between the various processes might reach eightfold factor. It should be noted, however, that the less sophisticated treatment is less controlled and at times might produce environmental hazards such as odors and dissemination of airborne pathogens, while the more sophisticated treatment is usually more controlled and compact and increases the flexibility and the spectrum of potential reuses.

VI. CONCLUSIONS

The coming few decades in Israel will be marked with an intensive effort to reuse wastewater effluents for a multitude of purposes and through various types of treatment and reclamation practices. In general, the possibility of turning wastewater into a new utilizable resource of water enhance, to a great extent, the allotment of financial resources into constructing and operating more sophisticated wastewater treatment facilities, which improves the general sanitary and environmental conditions of the wastewater-producing communities.

It is estimated that the increasing demand for new water resources in Israel and elsewhere, together with the trend toward a higher degree of treatment due to environmental protection measures, will accelerate the complementing cycle of wastewater treatment and wastewater reuse.

REFERENCES

Amir, Y., and Idelovitch, E. (1974). Reuse of Municipal Wastewater-Dan Region Project. *Proc. FAO Symp. Achievement Efficiency Use Reuse Water, 1974* p. 1–19.

Balashe-Yalon, Consulting Engineers Ltd. (1973). "Utilization of Jerusalem Wastewater Effluent in the Gazaza Region" (in Hebrew), Rep. No. PM-512.

Channabasappa, K. C. (1972). Application of membrane processes for water reuse. *In* "Applications of New Concept of Physical-Chemical Wastewater Treatment" (W. W. Eckenfelder and L. K. Cecil, eds.), pp. 1–19. Pergamon, Oxford.

Clayton, A. J., and Pybus, P. J. (1972). Windhoek reclaiming sewage for drinking water. *Civil Eng., Amer. Soc. Civil Eng.* **9**, 103–107.

Culp, R. L. (1968). Wastewater reclamation at South Tahoe Public Utilities District. *J. Amer. Water Works Ass.* **60**, No. 84, 84–102.

Frankel, R. J. (1971). "A Model for Multiple Reuse of Municipal Wastewaters in Israel." Water Commissioner Office, Tel Aviv, Israel.

Gottesman, A., Engel, G., Rebhun, M., and Kishony, S. (1973). Advanced treatment and reuse of industrial and municipal wastewater in the oil refineries at Haifa. *Proc. Sci. Conf. Isr. Ecol. Soc., 4th, 1973* pp. A115–A137.

Kneese, A. V., and Bower, B. T. (1968). "Managing Water Quality: Economics, Technology, Institutions," pp. 75–129. Johns Hopkins Press, Baltimore, Maryland.

Kott, Y. (1970). Chlorination of sewage oxidation pond effluents. *In* "Developments in Water Quality Research" (H. I. Shuval, ed.), pp. 189–197. Ann Arbor (Humphrey), Sci. Publ. Ann Arbor, Michigan.

Sachs, S. B. (1972). "Renovation of Municipal Effluents by Ultrafiltration," Res. Rep. Israel Desalination Engineering, Ltd., Tel Baruch, Israel.

Shelef, G. (1973). "Drinking Water Quality Standards," Report of National Committee, Ministry of Health, Jerusalem, Israel.

Shelef, G. (1975). "Draft Regulations on Quality of Effluent for Irrigation of Agricultural Crops." Ministry of Health, Jerusalem, Israel.

Shelef, G., Matz, R., and Schwarz, M. (1972). Ultrafiltration and microfiltration membrane processes for treatment and reclamation of pond effluents in Israel. *In* "Applications of New Concepts of Physical-Chemical Wastewater Treatment" (W. W. Eckenfelder and L. K. Cecil, eds.), pp. 335–345. Pergamon, Oxford.

Tahal. (1972). "National Plan for Reuse of Wastewater," (in Hebrew), Rep. No. HR/72/106. Water Planning for Israel.

Water Commissioner Office. (1971). "Survey of Wastewater Collection, Treatment and Reuse." WCO, Jerusalem, Israel.

Water Commissioner Office. (1975). "Report to the Government on Water Demand and Availability of Water Resources." WCO, Tel-Aviv, Israel.

12

Wastewater Reuse in Japan

Takeshi Kubo and Akinori Sugiki

I. THE CHARACTERISTICS OF WATER RESOURCES IN JAPAN

Japan owes most of its precipitation to seasonal winds, rains in the wet season, and typhoons. The average annual rainfall is approximately 1750 mm, but each district throughout Japan has its own climatic pattern. In the northeastern and Hokuriku districts, for instance, the peak of precipitation comes in winter, while on the Pacific coast, most of the rainfall is brought by typhoons and the rainy spell in early summer.

Japan depends for its water resources mostly upon rivers whose flux varies with the rainfall and is naturally subject to wide fluctuations.

In the case of Tokyo, 20% of the total annual rainfall is brought by typhoons, and another 20% by rains in the wet season. This means that rains are apt to occur on special days and weeks, and that creates difficulties in utilizing water resources effectively.

It is estimated that the total annual precipitation of Japan is 650×10^9 m^3. Of this total rainfall, 46×10^9 m^3 are used for irrigation, 25×10^9 m^3 for industrial purposes, and 9×10^9 m^3 for municipal water. In other words, only 12% of the total precipitation is being utilized, and as much as 200×10^9 m^3 are lost in floods, 170×10^9 m^3 just flow into the sea, and 200×10^9 m^3 evaporate.

A. The Water Use in Japan

In Japan, water resources are utilized mainly for the following purposes.

1. MUNICIPAL WATER

According to statistics, in 1968, 65.5% of the total population are served with public water supply, and, on average, each of them is supplied with 322 liters a day. Shown in Table I is the transition of water consumption per capita per day in the big six cities in Japan for the past 30-years. It suggests that in cities with higher living standards, there is a greater demand for water.

2. INDUSTRIAL AND AGRICULTURAL WATER

Water is used in various fields for industrial purposes such as agriculture, fishery, transportation, and for power generation, cooling, and processing. Discussed here, however, is water used for irrigation and industry only.

a. Water for Irrigation Use. Rice culture is one of the most important parts of Japanese agriculture, and the better part of the current riparian water use (90% of the total water right) is for rice fields.

The consumption of irrigated water is generally measured by the decrease of water depth due to permeation and evaporation, and the average daily decrease of the water depth of Japanese paddy fields is estimated to be 15 to 25 mm. Supposing that the average decrease of water depth is 20 mm, the daily consumption of irrigated water to cover the total rice field acreage (3,350,000 ha) in Japan may be calculated as 670×10^6 m^3 and the annual consumption as approximately 60×10^9 m^3 on the assumption that irrigation continues for 90 days/yr. Of these, only 46×10^9 m^3 are actually supplied from rivers, irrigation ponds, and groundwater, and 14×10^9 m^3 may be from rains that fall over the whole extent of paddy fields during the irrigation term.

b. Water for Industrial Use. As shown in Fig. 1, demand for industrial water

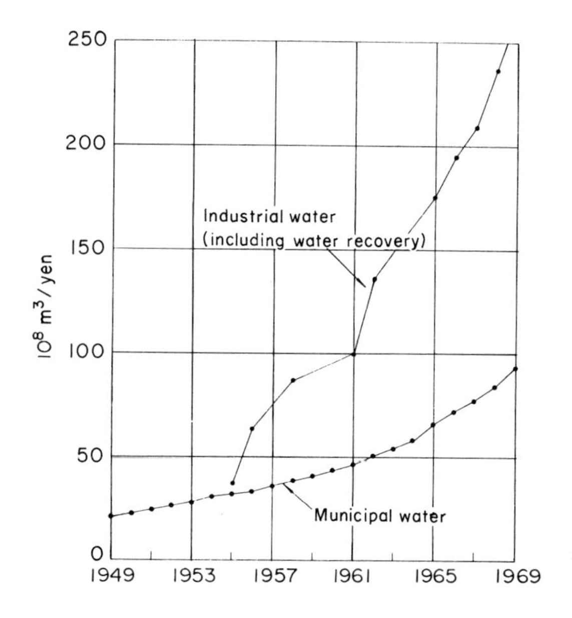

Fig. 1. Industrial and municipal water consumption.

consumption in Japan has rapidly increased since the late 1950s due to the progress of heavy industry which requires a large quantity of water. Table II shows the total consumption of water for industrial use in 1958 and 1968.

According to the Table II, the percentage of fresh water used for cooling and air-conditioning purposes has increased from 43.7% (1958) to 62% (1968).

As previously mentioned, most of Japanese riparian water is used for irrigation. This means that if the demand for municipal water and industrial water, is increased, a new source of supply must be developed.

The development of new water resources, however, has become more and more difficult, and securing of water for industrial use without rationing has now become unavoidable. This is where the problem of reusing fresh water comes in. Fig. 2 shows that fresh water required per unit production has been rapidly decreased in relation to the increased rate of reuse of fresh water in every type of industry.

B. Future Demand for Water and Reuse

The Ministry of Construction has predicted demand for water in 1985 based upon economic activities in 1965. The prediction shows that total national

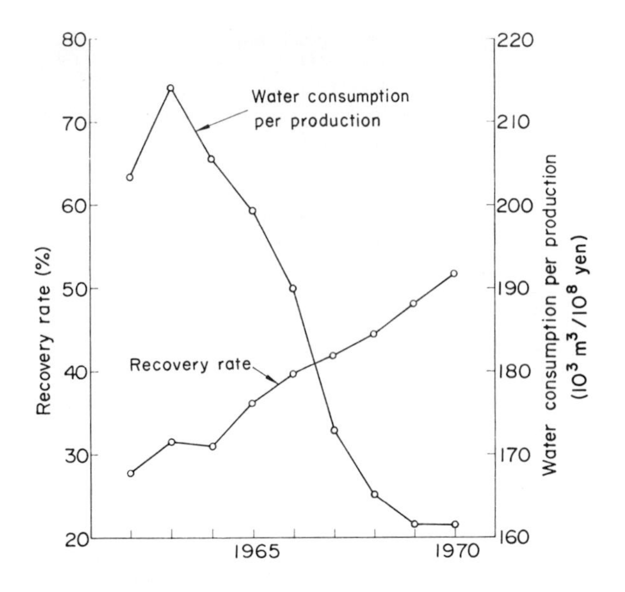

Fig. 2. Water consumption per production and recovery rate.

demand for water will be estimated at $117 \times 10^9 \, m^3$ ($20 \times 10^9 \, m^3$ for municipal water, $58 \times 10^9 \, m^3$ for agricultural use, and $39 \times 10^9 \, m^3$ for industrial use). Supposing that demand in 1965 is 1, then the estimated demand for water in 1895 will be 2.9 for municipal water, 1.2 for water for agricultural use, and 3.1 for water for industrial use. This estimate suggests that increased demand, especially for industrial water use, is expected in the near future.

On the other hand, the Kanto (Tokyo metropolitan area inclusive) and Kinki (Osaka, Kobe, and Kyoto area inclusive) districts will find it very difficult to meet the future demand for water to support the dense population and a great deal of industry.

The population of Tokyo metropolitan area is expected to increase from 30,260,000 (1970) to 36,500,000 (1985), and the industrial output from 2,420,000,000 yen to 5,992,000,000 yen. This estimated growth demands at least $788.1 \times 10^9 \, m^3$ of new water resources. Of this quantity, $591 \times 10^9 \, m^3$ are planned to be supplied from newly constructed dams, but the remaining $196.9 \times 10^9 \, m^3$ are left unsettled.

In the case of such coastal cities as Tokyo and Osaka, repeated reuse of sewage is practically impossible, while used water of inland cities discharged into rivers can be used repeatedly by other cities below, so long as the sewage treatment is adaquate. In the future coastal cities, therefore, one must consider measures to convert final effluents from sewage treatment plants into industrial water, as well as ways to ensure maximum utilization of municipal water and industrial water.

1. DIRECT REUSE OF TREATED SEWAGE FOR MUNICIPAL WATER SUPPLY

Some large-scale public housing areas and business centers have already started investigations into the possibility of diminishing the absolute quantity of water consumption by utilizing treated sewage for miscellaneous purposes. Shown in Table III is the average consumption of municipal water classified according to purpose on four housing areas in Tokyo and Osaka.

In an office building, for instance, each office worker requires 100 to 200 liters/day, and 50 to 60% of this quantity is used for the sanitary purpose of flushing toilets. They are now planning to reuse treated sewage for this purpose.

In some public housing areas in Osaka, Kobe, and Tokyo, pilot sewage treatment plants are already in operation.

On the other hand, reusing treated sewage for sanitary purposes has been included in the urban redevelopment plan for Koto Ward in Tokyo, which is now under consideration.

2. RECOVER OF INDUSTRIAL WATER AND REUSE OF TREATED SEWAGE

One of the most crucial restrictive factors of recent industrial development is water supply, and economy of water has now become an urgent problem.

Table IV shows the estimated rate of reuse of industrial waste water in 1985 compared with that in 1970, based upon a survey by the Ministry of International Trade and Industry.

In some particular fields of industry, such as iron and steel, oil, coal, and chemical industry, which utilize cooling water at a relatively higher rate, the rate of reuse is generally higher than that of paper pulp industry whose rate of reuse is 21%.

This rate of waste water reuse is expected to rise in accordance with the increased demand for industrial water use in the future.

The next section includes examples of conversion of treated sewage into industrial water already in practical use in some big cities such as Tokyo, Nagoya, Kawasaki, and Osaka.

II. REUSE OF TREATED SEWAGE FOR INDUSTRY

Tokyo and four other big cities in Japan have already started their reuse of treated sewage for industrial use.

For the time being, sewage is supplied to each factory after chemical treatment with coagulants and rapid sand filtration, followed by biological treatment. A few problems, however, still remain unsolved; further study on ways to improve quality by using active carbon treatment is now in progress.

A. Examples in Tokyo, Nagoya, and Kawasaki

1. TOKYO

In Tokyo, reuse of effluents from sewage treatment plants for industrial use was first planned to prevent depletion of groundwater supplies in Koto Ward in Tokyo due to excessive pumping of groundwater for factories there. The current facilities are now treating 138,000 m^3/day of final effluent from the Mikawajima Sewage Treatment Plant and supplying a total of 340 factories in the Koto Ward with 110,000 m^3/day of industrial water.

In this plant, effluent from the biological process undergoes prechlorination prior to coagulation. Then the rapid sand filtration is applied and the effluent is supplied to each factory after postchlorination.

Table V shows a list of factories that use treated sewage and the type of utilization.

According to Table V, approximately 50% of this treated sewage is being used for cooling water and 30% for washing purposes.

Table VI outlines the general standards for water quality for industrial use which is authorized by the Japanese Society for Industrial Water Supply.

The standard for the industrial water supply in Tokyo, as shown in table VII, also covers the treated sewage for cooling and washing purposes previously mentioned.

The recent improvement of sewerage systems has eliminated the intrusion of seawater and diminished the salt content in sewage to 1000 ppm or less.

Table VIII shows the quality of final effluents from secondary treatment recorded from February, 1967, to December, 1969.

Effluents from sewage treatment plants are usually treated as follows. First, prechlorination (3 to 7 ppm) is applied to oxidize organic matter and remove offensive odors. Then, chlorine is injected at an average rate of 15 ppm to keep the turbidity under 15 Jackson units. The following tabulation includes specifications for the rapid sand filters now in use:

Depth of sand layer	60 cm
Depth of gravel layer	50 cm
Effective diameter of filter sand	0.05 mm
Uniform coefficient of filter sand	1.3
Average rate of filtration	90 m/day
Turbidity removal	40–60%
Period of filtration	100 hr

For both disinfection and slime removal, an average dosage rate of 1 to 4 ppm of chlorine is added to the filtrate so that the residual chlorine may be kept 0.1 to 0.5 ppm, both in the distributing reservoir and distributing pipes.

2. NAGOYA

In Nagoya, effluents from sewage treatment plants have been supplied since 1966 to a total of 12 factories in the southern part of the city for various industrial purposes after treatment by coagulation and rapid sand filtration.

In 1968, average daily supply was 20,240 m^3. Treated sewage is mixed with purified surface water before being sent to each factory. This mixed water is mainly used for cooling purposes.

Table IX shows the results of seasonal water quality in the Chitose Sewage Treatment Plant. The quality standard of this reclaimed wastewater after the treatment by coagulation and rapid sand filtration is given in Table X, and the quality of wastewater effluent actually supplied during the year 1968 is shown in Table XI.

A qualitative presentation of the current problems can be summarized as follows: (a) an increase in chloride ions due to the intrusion of seawater causes troublesome scale and metallic corrosion; (b) ammonia included (usually 8–10 ppm as $NH4-N$) causes corrosion of alloyed copper pipes.

3. KAWASAKI

In Kawasaki, treated sewage is being supplied directly from the Iriezaki Sewage Treatment Plant to the Mizue Iron Foundry (26,000 m^3/day) of Nippon Kokan K. K. and the factories of Shin-Toyo Glass Industry Co., Ltd. (3000 m^3/day).

The effluent quality of the biologically treated sewage supplied during the year 1965 is shown in Table XII.

Nippon Kokan K.K. has its own facilities for rapid sand filtration and is filtering all effluent from sewage treatment plants before distribution to each factory, while Shin-Toyo Glass is using it without any further treatment. Thus, it may be quite natural that Shin Toyo Glass Co. faced problems of corrosion of cooling pipes and scale and slime forming in the supply pipes. To cope with this situation, they took the following countermeasures:

(a) Stop using reclaimed water for those facilities which allow no break in operations and no replacement, or to mix the treated effluent with high quality industrial water
(b) Replace old pipes with new galvanized ones
(c) Change the system from a circulating system to an overflowing system
(d) Inject chlorine to remove the slime growth
(e) Add anticorrosion chemicals if necessary

The average daily supply of this reclaimed wastewater during the year 1970 is reported as 154,000 m^3.

III. NIGHT-SOIL TREATMENT BY PHOTOSYNTHETIC BACTERIA AND *CHLORELLA*

A. Introduction

As Table XIII indicates, organic wastes such as night soil with BOD as high as 10,000 ppm usually include more than 1500 ppm of BOD even after chemical treatment or commonly used digestion methods (aerobic or anaerobic). This means that such wastes require further dilution and more extensive treatment, such as the activated sludge process.

The method of night-soil treatment using photosynthetic bacteria and *Chlorella,* that was developed in Japan, is said to be able to reduce BOD to approximately 30–50 ppm. Another advantage of this method is that the cultivated photosynthetic bacteria and *Chlorella* can also be reused for feed for livestock and fish.

B. Mechanism and Characteristics of Night-soil Treatment

As shown in Table XIII, the BOD value of both crude and treated night soil is very high because it includes much lower fatty acids (acetic acid, 60–70%; propionic acid, 10–15%; butyric acid, 5–10%; etc.). High concentrations of ammonium nitrogen and phosphorus also characterize night soil.

Dr. Kobayashi (1970), and Dr. Kitamura (1972) have made a systematic study of night-soil treatment by bacteria and showed that the kinds of bacteria that can grow vary with the strength of night soil. They have demonstrated that heterotrophic bacteria can grow in night soil of high strength and photosynthetic bacteria can grow only in medium strength night soil, while green algae can develop only in relatively dilute solutions. Therefore, useful by-products can be produced by night-soil treatment if an effective method is developed to cultivate the most suitable bacteria for each grade of strength.

As shown in Table XIV, photosynthetic bacteria are classified roughly into three groups, green sulfur bacteria, purple sulfur bacteria, and nonsulfur-purple bacteria.

According to Table XV, both photosynthetic bacteria and *Chlorella* are rich in albuminous substances and can provide a promising source of protein for animal feed. Photosynthetic bacteria can also be used as an addition to feed because they include more methionine and vitamin B_{12} than *Chlorella* and yeast. (See Table XVI and Table XVII.)

C. Pilot Plant Experimental Works in Kiryū City

As outlined in Fig. 3, crude night soil enters the first-stage tank for aerobic decomposition for 10–20 hr so organic solids can be reduced by sedimentation

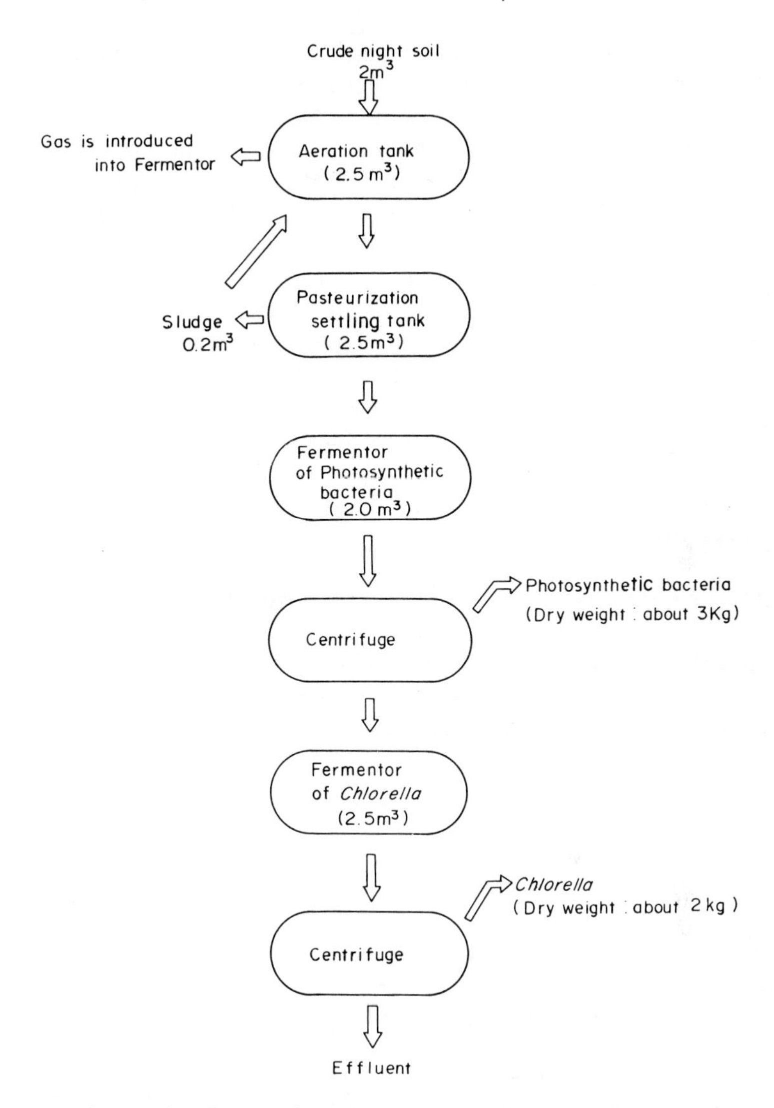

Fig. 3. Flow diagram of test plant at Kiryū city for night-soil treatment.

and by decomposition with aerobic bacteria. Then the supernatant liquor is sent to the photosynthetic bacteria tank. Treatment with UV light, electromagnetic waves, and sonic waves can help control the growth of anaerobic bacteria and kill microorganisms. In the first-stage tank of the three photosynthetic bacteria culture tanks the night soil is inoculated with (20% (v/v)) a culture of photosynthetic bacteria. Several thousand luxes of light are applied to the surface of treated night soil in the transparent plastic case. The tank requires aeration to suppress the propagation of anaerobic heterotrophic bacteria and to cultivate aerobic bacteria. This also serves the purpose of stirring the liquor. The required retention time for cultivation in each tank is 24 hours, and the cultivation of photosynthetic bacteria continues for 3 days from the first to the third tank. The cultivated photosynthetic bacteria are collected by the centrifugal separator. Night-soil liquor deprived of photosynthetic bacteria still contains organic matter such as amino acid, nucleic acid, vitamin B_{12} in concentrations suitable for growth of green algae. Night-soil liquor, inoculated with 20% *Chlorella*, undergoes a further 36 hours of retentive cultivation in each lighted aerobic cultivation tank. This is necessary since *Chlorella* requires a greater retention time for cultivation than photosynthetic bacteria.

The quality of treated night-soil liquor in the pilot plant experiment recorded from September to October of 1967 is shown in Table XVIII. According to Table XVIII, the average content of BOD in treated night soil is 15 ppm and for ammonia nitrogen is 31 ppm.

D. Construction of Model Plants and Their Operation

As just stated, night-soil treatment by photosynthetic bacteria and *Chlorella* in pilot plants has proved succesful and large-scale model plants have been put into operation. The layout of one such plant is given in Fig. 4.

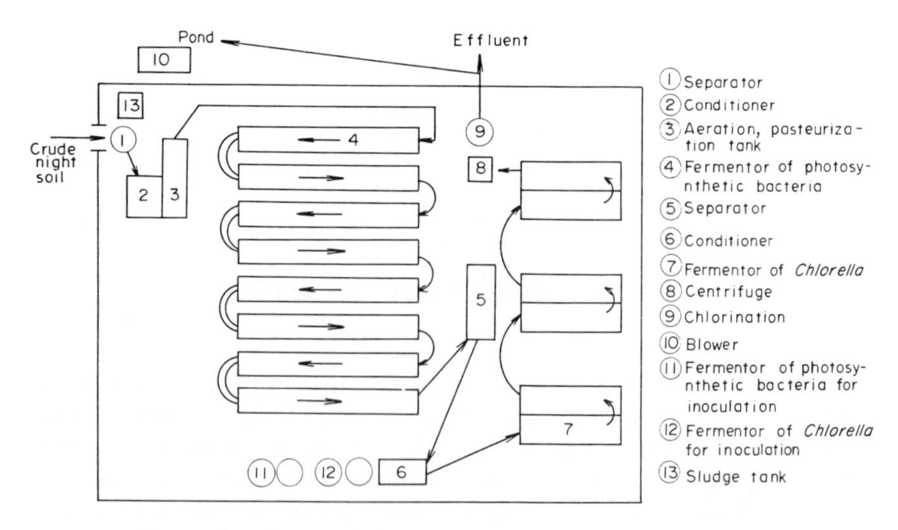

Fig. 4. Flow diagram of pilot plant (Kiryū type) for night-soil treatment.

Table I
Water Use Increase per Capita per Day (lpcpd) in Large Cities in Japan

	1921		1929		1935		1940		1946		1950		1955		1960		1965		1968	
Water use	Max	Mean	Max	Mean	Max	Mean	Max	Mean	Max	Mean	Max	Mean	Max	Mean	Max	Mean	Max	Mean	Max	Mean
Location																				
Tokyo	172	129	228	180	231	178	182	135	—	—	363	321	341	302	371	331	471	395	521	448
Osaka	189	144	196	144	213	169	238	190	641	549	466	400	421	356	481	394	651	519	750	604
Nagoya	164	126	161	104	179	135	239	196	—	559	460	376	367	280	433	337	493	358	519	391
Kyoto	212	141	176	123	214	165	217	175	363	320	329	285	262	219	358	291	423	322	450	358
Yokohama	258	204	235	174	262	215	309	261	752	688	444	398	479	409	459	399	464	370	496	398
Kobe	218	161	219	149	231	177	200	160	—	429	367	330	387	321	411	337	414	335	440	369

Table II
Industrial Water Use for Various Purposes

Year	Boiler feed water	Process[a] water	Washwater and process water	Cooling	Cooling (seawater)	Air-conditioning	Miscellaneous	Total[b]
1958								
1000 m^3/day	961	1,306	9,093	9,275	24,956	1,289	2,006	23,930
Percent	4.0	5.5	38	38.3	—	5.4	8.8	100
1968								
1000 m^3/day	1,595	530	18,549	36,729	26,297	3,871	3,664	64,938
Percent	2.6	0.8	28.6	56.6	—	6.0	5.4	100
Rate of increase 1958–1968								
Percent	166.0	40.6	204.0	396.0	105.4	300.3	182.7	271.4

[a]Water contained in products.
[b]Only fresh water not including seawater.

Table III
Water Consumption for Various Domestic Purposes

Items	Consumption (Ipcpd)[a]	Ratio (%)
Culinary	45	18
Bath	50	20
Washing	60	24
Sanitary	40	16
Cleaning	18	7
Miscellaneous	37	15
	250	100

[a]Ipcpd, Increase per capita per day.

Table IV
Future Estimation of Water Consumption per Unit Production and Ratio of Water Recovery[a]

Industry group	1970			1985		
	Production	Water consumption per production	Recovery rate	Production	Water consumption per production	Recovery rate
	(10^9 yen)	(10^3 $m^3/10^8$ yen)	(%)	(10 yen)	(10^3 $m^3/10^8$ yen)	(%)
Food and kindred products	1,874	0.095	17	14,120	0.08	30
Textile mill products	2,354	0.152	6	7,008	0.11	30
Cloth products	431	0.010	0	1,620	0.01	10
Lumber and wood products	838	0.016	17	3,021	0.02	40
Furniture products	426	0.014	5	4,403	0.02	20
Paper and allied products	1,705	0.792	27	5,931	0.36	40
Printing and allied products	1,320	0.016	39	7,165	0.02	70
Chemicals and allied products	5,093	0.521	65	27,911	0.44	90
Petroleum and coal products	1,758	0.152	70	5,861	0.15	90
Rubber and allied products	585	0.166	27	1,862	0.08	70
Leather and leather products	133	0.042	13	398	0.03	10
Ceramic products	1,621	0.168	48	3,965	0.10	60
Iron and steel products	5,545	0.309	75	21,515	0.34	90
Nonferrus metal products	2,292	0.132	57	9,358	0.10	80
Metal products	2,172	0.029	5	7,948	0.02	20
Machine products	5,658	0.016	19	31,264	0.01	40
Electric machine products	6,695	0.020	26	80,540	0.01	50
Transportation machine products	6,793	0.013	54	20,548	0.04	80
Precision machine products	727	0.016	7	1,827	0.01	20
Arms products	3			41		
Miscellaneous	1,551	0.055	16	16,211	0.01	20
	53,701	0.161	52	273,318	—	—

[a]Note: Based on 1965 prices.

Table V
Water Consumption for Each Industry Group (1970)

			Purpose cm³ m/day						
Industry group	Number	Consumption (m^3/day)	Cooling	Washing	Processing	Process[a] water	Air-conditioning	Boiler	Miscel-laneous
Chemicals and allied products	45	39,915	25,987	12,320	15	100	638	0	900
Iron and steel products	30	20,621	15,651	2,515	1,538	200	410	0	307
Metal products	31	11,495	5,722	4,858	32	19	75	2	787
Food and kindred products	12	3,630	2,867	593	60	0	0	60	50
Rubber products	19	3,273	2,683	535	0	0	0	0	55
Leather and leather products	163	12,456	0	12,456	0	0	0	0	0
Paper and allied products	4	9,149	200	4,512	2,750	1,677	0	0	0
Textile mill products	15	2,690	405	1,630	350	10	0	0	295
Gas supply	2	2,500	1,500	700	200	0	0	0	100
Miscellaneous	18	2,677	410	1,698	150	100	17	0	302
	339	108,396	55,335	41,817	5,095	2,106	1,185	62	2,796
Ratio (%)		100	51.0	38.6	4.7	4.7	1.1	0.1	2.8

[a]Water contained in product.

Table VI
Water Quality Criteria for Industrial Water Supply

Characteristic	Turbidity (ppm)	pH	Alkalinity (ppm)	Hardness (ppm)	Total solids (ppm)	Chloride (ppm)	Iron (ppm)	Manganese (ppm)
Industrial Water Supply	20	6.5–8.0	75	120	250	80	0.3	0.2
Public Water Supply	2	5.8–8.6	—	300	500	200	0.3	0.3

Table VII
Water Quality Criteria in Tokyo Industrial Water Supply

Characteristic	Criteria
Temperature (°C)	less than 27
Turbidity (ppm)	less than 15
pH	5.8–8.6
Chloride (ppm)	less than 1,500
Iron (ppm)	less than 0.7

Table VIII
Water Quality of Intake Water (1967–1969)

Characteristics	Max	Min	Mean
Temperature (°C)	32.0	9.8	20.4
Turbidity (ppm)	134.0	2.5	20.0
Color (ppm)	80	14	29
pH	7.3	6.3	6.9
Alkalinity (ppm)	144	53	115
Chloride (ppm)	1205	36	267
Conductivity (μv/cm)	4275	337	1064
Iron (ppm)	3.6	0.12	0.72
Hardness (ppm)	460	106	272
Total solids (ppm)	1957	328	914
ABS (ppm)	8.0	0.1	2.8
COD (ppm)	37.2	6.4	18.5
N H $^{-}$N	48.8	7.0	21.3

Table IX
Seasonal Water Quality in Chitose Sewage Treatment Plant, Nagoya (1968)

Characteristic	March–May	June–August	September–November	December–February
Transparency (cm)				
Crude sewage	2.3	2.5	2.7	2.4
Final effluent	30 <	30 <	30 <	13 <
pH				
Crude sewage	6.9	7.1	6.7	6.9
Final effluent	7.2	6.8	6.9	7.9
Total solids (ppm)				
Crude sewage	1770	1196	1060	1041
Final effluent	1448	984	894	832
Suspended solids (ppm)				
Crude sewage	192	242	138	206
Final effluent	30	22	30	20
BOD (ppm)				
Crude sewage	120	50	128	86.5
Final effluent	22	16.2	18.6	12
COD (ppm)				
Crude sewage	59	38	58	67
Final effluent	14	7	24	15

Table X
Water Quality Criteria in Nagoya Industrial Water Supply

Characteristic	Criteria
Temperature (°C)	Less than 27
Turbidity (ppm)	Less than 15
pH	6.0–7.5

Table XI
Water Quality of Reuse for Industrial Water in Nagoya Sewage Works (1968)

	Effluent from sewage treatment			Reuse water		
Characteristics	Max	Min	Mean	Max	Min	Mean
Temperature (°C)	26.5	11.5	19.3	27.0	12.2	19.8
Turbidity (ppm)	36.0	4.0	11.2	4.0	0.1	0.8
Color (ppm)	50.0	10.0	24.4	20.0	7.0	10.5
pH	7.1	6.7	6.9	7.1	6.6	6.7
Alkalinity (ppm)	156	52	109.8	108	71	89.1
COD (ppm)	22.5	68.2	12.9	11.1	3.9	11.1
Conductivity (μv/cm)	1920	770	1590	2280	1100	1713
Hardness (ppm)	320	118	222.2	362.5	121	213.6
Total solids (ppm)	1118	451	761	1252	572	916.4
Chloride (ppm)	494	140	353.8	648	209	379.5
Iron (ppm)	3.54	0.43	1.25	0.50	0.06	0.15

Table XII
Water Quality of Reuse Water from Sewage Treatment (1964)

Characteristic	Max	Min	Mean
Temperature (°C)	27.5	9.0	18.9
Turbidity (ppm)	230	8	43.4
pH	7.0	5.8	6.5
Alkalinity (ppm)	122	58	86
Acidity (ppm)	39.0	13.2	27.1
Conductivity (μv/cm)	2460	1300	1286
Total solids (ppm)	1520	446	867
Iron (ppm)	2.32	0.10	0.74
Chloride (ppm)	2318	188.4	547.3
Sulfate (ppm)	105.0	54.0	82.6
Nitrite (ppm)	0.10	0.00	0.02
Ammonia (ppm)	7.09	0.06	1.81
Silica (ppm)	40.0	16.6	29.7
Calcium (ppm)	85.44	37.33	49.20
Magnesium (ppm)	40.01	12.67	24.20
Hardness (ppm)	285.3	136.3	213.3
DO (ppm)	7.21	2.17	5.68
Free carbon dioxide (ppm)	34.32	11.62	23.85
COD (ppm)	12.87	3.64	10.79

Table XIII
Analytical Results for Each Night-Soil Treatment

Characteristic	Crude night soil	Supernatant liquor from chemical treatment	Supernatant liquor from anaerobic digestion (30 days)	Supernatant from aerobic digestion	Effluent from contact stabilization (5 days)
BOD (ppm)	12,000	6000	2000	1500	4000
COD (ppm)	3000	1500	2500	2000	1600
Volatile acid, as acetic acid (ppm)	6000	5000	1000	1000	2000
NH_4–N (ppm)	4500	4000	3000	600	3000
PO_4–P (ppm)	1000	Trace	300	—	300
pH	7–9	12	7–9	—	7–9

Table XIV
Biological Classification of Photosynthetic Bacteria and Their Nature

Photosynthetic bacteria	Species	Major hydrogen supplier	Major source of carbon	Requirement of growth factor	Nitrogen fixation	Condition of growth
Chlorobacteriaceae		H_2S, $H_2S_2O_3$, H_2	CO_2	–	+	Anaerobic, light
Green sulfur bacteria	*Chlorobium*					
Thiorhodaceae		H_2S, $H_2S_2O_3$, H_2	CO_2	–	+	Anaerobic, light
Purple sulfur bacteria	*Chlomatium*	Organic matter	Organic compound			
Athiorhodaceae		H_2	CO_2	+	+	Anaerobic, light
Nonsulfur purple bacteria	*Rhodospirillum*					
	Rhodopseudomonas	Organic matter	Organic compound			Aerobic, dark

Table XV
Composition of Photosynthetic Bacteria, *Chlorella*, etc.

Item	Crude protein (%)	Crude fats (%)	Soluble sugar (%)	Crude fibrous (%)	Material ash (%)
Photosynthetic bacteria	57.95	7.91	20.83	2.93	4.40
Chlorella	53.76	6.31	19.28	10.33	1.52
Rice	7.48	0.94	90.60	0.35	0.72
Soybean	38.99	19.33	30.93	5.11	5.68

Table XVI
Amino Acid Composition in Photosynthetic Bacteria, *Chlorella*, and Yeast (gm/100 gm dry wt.)

Item	Photosynthetic bacteria	*Chlorella*	Yeast
Lysine	2.86	2.71	3.76
Histidine	1.25	1.06	0.90
Arginine	3.34	3.24	2.50
Aspartic acid	4.56	4.74	3.11
Threonine	2.70	2.28	2.65
Serine	1.68	2.12	2.75
Glutamic acid	5.34	4.62	6.21
Proline	2.80	2.12	1.77
Glycine	2.41	2.28	2.18
Alanine	4.65	2.98	2.86
Valine	3.51	3.02	3.20
Methionine	1.58	0.27	0.51
Isoleucine	2.64	2.44	2.63
Leucine	4.50	4.46	3.54
Tyrosine	1.71	0.96	1.30
Phenylalanine	2.60	2.65	2.20
Tryptophan	1.09	0.64	0.66
Ammonia	4.01	2.58	5.30

Table XVII
Vitamin B Content in Photosynthetic Bacteria

Vitamin B Group	Rps. capsulatus (γ/gm)	Brewers yeast (γ/gm)
B_1	12	50 ~ 360
B_2	50	36 ~ 42
B_6	5	25 ~ 100
B_{12}	21	—
Nicotinic acid	125	310 ~ 1000
Pantothenic acid	30	100
Folic acid	60	3
Biotin	65	—

Table XVIII

Experimental Results of Night-Soil Treatment by Pilot Plant (September 3, 1967 to October 1, 1967)

Characteristics	Crude night soil	Aeration tank	Pasteurization tank	Photosynthetic bacteria culture harvesting		*Chlorella* culture harvesting		Final effluent
				first stage tank	third stage tank	first stage tank	third stage tank	
Water temperature (°C)	15	38	28	27	28	24	24	18
pH	8.7	8.4	8.2	7.5	8.5	6.8	7.2	7.2
Detention time (hr)	—	24	12	← 72 →		← 72 →		—
COD (ppm)	3824	1570	1410	760	231	134	15	10
BOD (ppm)	9740	2325	2032	1225	297	203	15	15
NH_4–H (ppm)	4126	1718	1326	853	352	182	35	31
Albuminoid–N (ppm)	783	568	436	214	83	46	12	7
Volatile acid (ppm)	2620	6215	6852	3726	312	152	28	20
Cl^- (ppm)	5230	4257	4027	2180	473	238	105	75
Weight of cells (gm/liter) by centrifuge	—	—	—	0.5	7.3	0.5	4.8	—

REFERENCES

Kanto Regional Office (1972). "New Concepts on Water Resources Planning in Kanto Metropolitan Area." Kanto Regional Office, Ministry of Construction. Kanto, Japan.

Kitamura, H. (1972a). Treatment of organic wastewater by photosynthetic bacteria and green algae and their application (I). *J. Ferment. Technol.* **30**, No. 2, 76.

Kitamura, H. (1972b). Treatment of organic wastewater by photosynthetic bacteria and green algae and their application (II). *J. Ferment. Technol* **30**, No. 4, 13.

Kobayashi, T. (1970). Treatment of wastewater by photosynthetic bacteria. *Chem. Biol.* **8**, No. 10, 604.

Ministry of Construction. (1972). "Report on Water Pollution Control Survey in Tokyo Bay." Ministry of Construction. Tokyo.

Editorial Committees of Handbook on Water and Wastewater Treatment. (1973). "Handbook on Water and Wastewater Treatment," 2nd ed. Maruzen, Tokyo.

Japan Industrial Water Association, (1971–1973). "Reports of Experimental Research Works on Reuse of Industrial Water and Reclamation Water of Sewage Treatment Effluent." Japan Industrial Water Association.

13

Water Reuse in South Africa

Oliver O. Hart and Lucas R. J. van Vuuren

I. INTRODUCTION

South Africa's rainfall is erratic, and prolonged droughts are common. Storage reservoirs with capacities exceeding the mean annual runoff are therefore required to provide for subnormal conditions. Water supplies can be augmented by improved water transmission systems, evaporation reduction measures, etc., but this chapter is primarily concerned with the recovery, treatment, and reuse of industrial and domestic effluents, increased efficiency in irrigation practices, and wastewater usage by power stations.

The reuse of effluent for agricultural, industrial, recreational, and indirect or direct municipal purposes already forms an integral part of the country's overall water management practice.

A. Water Pollution Control Legislation

The development of water pollution control legislation in South Africa has been marked by two major acts, viz., the Irrigation Act (Act 8 of 1912) and the Water Act (Act 54 of 1956). The Irrigation Act remained the principal water act for 44 years during which period the country's economy changed from predominantly rural to industrial/agricultural (Bredenkamp, 1964).

The English concept of riparian ownership was introduced in South Africa in 1856 and was confirmed in the Irrigation Act. In this concept, a public stream is defined as "a natural stream of water which when it flows, flows in a known and defined channel, if the water is capable of being applied to the common use of the riparian owners *for the purpose of irrigation.*" Furthermore, water use from a public stream was divided into three categories: (i) primary use—for cattle and domestic use; (ii) secondary use—for irrigation, and (iii) tertiary use—for mechanical and industrial use.

The requirements of the regulations under this act allowed some measure of river pollution as long as the water was not used for power, industry, or mining. It allowed power, industry, and mining to return used water to the stream only when its quality was not less than when it was withdrawn from the stream. These regulations contributed toward the practice of wastewater reuse for irrigation purposes. By far the most important reason why local authorities established extensive irrigation schemes prior to the Water Act of 1956 was that the powers of the Public Health Act of 1936 were used to make the irrigation of effluents compulsory.

The Water Act (Act 54 of 1956) was aimed at the consolidation and amendment of the laws in force relating to the control, conservation, and use of water for domestic, agricultural, urban, and industrial purposes. This act included the following provisions for water pollution abatement:

1. Section 12 deals with industrial use of water and requires that a permit be obtained from the Secretary of Water Affairs for abstraction exceeding 270 m^3 on any one day or 230 m^3/day on a monthly average.

2. Section 21 deals with purification and discharge of industrial effluent. It stipulates that the purification of the effluent to certain standards shall be regarded as an integral part of any industrial process. Wastewater thus purified

must be returned to a natural watercourse unless a permit of exemption is secured. Standards for quality of effluent are regulations which were promulgated in terms of the act. These standards are relatively stringent, e.g., OA* $<$ 10 mg/liter, COD $<$ 75 mg/liter. However, the administration of the standards is given flexibility by the minister having the right to grant relaxation from the gazetted standards, subject to such conditions as he may think fit.

3. Section 22 provides for reuse of purified wastewater, or return thereof to a public stream, subject to the permission of the minister.

4. The Water Act allows some measure of local administration in the form of irrigation boards, but these boards deal solely with irrigation matters in the irrigable area for which they were constituted, and water pollution abatement and control are dealt with by the government through the Department of Water Affairs.

5. The Water Act allows for differential application of effluent standards, with a special (high-quality) standard being applicable to certain upland river catchment areas and a general standard to other areas. This is currently the only classification of rivers.

One of the main reasons for central control by the government is that in a country which receives an average annual rainfall of only 487 mm (410 mm with South West Africa included), water matters must be seen in the broader national interest. The disposition of the country is such that utilization of its water resources requires careful long-term planning, large civil engineering works, and the investment of great sums of public money. The country has therefore enjoyed the advantage of central control of all aspects of the water cycle for some 60 years.

B. Water Balance

The country's average annual rainfall of 487 mm is theoretically equivalent to 1.63×10^9 m^3/day (Menné 1970). Ninety-one percent of the rainfall is lost to the atmosphere by evaporation and transpiration and only 9% reaches the rivers. The runoff of South African rivers is comparatively small and the total runoff of all the rivers is in fact less than the runoff of any one of the major rivers of Africa, e.g., the Zambesi River (du Toit Viljoen, 1970). The assured runoff which can be made available by prodiving storage is 57.3×10^6 m^3/day. Underground supplies are estimated to be 3.1×10^6 m^3/day, giving a total supply of 60.4×10^6 m^3/day.

Irrigation demands can be expected to reach a total of 34.8×10^6 m^3/day by the end of the century. At an estimated rate of increase of 7%/annum, the demand for urban and industrial use is estimated at 45.5×10^6 m^3/day by the

*OA = Oxygen absorbed by wastewater from acid N/80 potassium permanganate in 4 hours at 27°C.

year 2000, giving a total demand of 80.3 × 10^6 m^3/day as against the assured supply of 60.4 × 10^6 m^3/day. This implies a deficit of 20 × 10^6 m^3/day, i.e., some 25% of the potential demand. The assured supply can, however, be increased to 75 × 10^6 m^3/day by increasing the net assured yield of dams and the proportion of the runoff that is conserved. In this case, the deficit would be reduced to 6.6%.

These figures indicate that on the basis of current usage trends, South Africa is heading for a substantial water shortage (Stander and Clayton, 1971). On a regional basis, the problem may become more accentuated and, in certain industrialized areas of the country, further progress is already prejudiced by water shortages. Consequently, the most constructive course of action calls for a critical appraisal of present philosophies and technologies of water resources development and utilization.

The solution to the problem must be sought in better utilization of the available water in which reuse must play a vital part. Thus, if a two-cycle reuse of all municipal effluents could be practiced and if the efficiency of water utilization in irrigation could be increased by 25%, then a 54% surplus in the available supply would be obtained.

The challenges posed by this situation have forced the acceptance of the inevitable fact that water supply, wastewater reuse, and control of pollution are inseparable components in a broad water conservation plan for every metropolitan area, as well as for the country as a whole.

C. Wastewater Technology in South Africa

The reuse concept has led to critical appraisal of water quality criteria and the reliability and accuracy of measuring quality parameters. As a consequence, water quality criteria and standards have been developed to safeguard public health, lake and stream life, and the specific needs of public and industrial water supply. It has also focused attention on the efficiency of conventional water and wastewater reclamation plants and the extent of pollution in the water environment.

Technological progress is continually contributing toward the complexity of water pollution, especially in regard to complex chemical compounds and micropollutants, many of which are biologically and physically intractable with possible harmful physiological effects (Hart and Stander, 1972). The possible effects of these pollutants on public health and aquatic life together with growing concern regarding the effect of plant and plankton nutrient have stimulated research in the development of unit processes to cope with a diversity of pollutants.

For the direct reclamation of domestic wastewater, the research conducted in South Africa was focused on the most economic and efficient unit process for a particular situation. Current research is being conducted on pilot and full-scale facilities at the Daspoort sewage works, Pretoria (Stander and van Vuuren, 1969), where various combinations of biological and physico-chemical unit processes such as the following are under investigation:

Biological nitrification and denitrification
Phosphorus removal using various flocculants
Ammonia stripping with surface aeration and induced draught
Foam fractionation
Recarbonation
Active carbon adsorption
Disinfection using chlorine, ozone, and UV

A full-scale demonstration plant at Daspoort with a capacity of 4500 m^3/day which uses some of these unit processes is shown in Fig. 1.

Design criteria derived from research conducted on the Daspoort plant are currently being implemented in other parts of the country, notably Windhoek and the western Cape.

Fig. 1. Daspoort wastewater reclamation demonstration plant at Pretoria includes combina-ation of biological and physicochemical unit processes.

Table I

Daily Volumes of Sewage Effluents Available from Some Major South African Towns and Industries and Their Usage

Town	Population (1971)	Total effluent volume (m^3/day)	Power station cooling (m^3/day)	Irrigation (m^3/day)	Industrial (m^3/day)
Johannesburg[a]	1,181,321 }	317,800	54,500	113,500	—
Roodepoort[a]	132,970 }				
Durban	716,585	229,300	—	—	10,200
Cape Town	651,090	143,000	9,100	500	—
Pretoria[a]	518,314	86,000	37,600	16,500	—
Port Elizabeth	390,982	88,000	—	6,800	—
Germiston[a]	253,500	52,600	—	8,000	—
East London	205,789	3,300	—	—	—
Bloemfontein	160,000	25,000	6,500	300	—
Benoni[a]	142,630	20,400	—	2,000	2,000
Springs[a]	141,690	29,500	—	—	29,500
Welkom	135,700	40,400	—	—	13,200
Pietermaritzburg	123,031	27,300	—	6,800	—
Kimberley	115,200	8,000	—	1,600	1,100
Carletonville[a]	103,500	4,500	—	4,500	—
Boksburg[a]	95,950	27,300	—	—	—
Vereeniging[a]	93,090	10,200	—	—	—
Krugersdorp[a]	91,100	17,000	—	13,600	—
Brakpan[a]	85,702	7,000	—	—	—
Windhoek[b]	79,000	6,200	—	—	3,000
Kempton Park[a]	71,160	25,500	—	—	—
Klerksdorp[a]	65,050	5,800	—	5,800	—
Randfontein[a]	50,398	4,500	—	1,400	—
Bellville	47,700	9,100	—	—	1,100
Grahamstown	41,375	2,300	—	2,300	—
Worcester	40,590	5,900	—	5,900	—
Westonaria[a]	40,027	6,800	—	6,800	—
Parys[a]	17,357	1,400	—	—	—
Sasol[a]	30,230	23,000	—	—	23,000
Slurry	1,000	2,100	—	—	2,100
Ulco	2,500	100	—	—	100
King William's Town	16,600	2,300	—	2,300	—
Modderfontein	6,400	1,900	—	—	1,900
Total	5,845,581	1,233,500	107,700	198,600	87,200
Percentage			8.7	16.1	7.1
Vaal River triangle					
Total	3,113,989	641,200	92,100	172,100	56,400
Percentage			14.4	26.8	8.8

[a]Situated in the Vaal River triangle.
[b]Average volume reclaimed for domestic use.

Performance data and cost analyses available to date clearly indicate the economic potential of reuse and its rewards in counteracting the deficit in the country's water balance, as well as in preventing pollution (van Vuuren and Henzen, 1972).

D. Availability of Wastewater for Reuse

In South Africa, wastewaters are currently being used for a diversity of applications, but only to a limited extent. Table I shows the average daily volumes of sewage effluent available and their usage in the 20 major cities and towns in the Republic of South Africa, as well as in some other minor towns and industries where sewage effluents are being reused.

It can be seen from Table I that these sources, representing a population of 5.8 million people, produce an average 1,230,000 m^3 of treated sewage effluent per day or 210 liters per capita per day. At present, 31.9% of this effluent is reused: 8.7% for power station cooling; 16.1% for irrigation of crops, parks, trees, and sportsfields; and 7.1% for industrial purposes. This last figure includes the utilization of an average 3000 m^3/day for domestic consumption at Windhoek (van Vuuren and Henzen, 1972).

In the case of the Vaal River triangle, often called the industrial powerhouse of South Africa, the reuse pattern changes drastically, mainly because the water resources for this area are limited. Of the 640,000 m^3 sewage effluent available daily, 50% are being reused; 14.4% for power station cooling, 26.8% for irrigation, and 8.8% for industrial purposes.

II. AGRICULTURAL REUSE

The use of treated sewage effluents for irrigation has been practiced in South Africa for many years without any apparent deleterious effect on crops and soil. There are, however, certain factors that must be considered before embarking on irrigation schemes with sewage effluent:

1. Quality requirements of plants, crops, and livestock with respect to dissolved minerals in the irrigation water
2. The sodium adsorption ratio (SAR) and salinity hazard of the irrigant
3. Soil structure and drainage characteristics
4. Mineralization of seepage and its disposal
5. Maintenance of equilibrium in total dissolve solids in effluents due to bleedoff for irrigation purposes
6. Presence of human or animal pathogens in sewage effluent, in relation

to the system of irrigation practiced, and the barrier effect required against transmissible human disease or zoonosis

7. Pollution of underground waters

To ensure optimum productivity in modern scientific agricultural and livestock farming practices, we need to know the sensitivity of plants and crops to the dissolved minerals in the irrigation water and the water quality requirements of livestock. Table II is a suggested classification of mineral water quality requirements for agricultural use (Henzen, 1970).

The use of mineralized waters for irrigation purposes can lead to excessive water requirements. According to Klintworth (1952), the amount of water required to produce the same amount of crop increases fivefold if the concentration of the total dissolved solids in the irrigation water increases from 1000 to 4000 mg/liter. The additional amount of water is required to leach out the accumulated salts and thereby maintain the threshold concentration total dissolved solids in the root zone of the plants (Fig. 2).

For salt-intolerant crops, such as citrus, an increase of the total dissolved solids in the irrigation water from about 750 to 1450 mg/liter will entail an increase of 28% in the amount of irrigation water required to effect the necessary leaching (Fig. 3).

The sodium adsorption ratio (SAR) is an important parameter for the appraisal of waters for irrigation purposes and represents the sodium hazard of the irrigant. Sodium in effluents used as irrigation water may increase the salinity and thereby the osmotic concentration to a level that will affect plant growth. It may also replace the calcium and magnesium in the soil, which may result in a breakdown of the granular soil structure that will impair the permeability and ultimately result in the formation of alkali soils.

Table II

Mineral Quality Requirements for Agriculture Use

Agricultural use	Total dissolved solids mg/liter	Remarks
Crops	< 500	Satisfactory for all crops
	500–1000	Water that can have detrimental effects on sensitive crops; additional water may be required for leaching
	1000–2000	Water that may have adverse effects on many crops and thus requires careful application practice
	2000–5000	Water that can be used for tolerant plants on permeable soils with careful management practice
Livestock	2500*–5000†	*Threshold value at which poultry or sensitive animals might show slight effects from prolonged use †Animals in lactation or production might show definite adverse reaction

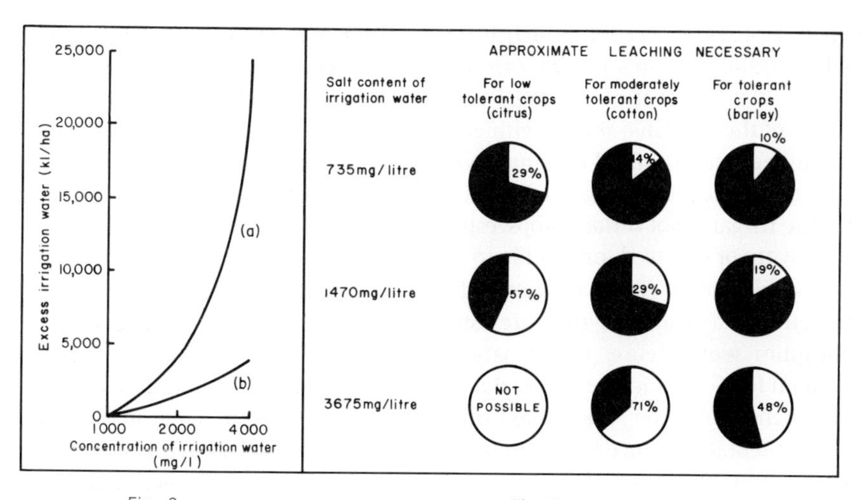

Fig. 2. Additional water (m³/ha) of varying salinity required for leaching to maintain soil solution at 5000 mg/liter (curve a) and at 10,000 mg/liter (curve b).

Fig. 3. Approximate leaching requirements—percentage of irrigation water (white area) that must pass below the root zone to prevent an appreciable reduction in crop yields because of a salt accumulation in the soil.

Good drainage of the soil may be a more important factor for crop growth than the concentration of salts in the irrigation water. Even with excellent water, poorly drained land may reduce crop production, whereas saline waters may often be used on open well-drained soils. When considering sewage effluent for irrigation, good drainage is of paramount importance because the effluent must be irrigated throughout the year, even during periods of excessive rainfall.

In addition to the effect of total salinity, individual constituents in irrigation water may have varying effects on plant growth. The accepted threshold limits for trace elements must also be considered in the appraisal of the quality of effluents for irrigation purposes.

The large variety of methods for demineralization of water are at present still unacceptable economically for irrigation return waters. The most positive approach, when the volume of irrigation return is substantial, is to devise means whereby seepage from irrigation can be isolated from fresh water sources by the creation of saltwater impoundments or by separate drains to the sea.

Irrigation of sewage effluents is an important adjunct of unrestricted reuse, since it provides bleedoff which is essential for maintaining a total dissolved solids equilibrium in a partially closed system. Henzen *et al.* (1970) showed that an equilibrium level of 500 mg/liter total dissolved solids can be maintained if 80% of the available sewage effluent in the Johannesburg metropolitan area is reclaimed and 20% used consumptively for cooling and irrigation purposes.

The presence of pathogenic organisms in sewage effluents precludes their use for the irrigation of certain crops, especially those which may be consumed raw.

"Aware of the need for a guide to water pollution abatement and reuse based on established scientific operational and experimental data, epidemiological patterns, practicability, the interests of public health and the necessity for optimum water reuse, the Department of Health, in consultation with the Council for Scientific and Industrial Research, the Bureau of Standards and the Department of Water Affairs, has formulated certain recommendations for interdepartmental guidance in regard to quality standards and the reuse of effluents. The recommendations, shown in Table III, are not comprehensive, but do serve as a starting-point for future epidemiological studies" (Smith, 1969). It must be emphasized that these standards are recommended criteria only. In practice these standards are seldom enforced except with regard to 1 (i) and 3 (iii) which is strictly observed.

Irrigation with sewage effluents in dolomitic areas is not recommended because underground waters may become polluted. This practice is especially

Table III
Abridged Guide to Recommended Bacteriological Standards for Effluents

Purpose of effluents used	*Escherichia coli* (per 100 ml)
1. *Irrigation*	
(i) Crops eaten raw	Use of effluent not permitted
(ii) Crops unlikely to be eaten raw	1000
(iii) Fruit trees, trellised vines	1000
(iv) Golf courses, sportsfields, etc.	1000
(v) Pasturage grazing	1000
(vi) Pasturage nongrazing	No standard
2. Industrial use	Zero
3. Discharge	
(i) River, stream, etc.	1000
(ii) Sea wave zone	
(a) Bathing beach	1000
(b) Remote from bathing beach	Free of floatable material
(c) Intermediate between (a) and (b)	Free of floatable material
(iii) Beyond wave zone	Each case considered on its merits

dangerous where domestic water supplies are extracted from wells in the vicinity or downstream from the point of application.

The reuse of sewage effluents for irrigation purposes in South Africa has found widespread application throughout the country. Table IV shows the average daily volumes of sewage effluents available and reused at those cities, towns, and industries that utilize some or all of their sewage effluent for irrigation purposes.

It can also be seen that the major portion of sewage effluent reused for irrigation purposes is being applied for pasturage irrigation. The most important of these schemes is that of the City Council of Johannesburg which is dealt with here in some detail.

A. Irrigation Farming—City of Johannesburg (Bolitho, 1970)

The Johannesburg City Council has been using sewage effluent for irrigation purposes since 1914, mainly for the purpose of disposal, because irrigation was cheaper than the installation of purification plants. Today, the Council owns one of the major cattle herds in the republic—totaling over 7500 beef animals of which 3800 are breeding cows. Johannesburg has a northern farm in the Jukskei River catchment and a southern farm in the Klip River catchment. The total area of the farms exceed 6000 ha of which about 1800 ha is under irrigation. A total volume of 113,500 m^3/day sewage effluent are being applied at a rate of 230 cm/annum. This is a rather high application rate with the result that between 40 and 50% of the effluent applied find its way back to the river either as seepage or as runoff, particularly during wet weather. At this application rate, the potential fertilizing value of the effluent (34 mg/liter as N and 23 mg/liter as PO_4) is equivalent to applications of 730 kg of N and 540 kg of phosphorus (as PO_4)/year/ha.

1. CROP PRODUCTION

The irrigated land is divided approximately according to the following four uses:

a. Winter Grazing. Winter pasture consists of Italian ryegrass, tetraploid ryegrass, fescue, and clover. Grazing is controlled by an electric fence, with the intake of each cow strictly controlled to a given weight per day according to the reproductive status of the group. Ryegrass pasture provides the principal source of protein intake, especially in the winter. Cows are grazed at a net intensity of from 15 to 25 animals/ha on such pastures throughout the winter, but the

Table IV
Daily Volumes of Sewage Effluents Reused for Irrigation in South Africa

Locality	Total vol. available (m^3/day)	Vol. reused (m^3/day)	% of total reused	Pasture m^3/day	Pasture area (ha)	Crops (m^3/day)	Parks (m^3/day)	Trees (m^3/day)	Sports fields (m^3/day)	Disposal rate (m^3/day/ha)
Johannesburg	317,800	113,500	35.7	113,500	1,820					62
Cape Town	143,000	500	0.3						500	
Pretoria	86,000	16,500	19.2				200	16,300		
Port Elizabeth	88,000	6,800	7.7						6,800	
Germiston	52,600	8,000	15.2			8,000				
Bloemfontein	25,000	300	1.2			200			100	
Benoni	20,400	2,000	1.0			400	1,000		600	
Pietermaritzburg	27,300	6,800	24.9					6,800		
Kimberley	8,000	1,600	20.0						1,600	
Carletonville	4,500	4,500	100.0					4,500		
Krugersdorp	17,000	13,600	80.0	13,600	100					
Klerksdorp	5,800	5,800	100.0			5,600			200	
Randfontein	4,500	1,400	31.1			1,400				
Grahamstown	2,300	2,300	100.0			2,300				
Worcester	5,900	5,900	100.0	5,900	150					39
Westonaria	6,800	6,800	100.0			6,800				
King William's Town	2,300	2,300	100.0	1,400	77				900	18
Total	817,200	198,600	24.3	134,400		24,700	1,200	27,600	10,700	
Percentage				67.7		12.4	0.6	13.9	5.4	
Industries										
Ngodwana Pulp and Paper		1,300		1,300	54					24
AE & CI Modderfontein		1,100			1,028					1

pastures naturally supply only part (essentially the protein segment) of the nutrient intake.

b. Summer Grazing. Established pastures of an indigenous South African grass *Eragrostis curvula*, with a crude protein as high as 13% when cut before seeding, are mainly used for summer grazing.

c. Summer Hay Production. Summer hay production is also based on *Eragrostis curvula* because of its high response to irrigation and nitrogen. Hay yields of as high as 15 to 23 tonnes/ha are obtained.

d. Maize Silage. Maize silage is grown during the summer on lands which are partly irrigated with sewage effluent and partly with liquid digested sewage sludge. To overcome the problem of high ammonia in the liquid sludge, more pure effluent is applied. Yields are about 78 tonnes/ha.

The total annual crop production on both farms for 1970 was 9000 tonnes hay and 7000 tonnes silage.

The overall carrying capacity of the farms averages at 4.2 animals/ha, including calves.

2. VETERINARY PROBLEMS

Many important veterinary problems arise from the high degree of intensification. When cows and their calves are kept in permanent sleeping camps with an area of 2 to 4 ha every night, considerable risk of the spread of infective bacterial disease and parasite infestation can be expected.

Tickborne diseases, especially redwater (babesiosis), are controlled by weekly passage of all animals through spray races for external parasites, in addition to inoculation. In general, the moist warm conditions not only favor the bacterial calf diseases referred to, but also nematode, cestode, and trematode infestation. These infestations with gastrointestinal roundworms, lung worms, and liver fluke are controlled by regular antihelminthic treatment. The most serious disease has proved to be fascioliasis which is widespread on the farm as a result, principally, of the presence of limnea snail species, especially *Natalensis* and *Truncatula*.

Continual investigations are carried out (a) into all deaths that occur on the farms, and on the slaughter floor, to gather information on the possibility of the occurrence of Johnes disease (to date negative); (b) into fertility problems experienced on the farm by the examination of all genitalia; (c) into fascioliasis; and (d) into trace element surveys of the livers of all culled cows and heifers. In addition continuous investigations are made into the cause of abortions, stillbirths, and prenatal deaths which account for 5.3% of the total births.

B. Irrigation of Industrial Effluents

Following municipal practice, certain industries started to dispose of their effluents by irrigation on established kikuyu grass (*Pennisetum clandestinum*). It was found that this grass not only survived the devastating effect of saline waters, but flourished to such an extent that it has to be grazed by cattle and sheep in order to cope with the prolific growth. This phenomenon led to a scientific investigation by the National Institute for Water Research into the disposal of industrial effluents by irrigation (Murray, 1973).

A series of 48 experimental plots (3.05 m × 1.93 m) were laid out on a fairly level area consisting of a medium-textured soil at least 2 m deep. The irrigant used throughout the plot experiments, which lasted 5 years, was prepared for the purpose in a pulp and paper mill. The consistency of the vàrious batches of effluent varied considerably as shown in Table V, and apart from the mineral content, the effluents had considerable amounts of organic matter in solution.

On 24 of the experimental plots, 8 types of grass and fodder pastures were sown or planted and subjected to three levels of fertilizer treatment, viz. 300 kg superphosphate, 150 kg urea, and 100 kg potassium chloride per ha per year as full treatment. Half treatment consisted of exactly half of this application. The yields obtained over 5 cropping seasons per unit area are presented in Table VI.

The plant that gave the most consistent yields was lucerne and was second in total yield only to antelope, a high-growing reedy grass. The other grasses yielded substantial crops despite the highly saline irrigant. Over the 5 cropping seasons, the total yields for no treatment, half treatment, and full treatment were 197, 242, and 244 kg dry mass, respectively. The full treatment therefore has no significant advantage over the half treatment.

Two soil amendments, gypsum and ferrous sulfate, were applied at various rates with no effect on crop response. The chemical analysis of the soils over 5 years shows, however, that the soil amendments do have a beneficial effect specifically with respect to the SAR of the soil. Of the two amendments, gypsum appears to have a more beneficial effect.

Table V

Analyses of Industrial Effluents Used for Irrigation Plots[a]

Measurement	TDS	Na	K	Ca	Mg	Cl	SO_4	CO_3	pH	SAR
Maximum	5960	1540	38	65	30	1710	320	297	7.3	39.5
Minimum	2020	561	15	77	13	690	146	243	8.3	15.6
Average	3181	823	16	119	26	1070	200	231	7.5	17.7

[a]Measured in mg/liter excepting pH and SAR.

After 5 years of irrigation with the saline effluent, none of the soils, whether treated with amendments or not, showed any signs of structural degeneration or impaired drainage.

1. SOUTH AFRICAN PULP AND PAPER INDUSTRIES—NGODWANA MILL (MURRAY AND HAYMAN, 1972)

The South African Pulp and Paper Industries was granted permission by the Department of Water Affairs to draw more water for their mill at Ngodwana, eastern Transvaal, on condition that no effluent found its way back to the river. The only way by which this restriction could be met was by disposal of the effluent by irrigation. The volumes and composition of the mill effluents are shown in Table VII.

To ensure complete utilization of the irrigated effluents, the Department of Water Affairs advocates a disposal rate of 24 m^3/day/ha. A total disposal area of 54.2 ha is available which could receive 1300 m^3/day of irrigated water according to this specification. The soil is of alluvial origin and consists mainly of brown to reddish-brown clay loam in the surface layer. The soil profile is fairly permeable to water and exhibits a good water-holding capacity. A storage dam with a capacity of 2100 m^3 at the lowest point of the land is capable of impounding 3 days effluent flow.

At the contemplated rate of effluent application on lands established to kikuyu grass, there is little likelihood of a building up of a water table in the soil. The records of analysis of the water of the Elands River upstream and downstream from the sphere of influence of the Ngodwana mill indicate that no pollution of the river has occurred during 4 years of effluent disposal by irrigation on

Table VI
Yields (Dry Mass in kg) of Different Grasses/Unit Area (16.9 m^2) over a 5-Year Period

Type of grass or pasture[a]	1966–1967	1967–1968	1968–1969	1969–1970	1970–1971	Total
Antelope (*Echinochloa pyramidalis*)	27.23	28.44	21.44	44.62	33.04	154.77
Lucerne (*Medicago sativa*)	2.36	17.15	28.03	42.99	49.42	139.95
Nile (*Acroceras macrum*)	2.30	6.70	17.83	25.02	30.78	82.63
Panicum maximum	21.00	8.30	12.43	20.96	13.87	76.56
Eragrostis curvula	5.35	4.04	20.14	16.69	22.31	68.53
Mixed pastures (ryes, clovers, etc.)	1.36	12.20	14.61	18.20	18.05	64.42
Rhodes (*Chloris gayana*)	0.54	3.49	14.79	19.05	18.94	56.81
Kikuyu (*Pennisetum clandestinum*)	9.51	8.51	9.45	10.83	17.50	55.80

[a]All the types except lucerne are frost sensitive or only summer producers. Lucerne can only be grown successfully in temperate climates that are not too hot and when soil depth and drainage are good.

grasslands. It must, however, be noted that this is a frost-free area in which the grass can utilize the irrigation throughout the year.

2. AFRICAN EXPLOSIVES AND CHEMICAL INDUSTRIES—MODDERFONTEIN FACTORY

The nature of the strong effluent (see Section III,D) from the Modderfontein factory of African Explosives and Chemical Industries is such that it is suitable for use as a liquid fertilizer. The liquid has a total nitrogen content of about 3000 mg/liter and is sprayed onto grassland at a rate of 637 kg/ha in four equal amounts of ca. 160 kg (Hyam, 1969). The total amount of liquid applied to the soil per annum is about 390,000 m^3 or about 1000 m^3/day. The scheme is therefore not one of irrigation, but of applying a dilute liquid fertilizer to grassland grazed by cattle. The total area of land treated is ca. 1028 ha, consisting of natural grassland, *Eragrostis curvula*, and kikuyu grass (*Pennisetum clandestinum*).

The carrying capacities obtained from the different types of grasslands for a given period are shown in the following tabulation.

Grassland	Period (months)	Head of cattle/ha
Natural veld	17	1.7
Eragrostis curvula	3	6.8
kikuyu	6	10

The soils on which the effluent disposal scheme is established are essentially sands to sandy loams of granite origin.

The high application rate of nitrogen made it imperative that the balance of elements be maintained in the soil. Factorial experiments showed that significant responses were obtained with phosphorus addition, but none with

Table VII
Average Analysis (April 1972– March 1973) and Volumes of Effluents from the Ngodwana Pulp and Paper Mill (Analysis in mg/liter)

Criterion	"Black" effluent	"White" effluent
m^3/day	592	151
COD	5059	986
OA	936	237
TDS	3625	1201
Suspended solids	551	333
Sodium as Na	559	167

potassium on the natural veld. *Eragrostis curvula*, however, showed an increase in yield of 52% by the addition of 1.3 tonnes/ha/annum of both phosphorus and potassium.

The effluent disposal scheme at Modderfontein Factory has now been in operation for more than 8 years and, as far as the veld is concerned, production is steadily increasing.

III. INDUSTRIAL REUSE

The population and industrial activity of the country is highly centralized. About 20% of the total water consumption in the republic is used by cities, towns, industry, mining, and power generation (Stander and Clayton, 1971). A large portion of this water, estimated to range from 40 to 70%, but excluding the water used by mining and power generation, is not used consumptively, but merely for washing purposes and the conveyance of waste products and sewage.

It is important to note that industry and mining alone produce 40% of the country's gross national product of which 81% is produced by the four main metropolitan centers (du Toit Viljoen, 1970). The most important of these centers is the Pretoria/Witwatersrand/Vereeniging complex, or the Vaal River triangle, which is responsible for about 45% of the industrial production of the country. However, natural water resources in this area are limited, with the result that interbasin transfers and effluent reuse will be inevitable to ensure sufficient water by the end of the century. At present, some 23% of the purified effluent available in the Vaal River triangle is reused directly in industry and for power generation.

Another important center of economic activity is the Western Cape where 11% of the purified sewage effluent available is presently being reused, mainly for power generation. With the exploitation of all the available sources and purified sewage effluent there would, however, be sufficient water for the foreseeable future for this area.

The other important metropolitan centers, viz., the Durban/Pietermaritzburg, and Port Elizabeth/Uitenhage complexes, have a reasonably assured supply of raw water for the future, although economics might swing the balance in favor of reuse of wastewater.

The strongest argument for the reuse of water in industry is the financial benefit that may be derived from reduced water intake and effluent discharge. Consequently, water use and effluent disposal must be considered technically and economically as integral parts of production costs. Water conservation in industry is, however, limited as is apparent from data on water supplied in the

Vaal River triangle in recent years (Laburn and Wells, 1969). In 1966, when water cuts of 25% and later 30% were imposed, a saving of 27% of the estimated unrestricted demand was achieved. Although restrictions were lifted in February 1967, consumption in 1968 was lower than in 1965 and totaled only 82% of the anticipated demand. This demonstrated that, having learned to make do with less water, consumers did not revert to wasteful practices once water became plentiful again. This limited the scope for further savings and when, in 1969, it again became necessary to impose restrictions on water use, the aim of a 25% saving could not be realized and a reduction of only 9% on estimated normal requirement was achieved.

Some aspects of in-plant water economy measures and direct reuse of wastewaters by South African industries are discussed in the following section.

A. Water Economy Measures

There are two basic methods by which industry can reduce their potable water intake; namely, internal reuse of process waters and wastewater reuse. Depending on the nature of the manufacturing plant, a reduction in water consumption of between 50 and 95% can be achieved by internal cascading and recycling as can be seen from the examples recorded in Table VIII (Stander and Clayton, 1971).

The extent to which water consumption is reduced by internal recirculation of process water by some major industries is shown in Table IX.

Table VIII
Water Savings Effected by Planned Reclamation of Process Waters for Internal Reuse

	Water requirements in m^3/tonne	
Industry	Without reclamation and reuse m^3/tonne	With reclamation and reuse m^3/tonne
Fruit and vegetable canning	11.2	5.4
Kraft paper pulp	201	4.0–11.2
Newsprint	116	27
Hardboard	67	33.5
Soap, oils, and fats	54	10.7
Steel	246	5.3–6.7
Glass containers	1.8	0.7

B. Wastewater Reuse for Cooling Purposes

About 67% of the total demand for industrial water, the mining industry excluded, is required for cooling purposes, while 4% is used for steam generation (Henzen and Funke, 1971). The largest consumers of water for cooling purposes are power stations, where cooling water constitutes 90 to 95% of the total water intake.

The extent to which purified sewage effluent is currently used as cooling water for power generation plants is reflected by the statistics in the following tabulation:

Location	Power station capacity MW	Consumption of reclaimed sewage effluent m^3/day
Capetown (Athlone)	150	9,100
Bloemfontein	50	6,500
Johannesburg (Orlando)	300	22,700
Johannesburg (Kelvin)	360	31,800
Pretoria West	250	19,800
Pretoria (Rooiwal)	120	17,800

In using treated sewage effluent for power station cooling, the primary problem is control of the pH-alkalinity and nitrogen-phosphorus relationships. High pH tends to deposit phosphate on the tubes of condensers; on the other hand, low pH associated with low alkalinity and low total hardness indicates potentially corrosive water.

Table IX
Percentage Ratio of Intake Water to Water in Recirculation for Some Major South African Industries

Product	Water recirculated (m^3/day)	Intake water (m^3/day)	Intake (%)
Steel (Iscor, Pretoria)	732,000	21,200	2.9
Thermal power[a] (Komati)	3,928,000	98,000	2.5
Oil from coal (Sasol)	823,000	63,600	7.8
Chemicals (AE & CI, Modderfontein)	491,000	17,200	3.5
	5,974,000	200,000	3.3

[a]All inland power stations employ natural draught wet cooling towers.

The recycling of cooling water is limited by progressive concentration of dissolved minerals. The maximum concentration of dissolved solids which can normally be tolerated in cooling water is about 1200 to 1500 mg/liter. With proper conditioning, however, a total dissolved solids concentration of 4000 mg/liter and higher can be tolerated without undue scale formation or corrosion hazard.

A further problem in using treated sewage effluent for power-station cooling systems is biological fouling as a result of excessive growth of bacteria, fungi, and algae. The growth of gelatinous slimes is stimulated by the presence of nutrients, or by intense aeration and elevated temperatures in cooling towers. Certain types of bacterial deposits may be corrosive and fungi will attack wooden structures, while slimes and deposits in condenser tubes will seriously reduce the rate of heat transfer.

A number of treatment methods and conditioning agents are available which will produce a "stable" cooling water that is neither corrosive nor scale-forming. The most commonly accepted indicators for the determination of the stability of cooling water are the Langelier Saturation Index, the Ryznar Stability Index, and the Phosphate Stability Index. Suitable stability can be maintained by judicious addition of acids. The deposition of scale can be prevented by treating the cooling water with polyphosphate compounds or amino compounds at concentrations between 0.5 and 5 mg/liter. These compounds strongly inhibit the crystallization of calcium carbonate from supersaturated solutions, even at these low concentrations, and are, therefore, capable of preventing the formation of scale. These compounds also serve as corrosion inhibitors by interfering with the corrosion mechanism, mainly by forming a film on all metal surfaces. Combined polyphosphate/tannin compounds are often more effective in preventing scale formation or corrosion than polyphosphate alone.

The stability of treated sewage effluents depends largely upon the alkalinity of the original water supply and the phosphate and nitrogen content of the sewage (Osborn, 1969). The expected concentrations of the latter two parameters are relatively constant and independent of the composition of the carriage water. The main variable is therefore the alkalinity of the water supply and the loss thereof during the purification process. The total loss of alkalinity is 7.14 mg/liter $CaCO_3$ for every 1 mg/liter of nitrogen oxidized to nitric acid. Since all nitrogen compounds are oxidized during aeration in cooling towers, it is clear that loss of alkalinity must occur except when denitrification is specifically provided for.

The incorporation of a large storage pond as part of the cooling circuit of a power station will ensure that unwanted precipitation occurs mainly in the pond. This is because algal activity in the constantly warm water continues at a satisfactory rate and so maintains a high pH level.

Excessive biological growth can be controlled by chlorination or addition of slimicides to the water circuit. Slimicides on the market which are most toxic to the organisms contain copper sulfate or chlorophenates or are mercury-organic compounds. Intermittent shock dosages of different types of slimicides are frequently more effective than maintaining a constant level of the same biocide. Where chlorination is the method used, a shock dosage of up to 25 mg/liter of chlorine may be applied, but the normal rate of dosing is of the order of 5 to 7 mg/liter or even less (Stander and Funke, 1967).

C. Wastewater Reuse in the Pulp and Paper Industry (van Vuuren et al., 1972)

The Enstra Mill of South African Pulp and Paper Industries Ltd., (SAPPI), near the city of Springs was the first large manufacturing industry in this country to utilize purified sewage effluent as the major part of its water supply. This reliance on effluent became a necessity since this mill was established in the Witwatersrand area on account of the favorable market situation, but also because it is an area inherently short in potable water. The present water usage is made up of 16,000 m^3/day from the Rand Water Board and a further 27,000 m^3/day of purified sewage from the Springs Municipality.

Initially, purified sewage from the Springs Municipal wastewater plant which supplied the factory with effluent for process water received only limited tertiary treatment consisting of sand filtration and low-level chlorination.

The demand for a higher brightness paper called for further refinement of the available purified effluent. A quality survey confirmed the presence of heavy metals, particularly iron, manganese, and copper, and organics, which are known to affect paper brightness. Research was then conducted using various adsorbents, oxidants, and flocculants, such as lime and aluminium sulfate, to improve the water quality. These studies culminated in the design and construction of a full-scale advanced treatment plant commissioned in July 1970.

The full-scale plant consists essentially of the following units (Fig. 4): flotation tank of 750 m^3 capacity; feederline (0.6 m) with booster pump; aeration vessel with high-speed disperser and air compressor operating at 10 psig; storage tanks and dosing equipment for aluminium sulfate, sodium hydroxide, and chlorine; and auxiliary equipment such as pH and flow-recorder controllers.

Aluminium sulfate is dosed at 75 mg/liter into the feed-line at a point succeeding the aeration stage. Approximately 1.4 mg/liter of polyelectrolyte is added for improved flocculation and flotation and sodium hydroxide (10 mg/liter) is dosed in the effluent launder to adjust pH value. After addition of 3 mg/liter of chlorine as a measure against algal growth, the effluent is passed through gravity sand filters. Typical operational results are shown in Table X.

The reclaimed water was of such a quality that it could be used in all sections

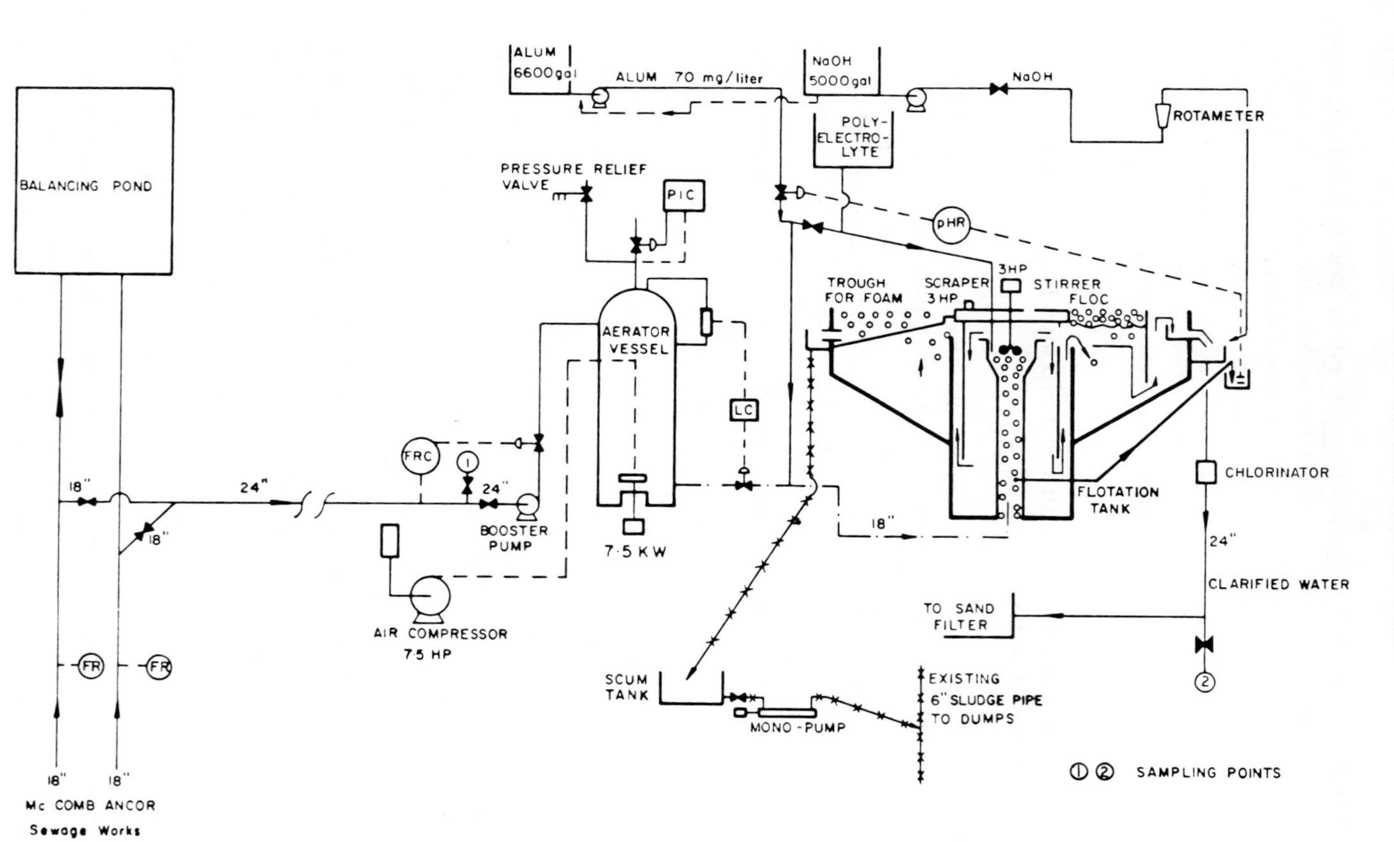

Fig. 4. SAPPI sewage water purification simplified flow scheme.

Table X
Quality of Purified and Reclaimed Sewage Effluent from Full-Scale Plant at SAPPI (mg/liter where Applicable)

Quality	Springs purified sewage	Reclaimed water
pH	7.2	6.7
Conductivity	1100	1200
Color (Hazen)	40	10
Chemical oxygen demand	75	40
Total phosphate (as P)	2.6–6.7	0.6–1.2
Methylene blue active substances	1.0–1.5	0.7–0.9
Iron (Fe)	0.26	0.06
Manganese (Mn)	0.55	0.50
Copper (Cu)	0.45	0.02
Relative paper brightness in Elrepho Units	76.9	84.4
Distilled water in Elrepho Units	85.6	
Rand Water Board water in Elrepho Units	82.4	

of the mill without deleterious effect on the quality of the product paper. Very low turbidities (0 to 1 mg/liter) were carried over from the flotation unit which has a hydraulic retention of less than 30 minutes.

1. COSTS

Average costs over almost 2 years of continuous operation are compared in Table XI, with costs for water as supplied by the Rand Water Board. The annual savings by using reclaimed water instead of Rand Water Board water amounts to R156,000,* which is of the same order as the initial capital expenditure.

The excellent performance record of this plant, the relatively low capital

Table XI
Treatment Costs at SAPPI's Water Reclamation Plant

Cost Item	Cents/m³
Rand Water Board water	3.52
Springs purified sewage	0.22
Operating costs (chemicals, maintenance, and supervision)	1.21
Capital Expenditure	
R159,000 at R14,000/annum, 25 Ml/day for 325 days/annum	0.17
Total cost reclaimed water	1.60

*Rand = 1.408 U.S. dollars (April, 1972).

expenditure and operating costs, and the production of a quality water suitable for pulp and paper production have confirmed the suitability of sewage effluent treated with selected processes to meet the target requirements. A cost saving of more than 50% in relation to conventional supply sources is achieved.

2. MONDI PAPER MILL

The Mondi Paper Company near Durban is another paper mill which uses secondary sewage effluent for process water. The total water intake to the mill at present amounts to about 1600 m^3/day of potable water and 10,600 m^3/day of treated sewage effluent from the Durban Corporation's southern sewage works.

The factory's advance treatment plant has a capacity of 11,400 m^3/day, consisting of chemical flocculation, foam fractionation, sand filtration, and activated carbon treatment. Spent carbon is regenerated on site.

Because this plant has only recently been commissioned, optimum dosing rates and cost figures are not yet available.

D. Wastewater Reuse in the Chemical Industry

The chemical industry is a large user of water in an industrialized country. One of South Africa's largest industrial consumers and also the largest dynamite factory in the world, African Explosives & Chemical Industries (Alexander, 1969), attacked their problem of water use, reuse, and effluent control in an admirable manner. Their Modderfontein factory near Johannesburg had a potable water consumption of approximately 500,000 m^3/month in 1968. These supplies are limited by permit and on two occasions since 1966 have been curtailed due to statutory-imposed restrictions during a period of drought. Intensive recycling and reuse of process waters considerably reduced the factory's total fresh water requirements to 430,000 m^3/month in 1973. Additional utilization of various classes of wastewaters, particularly for cooling purposes, assisted materially in making up further deficits in potable water supplies. Over the period 1962 to 1968, the flow of strong effluent from the factory has been reduced by more than 80%.

At the factory, two separate drainage systems have been constructed. One handles clear water containing low levels of dissolved solids and storm water from factory areas which drain to the fresh water dams I and III; the other is for strong effluents draining to effluent dams IV and V (Fig. 5).

Dams I and III with capacities of 477,750 m^3 and 750,750 m^3 respectively, receive the type of effluents listed in the following tabulation:

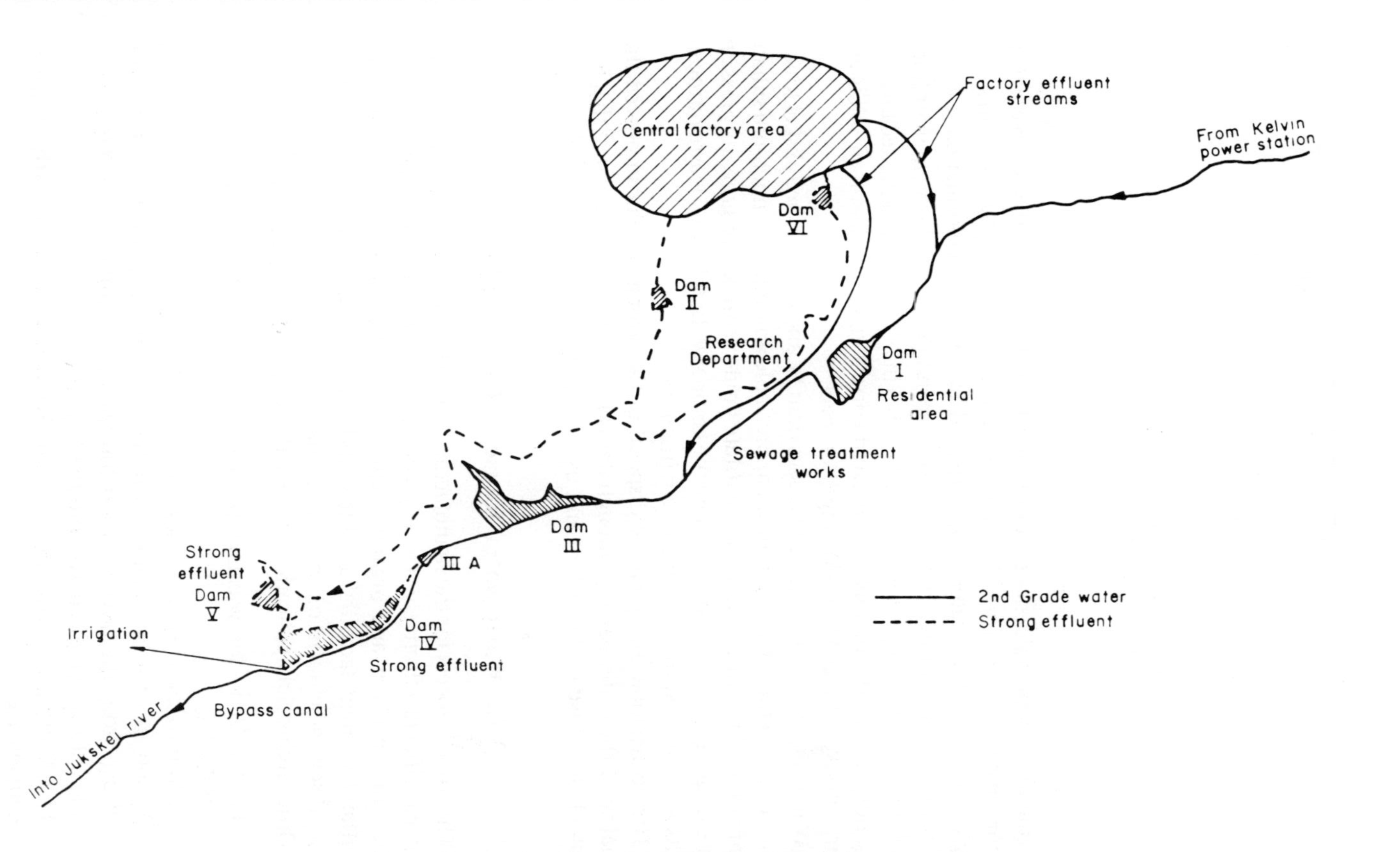

Fig. 5. Layout of effluent dams at Modderfontein.

Type of effluent	m^3/month
Industrial effluent (1 factory only)	1,600
Power Station (Cooling tower blowdown from a municipal power station using treated sewage for cooling)	220,000
Modderfontein sewage works	56,000
Modderfontein factory area	123,000
	400,600

Due to the long residence time of 3 months in dams I and III, some biological improvement occurs (principally microbial denitrification by the dam's ecological system). The water is then treated by marginal chlorination and pH adjustment and returned to the factory's second-grade water system mainly for cooling-water make-up. More than 18,000 m^3/hr of water is recirculated for cooling purposes at Modderfontein. To avoid corrosion, excess dissolved solids are purged out at the 2000 to 3000 mg/liter level.

Water released from dam III bypasses dam IV and flows to the Jukskei River. The stronger effluent, containing an appreciable amount of ammonium sulfate and ammonium nitrate, is used as a dilute liquid fertilizer for growing grass on an extensive area of land owned by the company. In this way, waste nitrogen salts are removed from the water environment.

The segregation and separate disposal of the strong effluent improved the quality of the water passing through dams I and III to such an extent that it is suitable for reuse in the factory and use by riparian owners.

E. Wastewater Reuse in Other Industries

The slurry process for the manufacture of Portland Cement can utilize treated sewage effluent effectively. The effluent quality need not be of a high standard since all the water and organic matter is volatilized in the rotary kilns. The Portland Cement factories at Slurry and Ulco together use 2,200 m^3/day of treated sewage effluent on their slurry processes.

Other industrial application of treated sewage effluents include

1100 m^3/day for washing purposes at the De Beers diamond mines, Kimberley

1100 m^3/day by South African Board Mills at Bellville

23,000 m^3/day by Sasol (the oil-from-coal industry) for the conveyance of ash from the power station followed by maturation pond treatment and subsequent discharge into the Vaal River

13,200 m^3/day by gold mines in the Welkom area, particularly for slurry conveyance

2000 m^3/day for cooling of plate metal and rollers in a steel mill at Benoni

In general it can be said that the major industries in the republic are already water conscious and that these industries are endeavoring to modernize their factories and to keep abreast with new developments to cut down effluent pollution and water consumption.

Unfortunately, water consciousness is largely lacking among industries which discharge their effluents into municipal sewers. This is no doubt inevitable so long as water remains relatively cheap (7.4 to 11.5 c/m^3* to industry). Regulations for the discharge of effluents into municipal sewers are currently still based on limiting the concentration of pollutants, which does not encourage water conservation. The use of reclaimed effluents by industry, therefore, probably constitutes the most effective means for conservation of water.

IV. DIRECT AND INDIRECT REUSE OF WASTEWATER

In regard to domestic use of reclaimed water, an objective appraisal of the question of direct and indirect reuse of wastewaters is very necessary. This is a vexing and controversial issue which, in some countries, has precipitated a stalemate in pollution control, in wastewater reuse, and in the upgrading of conventional water-purification facilities.

It is necessary to state emphatically that the intake water for potable reclamation should consist of predominantly domestic wastewater which has undergone effective oxidative biological stabilization by any of the recognized systems.

The reservations that have been expressed regarding wastewater reclamation for domestic reuse, namely, the risk of infections, chronic toxicity, carcinogenic effects, sex hormone effects, and radiological effects, may apply with equal force to many conventional water purification plants treating natural waters which may be polluted. A critical evaluation of existing conventional water treatment plants is therefore deemed necessary (see Chapter 2).

A. Direct Reuse

During the past decade, the importance of pollution control and water reuse has been strongly emphasized in scientific literature and numerous technological processes have been developed. It is evident that available technology

*100c = 1 Rand = 1.144 U.S. dollars (April, 1976).

can be harnessed successfully to solve pollution problems or to reclaim water at a cost which has to be assessed for a particular situation (Henzen *et al.*, 1972).

In South Africa, the ever-increasing demand for more water, rather than the pollution problem, stimulated extensive research into water reclamation. The first plant for the reclamation of sewage for direct potable reuse was put on stream in Windhoek, South West Africa, during the late sixties (Stander and van Vuuren, 1970). The original planning of and motivation for the Windhoek reclamation scheme, the preliminary investigations, including laboratory tests and pilot runs, were reported in several earlier publications (Cillié *et al.*, 1967; van Vuuren *et al.*, 1971). The plant was commissioned in October 1968, and up to the end of 1970, has produced 1.8 million m^3 of reclaimed water. The public's favorable acceptance of the reclaimed water is ascribed to the progressive disclosure of information via the press and invitations to visit the plant (see Fig. 6).

The reclaimable treated sewage effluent is derived from a conventional sewage treatment plant followed by further biological purification in nine maturation ponds in series with a total hydraulic retention of 14 days. Variations in flow from the maturation ponds are balanced out by utilizing the available

Fig. 6. Windhoek wastewater reclamation plant in South West Africa. The effluent has been used to supplement the municipal water supply.

freeboard capacity in the ponds and throttling the inlet point to the reclamation works. The reclamation plant has a nominal capacity of 4500 m^3/day and, essentially, it comprises additional facilities for algae flotation, foam fractionation, breakpoint chlorination, and adsorption of organics onto activated carbon.

A surface water supply undergoes purification in an existing conventional plant comprising clarification, filtration, and disinfection. The conventional plant is situated side by side with the reclamation plant. The reclaimed water, after carbon adsorption, is mixed with purified surface water and the admixed streams are then postchlorinated to a free residual of 0.2 mg/liter. A gravity-flow system is used throughout the plant except for the active carbon filters.

1. PERFORMANCE OF PLANT DURING FIRST SIX MONTHS OF OPERATION

a. Virological and Bacteriological Efficiency. During a strict monitoring program followed over the course of several months, an average of 20 $TCID^{*}_{50}$/ml virus was found in the settled sewage entering the sewage purification works. The cytopathic effect of the viruses isolated resembled those of enterovirus and reovirus. Reovirus were also isolated from the maturation pond effluent. Beyond the foam fractionation stage, no virus was ever detected in any of the hundreds of samples tested. This was also the case in samples taken from the distribution system. Control samples were sent by air to H. I. Shuval, of the Hebrew University, Jerusalem, for independent confirmation of results.

The multiple safety barriers (prefilter breakpoint chlorination, activated carbon adsorption, and post-chlorination) built into the reclamation plant effectively ensured that no viable virus remained in the fully treated water.

Escherichia coli (MPN *E. coli*) and membrane-filtered *E. coli* I were never recorded after breakpoint chlorination, whereas the total count was drastically reduced within acceptable low levels. On several occasions, however, a significant increase in total count was recorded after carbon adsorption, but postchlorination was sufficient to reduce total counts below 100 per ml. (Fig. 7).

b. Chemical and Biochemical Quality. The total dissolved solids of the reclaimed water increased from approximately 740 to 810 mg/liter, mainly as a result of chemical dosing. Due to changes in alkalinity and algal density, the aluminium sulfate dosing requirements varied between ca. 130–220 mg/liter. This obviously affected the increase in TDS of the reclaimed water. Depending on the ratio of admixture of reclaimed water with purified raw water having a TDS of about 200 mg/liter, the TDS of the blended water could be controlled well below 600 mg/liter.

*Tissue culture infectious doses.

The COD of the maturation pond effluent was reduced by some 70% up to the sand filter stage, with a substantial further reduction by carbon adsorption. After optimum operational conditions were established, a final quality reclaimed water could be produced with COD less than 10 mg/liter and ABS of less than 0.2 mg/liter (see Figs. 8 and 9) before blending with purified raw dam water.

In the case of permanganate values and BOD, the major reductions occurred in the flotation, sand filtration, and carbon adsorption stages. Under the most favorable operational conditions, an overall reduction of 93% and 98%, respectively, was achieved.

The carbon-chloroform-extractables of 0.13 mg/liter was within the acceptable standards.

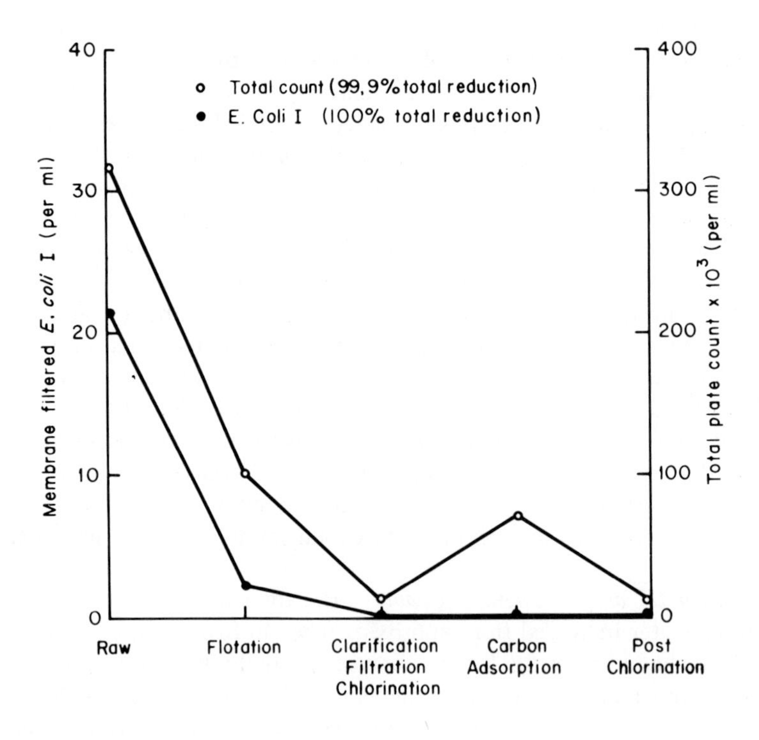

Fig. 7. Residual bacterial counts at various process stages of Windhoek reclamation plant.

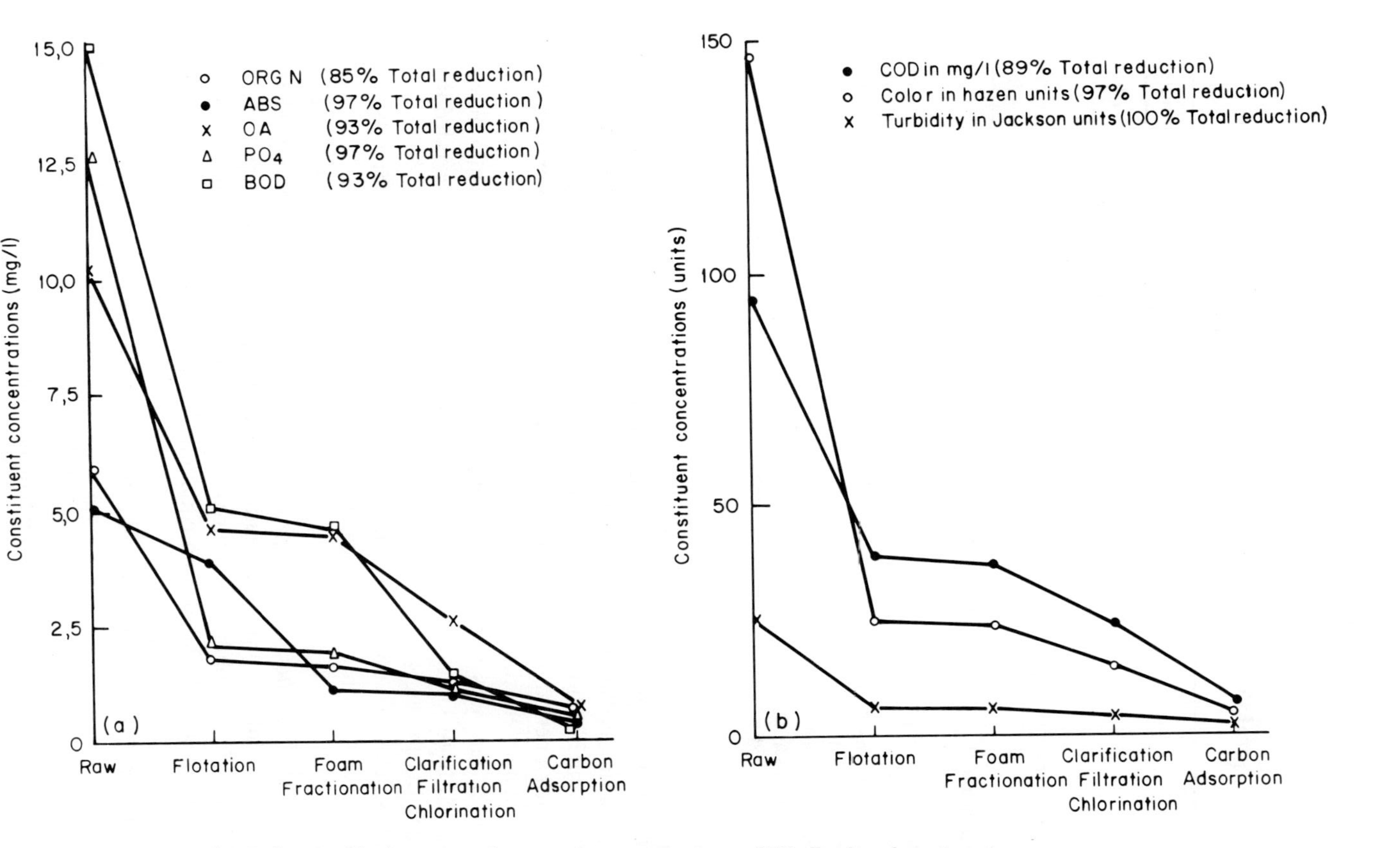

Fig. 8a,b. Residual concentration at various process stages of Windhoek reclamation plant.

Table XII
Typical Composition of Waters from Individual and Combined Sources at Windhoek (mg/liter)

Constituent	Treated dam water A	Reclaimed water B	Borehole water C	Dam and reclaimed water mixed in ratio 1:1 = (A + B)	Reticulated water: Before integration 1:1 = (A + C)	Reticulated water: After integration 1:1:1 = (A + B + C)
Total dissolved solids (TDS)	186	740–810	498	525–609	375	475
Phosphate as PO_4	0.1	0.5–1.2	0.03	0.4	0.4	0.6
Nitrate as N	0.3	18.0	0.0	9.0	0.9	3.3
Synthetic detergents as Manoxol O.T. (ABS)	—	0.05–0.40	0.0	0.10	—	0.1
Chemical oxygen demand as COD	18	7.5–28	7.0	7.5–34	8–12	11.0–14.5

The TDS in the distribution system increased from 375 to 475 mg/liter after integration (see Table XII). This increase, however, depended on the proportions of dam and borehole water present in the reticulated water. These figures coincide with a period when a high proportion of borehole water was extracted as a result of a severe shortage of dam water.

The nitrate was of the order 3 mg/liter after integration, while the COD changed from 11 to 14 mg/liter and no significant change was recorded in the 20-day BOD values.

Phosphates were increased from 0.4 to 0.6 mg/liter and a final ABS of 0.1 mg/liter was recorded after integration.

c. Mineral Buildup. The large consumptive use of water at Windhoek (65% of the total intake) provides a natural bleedoff to prevent a buildup in salinity.

The buildup of soluble substances (total dissolved solids) in such a semiclosed system can be calculated from equations for the material balance pertaining under equilibrium conditions.

Thus the TDS concentration of recycled sewage effluent is

$$x + y/(1 - a)$$

and the TDS concentration of the reticulated water is

$$x + ay/(1 - a)$$

where a = fraction of reticulated water returning as sewage, x = TDS concentration of existing supply, mg/liter, and y = salt pickup in sewage and treatment process, mg/liter.

By inserting applicable values in these expressions, it was calculated that the TDS concentration of the reticulated water would increase by 160 mg/liter when purified sewage effluent is recycled continuously. The observed buildup of 100 mg/liter supports the validity of this expression.

2. CONTINUED PERFORMANCE AND COST EVALUATION

Subsequent to the first 6 months of operation, research at Windhoek was directed toward a continuous monitoring program, particularly with respect to pathogenic and biochemical quality, as well as epidemiological surveys in relation to the reclaimed and other water supply sources in the area.

During the first 2 years of operation, the reclamation plant has contributed an average of 13.4% of the total water consumption (van Vuuren and Henzen, 1972).

The average inflow of raw sewage during this period was 6200 m^3/day, which is equivalent to a capacity utilization factor on the nominal design output of a little over 50%. The low utilization performance is attributed largely to interruptions in the operations caused by mechanical failures, process losses, recharging of carbon filters, and cessation of operations during the winter months due to high ammonia concentration levels at which the existing chlorination facilities were inadequate to ensure breakpoint chlorination.

The Windhoek water reclamation plant operated from October 1968 until the end of 1970. Toward the beginning of 1971, the load on the conventional sewage treatment works had increased to such an extent that the quality of the maturation pond effluent did not comply with the water quality specifications of the reclamation plant. Reclamation of purified sewage effluent was therefore stopped temporarily pending completion of expansions to the conventional sewage-treatment facilities. Modifications to the reclamation plant are currently being carried out with the view to achieving a higher ultimate utilization factor.

During the entire period of operation, the quality of the product water complied with World Health Organization Standards. Hence, the unit costs of reclaimed water were higher than would be the case for water of lower quality for nonpotable purposes.

The costs obtained for the first 2 years of operation are given in Table XIII. These costs do not include that of conventional sewage treatment and are based on the actual utilization factor of 50%. For an 80% utilization factor, the total unit cost would drop to 9.19 c/m^3. The latter figure compares favorably with the unit cost of 10.7 c/m^3 for conventional treatment of water from supply sources fed by natural runoff in this area. It should be taken into account that for geographic reasons unit costs for chemicals and powers are relatively high in the Windhoek area.

Table XIII
Costs for Reclaiming Potable Water from Maturation Pond Effluent at Windhoek

Cost item	Cents/m^3
Capital costs	3.88
Labor	0.89
Chemicals	2.87
Activated carbon	2.40
Specialized supervision	1.50
	11.54

The combination of the individual processes of algae flotation (using aluminium sulphate as flocculant), foam fractionation, breakpoint chlorination and granular activated carbon adsorption followed by postchlorination proved to be an economic system for the treatment of this type of effluent, provided the ammonia concentrations in the influent were maintained at levels below about 7 mg/liter. The low utilization factor referred to, as well as the inability of the conventional sewage treatment facilities and attendant maturation ponds to provide an effluent fit for reclamation at all times of the year, particularly from a chlorination point of view, impose a constraint on the cost evaluation.

The Windhoek plant has demonstrated that a high quality potable water can be produced by using maturation pond effluent as an intake source to supplement other supply sources.

Although the critical water supply situation at Windhoek did not allow a long and detailed epidemiological study before the introduction of reclaimed water, data were available from hospital and laboratory records on incidences of typhoid, salmonella, shigella, and infectious hepatitis cases as far back as 1964. To date, there has been no significant change in the disease pattern of the area.

On routine monitoring, no enterovirus or reovirus was isolated at any time from the final effluent (10-liter samples) or from the effluent of any of the units of the wastewater reclamation plant after breakpoint chlorination. Enterovirus was, however, isolated from the raw surface water supplies and from the product of the water purification plant treating these supplies. On one occasion, one enterovirus isolate (less than 1 $TCID_{50}$/liter) was found in the mixed water, before final chlorination at the reservoirs. It is to be noted that the mixed water received chlorination prior to distribution to the city and, at no time, was enterovirus found in the distributed water. Exhaustive evaluation of the aforementioned occurrence proved the necessity of installing breakpoint chlorination facilities in the water purification system which treated the surface water supply.

It also confirmed the provision of breakpoint chlorination facilities in the design and construction of the reclamation plant circuit before the settling tanks and the maintenance of a free residual chlorine of 0.5 mg/liter after sand filtration prior to the carbon filters. This particular aspect was further confirmed by exhaustive tests after the normal closing of the plant during the winter months, when the water demand was low and ammonia concentration in the maturation pond effluent too high for economic breakpoint chlorination. When the reclamation plant was operated with an ammonia load beyond its breakpoint capacity and the water bypassed to the river, enterovirus (1 $TCID_{50}$/liter) was isolated in one of the eight carbon filters, which proves the need for the maintenance of breakpoint chlorination and a free chlorine residual in units prior to carbon filtration.

The free residual chlorine, pH, and turbidity are furthermore regarded as critical parameters for effective breakpoint chlorination to destroy viruses.

Before wastewater can be reclaimed for domestic reuse in a particular type of advanced purification plant like that at Windhoek, it must undergo extended biological purification to yield an effluent of predetermined quality suitable for intake to the reclamation facility. In the case of Windhoek, special care was taken to separate abattoir effluent from the domestic sewers. (For a detailed discussion of the pros and cons of water reuse for domestic purposes see Chapter 2.)

B. Indirect Reuse

In South Africa several examples can be quoted where indirect reuse of domestic effluents is practiced. In the Witwatersrand/Pretoria area, a substantial flow (321,000 m^3/day) of purified sewage, which is not utilized for industrial or agricultural reuse, finds its way in natural streams and water courses and, thus, indirectly into domestic water supplies. The increasing signs of eutrophication and deterioration in chemical and biochemical quality in these catchments are a matter of grave concern and, in some instances, the conventional water purification plants have reached the stage where advanced techniques such as activated carbon treatment have become a dire need. The major problems encountered are the presence of algae and aquatic plants and the occurrence of dissolved organics, nitrogen, and phosphorus. The following case histories demonstrate typical examples of the extent to which treated sewage effluents are used indirectly.

1. HARTBEESPOORT DAM

This impoundment reservoir essentially serves as an irrigation supply source for an area of about 13,400 ha (Meiring and Liebenberg, 1970). It also serves as a recreational area for aquatic sport such as angling, boating, and skiing. The township of Brits (20,000) and several smaller communities obtain water for their domestic and industrial needs from this impoundment.

The average flow of the Crocodile River feeding this dam is 402,000 m^3/day. The Johannesburg metropolis within the catchment of the Crocodile is expanding rapidly and approximately 47,000 m^3/day of purified sewage is disposed of into this catchment. An average 10% of treated sewage enters the Hartbeespoort Dam. This percentage is, however, increasing rapidly and during the dry season can be as high as 50%. The nutrient load is reduced by storage in the dam, but approximately 1.7 mg/liter of P and 1.4 mg/liter of N still occur, on average, in the water extracted for irrigation and domestic purposes. These nutrients obviously do not pose a problem regarding the quality necessary for irrigation purposes. However, as a source for domestic water supply, these nutrients cause operational problems in the conventional treatment works and, on several occasions, public complaints have been received about the taste and odor of the domestic water supply.

The existing water treatment works at Brits, completed in 1958 with a capacity of 3500 m^3/day, has become inadequate to meet growing demands in this rapidly developing area. The works is based on a conventional design comprising clarification, filtration, and disinfection. The major problem encountered is that this plant cannot cope efficiently with the high incidence of algae blooms that is characteristic of the raw water for the greater part of the year. A new treatment plant is currently under construction where breakpoint chlorination and activated carbon units will be included in the overall treatment and where more efficient removal of algae is envisaged.

2. RIETVLEI DAM

This impoundment receives approximately 25,000 m^3/day of purified sewage from the township of Kempton Park together with an average 25,000 m^3/day runoff. Treated sewage effluent therefore constitutes an appreciable proportion of the inflow into the dam which contributes to the potable supply for the city of Pretoria. The proportion of treated sewage in this dam is therefore higher than that for the Hartbeespoort Dam, but indications are that the self-purification capacity of the river is higher than that of the Crocodile River.

Over the past years, the problem of increased eutrophication has also had a significant effect on the control and operation of the conventional water treatment plant (supplied by the Rietvlei Dam), which has a capacity of 25,000 m^3/day. Clogging of the rapid gravity sand filters by algae is a common occurrence that contributes to overall treatment costs.

Investigations are currently underway to improve algae separation by the addition of ferric salts and polyelectrolytes. Another study planned is on phosphorus removal from the purified sewage at source by using chemical addition. Quality data on the purified sewage and the water in the dam are depicted in Table XIV.

It is evident that very low levels of phosphorus can still cause eutrophication and affect the subsequent performance of conventional water-treatment facilities.

3. APIES RIVER AND BON ACCORD DAM

This is a relatively small stream of natural runoff which is heavily polluted along its course with purified sewage, agricultural seepage water, and industrial effluent. During the dry season, the upper reaches are essentially fed by purified sewage from the Daspoort sewage works (19,000 m^3/day) and effluent from a power station and steel foundry ($\pm$10,000 m^3/day). This water enters the Bon Accord Dam at a rate of about 45,000 m^3/day. The average retention in the dam is 4 to 5 months which allows for some degree of self-purification, however, this dam also shows marked signs of eutrophication.

During the dry season, a minimum of 3500 m^3/day of compensation water is released and during the wet season the dam spills over. About 10 km downstream of the dam, purified sewage from the Rooiwal sewage works, 15,000–20,000 m^3/day is allowed into the river. This river is the supply source for a purification works for a township of 18,000 inhabitants, at a point 20 km downstream.

In planning this water purification plant during 1964, account was taken of the inferior quality of the raw water, particularly the organic constituents.

Table XIV
Quality of Sewage Effluent from Kempton Park and Water in the Rietvlei Dam (mg/liter where Applicable)

Parameter	Rietvlei Dam	Kempton Park sewage effluent
pH	8.51	8.0
Total alk. ($CaCO_3$)	199	204
COD	25	51
NH_3(N)	0.5	4.9
Kjeldahl (N)	1.2	6.0
NO_2(N)	<0.1	0.2
NO_3(N)	0.2	0.4
PO_4(P)	0.2	7.6
Cl	35	63

Surveys indicated the presence of 1–3 mg/liter of detergents and COD of about 30–40 mg/liter. The presence of 5 mg/liter of P and 6 mg/liter of $NO_3(N)$ confirmed the presence of a substantial proportion of purified sewage.

The water-purification plant has a capacity of 4500 m^3/day and includes facilities for lime and alum flocculation, sedimentation, breakpoint chlorination, rapid gravity filtration, and active carbon-adsorption treatment.

The purified water is of acceptable potable quality, but chemical treatment costs are relatively high due to activated carbon treatment. Phosphorus is reduced below the 1 mg/liter level and COD below 15 mg/liter. Presumptive *E. coli* I has never been detected in the product water.

These case histories of indirect reuse of domestic effluents are typical of several other catchments in South Africa. In terms of quantity, these examples represent relatively small volumes of indirect reuse as compared with the Vaal Barrage which receives the major part of sewage effluent from the Johannesburg metropolis and Vaal Triangle. Because of the high dilution factor in the latter, the eutrophication effect is, however, less pronounced, although it is beginning to cause concern. The quality of runoff water in the Vaal Barrage reservoir has, however, deteriorated significantly over the past decade. Trends for future development have drawn attention to the need for higher quality standards for domestic effluents and more sophisticated sewage-treatment plants in order to reduce pollution loads in this highly industrialized area. The implementation of activated sludge systems based on the principle of nitrification and denitrification and the chemical precipitation of phosphorus are envisaged for future and existing treatment plants.

4. RECREATIONAL REUSE

Direct reuse of sewage for recreational purposes is not practiced in South Africa. The presence of sewage effluents in impoundment reservoirs and rivers, as just described, could perhaps be regarded as a mode of indirect recreational reuse. The reason for this approach in South Africa should be obvious in the light of the pressing need for more water for unrestricted reuse, rather than for a limited application such as recreation. Therefore, research in South Africa is primarily focused on direct wastewater reclamation for industrial and domestic purposes and, as a secondary objective, the prevention of pollution of the water environment.

V. FUTURE PLANNING

The water resources of the Republic of South Africa have hitherto been utilized primarily for irrigation. Up to 83.5% of the total supply is currently consumed by agriculture. According to present evidence, there is no urgent need for additional large-scale allocation of water for this use and the emphasis is on increased yields of the land already under cultivation. Therefore, it is anticipated that by the year 2000 the share of irrigation will be reduced to about 45%.

The position is different in regard to urban and industrial usage of water. With planned decentralization of economic activity and strong emphasis on optimum reuse, there should be sufficient water to meet the increased requirements of urban areas. Reuse is of particular importance to the southern Transvaal industrial complex which is responsible for 45% of the industrial production of the country. A second important center is the western Cape, where the consumption is expected to increase to 1.2 million m^3/day by the end of the century.

About 50% of the available effluents are already being reused in the southern Transvaal compared with only about 11% in the western Cape. The potential of reclaimed water is thus relatively unexploited in the western Cape. Statistics indicate that existing raw water supplies are already fully utilized and interbasin transfer is possible only at high costs. Current research is, therefore, intensively directed toward future wastewater reclamation systems for the area.

An interesting new approach being pursued is the utilization of the vast underground storage capacity of sand beds in the Cape Flats area as an evaporation-free underground reservoir (Cillié and Henzen, 1970). This area is ideal as a storage reservoir for reclaimed water because of its suitable hydrological characteristics. Research has reached the stage where a 4500 m^3/day reclamation and recovery plant is now being designed that will serve as a research and demonstration plant from which full-scale design criteria could be derived. Unit processes under study are essentially similar to those at Daspoort, Pretoria, with the addition of infiltration and extraction facilities.

In view of the urgent need to preserve a high quality of river water in South Africa, it has become imperative that advanced treatment, reuse, and pollution abatement at the source be considered seriously in planning for future development.

Optimum utilization, reclamation, and reuse of water can be effected most efficiently when town and regional planning is undertaken with a view to the principles involved in the former.

Future design of water supply works will have to embody improved purification techniques to take care of the increasing occurrence of trace toxicants in raw water supply. Fortunately, such techniques have already been developed for wastewater renovation. Therefore, it may be expected that water and sewage treatment systems will become even more akin in future.

VI. CONCLUSIONS

South Africa is not richly endowed with vast water supplies. Since the competitive demands for water will put an ever-increasing premium on its value, one can expect that technological research will provide new processes for the reclamation of water by industries and municipalities. For a country as short of water as South Africa, it is absolutely necessary to proceed with major water research efforts to promote wastewater reuse on a national basis. Furthermore, the state has established a National Water Research Commission to promote such work. It seems logical to conclude that the continuity in the current explosive socioeconomic progress in South Africa will only be ensured by effective water resources exploitation through reuse.

From the statistics quoted and case histories cited in this chapter, there appears to be much scope for a broader application of wastewater reuse in South Africa. This will not only relieve pressures on present and future demands on fresh water supplies, but, equally important, will also preserve the quality of our water environment.

ACKNOWLEDGMENTS

The authors wish to express their gratitude to all the authors listed in the references from whose publications subject matter was drawn in preparing this chapter. A special word of thanks is due to the adjudicating panel and other staff members of the National Institute for Water Research for advice and assistance. Special thanks to Mr. V. Bolitho for valuable comments and editing of the chapter.

Finally, the authors are indebted to the director of the institute for granting permission to prepare this chapter and for his guidance in its preparation.

REFERENCES

Alexander, A. T. (1969). Control of liquid effluent from African Explosives and Chemical Industries Limited, Modderfontein Factory. *Symp. Chem. Contr. Hum. Environ., 1969* Paper No. 39.

Bolitho, V. (1970). "Irrigation Farming," Rep. City Engineer's Department, City of Johannesburg.

Bredenkamp, C. S. (1964). "Water Pollution Abatement" (A literature survey of conditions in other countries and a short summary of the historical development in South Africa), Intern. Rep. Natl. Inst. Water Res., S. Afr. Counc. Sci. Ind. Res., Pretoria.

Cillié, G. G., and Henzen, M. R. (1970). Die hergebruik van water in die Kaapse Skiereiland. *Tegnikon* **19**, No. 4, pp. 15–24.

Cillié, G. G., van Vuuren, L. R. J., Stander, G. J., and Kolbe, F. F. (1967). The reclamation of sewage effluents for domestic use. *Proc. Int. Conf. Water Pollut. Res., 3rd, 1966* Vol. II, pp. 1–19.

du Toit Viljoen, S. P. (1970). Review of the future water requirements of the Republic of South Africa. *In* "Convention: Water for the Future, Water Year, 1970," pp. 205–209. Pretoria, Republic of South Africa.

Hart, O. O., and Stander, G. J. (1972). The effective utilisation of physical-chemically treated effluents. *Conf. Appl. New Concepts Phys. -Chem. Wastewater Treat., 1972* pp. 95–102.

Henzen, M. R. (1970). Mineral pollution: South Africa's greatest problem in water quality protection. *In* "Convention: Water for the Future, Water Year, 1970," pp. 417–425. Pretoria, Republic of South Africa.

Henzen, M. R., and Funke, J. W. (1971). The appraisal of some significant problems in the re-use of waste water. *J. Inst. Water Pollut. Contr.* **70**, 177–186.

Henzen, M. R., van Vuuren, L. R. J., and Stander, G. J. (1970). The current status of technological developments in water reclamation. *Symp. Dissolved Solids Loads Vaal Barrage Water Syst., 1970* Paper No. 7.

Henzen, M. R., Stander, G. J., and van Vuuren, L. R. J. (1972). Implications of the mineral load in the Vaal Barrage water system on waste water reclamation. *Proc. Int. Conf. Water Pollut. Res., 6th. Workshop Sess., 1972* Paper No. 23.

Hyam, G. F. S. (1969). Disposal of factory liquid effluent—an agricultural scheme. *Symp. Chem. Contr. Hum. Environ., 1969* Paper No. 89.

Klintworth, H. (1952). "Die Gebruik van Brakwater vir Besproeiing," Pam. No. 328. Union of South Africa, Department of Agriculture, Govt. Printer, Pretoria.

Laburn, R. J., and Wells, R. J. (1969). The supply and treatment of water for the Pretoria-Witwatersrand-Sasolburg region. *Symp. Chem. Contr. Hum. Environ., 1969* Paper No. 51.

Meiring, P. G. J., and Liebenberg, D. P. (1970). Beplanning van 'n nuwe waterwerke vir Brits met inagneming van die verryking van die water van die Hartbeespoortdam. *In* "Convention: Water for the Future, Water Year, 1970," pp. 591–598. Pretoria, Republic of South Africa.

Menné, T. C. (1970). Surface water potential of South Africa. *In* "Convention: Water for the Future, Water Year, 1970," pp. 233–242. Pretoria, Republic of South Africa.

Murray, G. (1973). "Irrigation of Industrial Effluents," Intern. Rep. W 6/454/1. Natl. Inst. Water Res., S. Afr. Counc. Sci. Ind. Res., Pretoria.

Murray, G., and Hayman, J. P. (1972). "Investigation of the Proposed Extension to the Effluent Disposal Area at the Ngodwana Pulp Mill of the South African Pulp and Paper Industries," Intern. Rep. W 13/218/3. Natl. Inst. Water Res., S. Afr. Counc. Sci. Ind. Res., Pretoria.

Osborn, D. W. (1969). Factors affecting the use of purified sewage effluent for cooling purposes. *J. Inst. Water Pollut. Contr.* **69**, 456–464.

Smith, L. S. (1969). Public health aspects of water pollution control. *J. Water Pollut. Contr. Fed.* **41**, 355–367.

Stander, G. J., and Clayton, A. J. (1971). Planning and construction of waste water reclamation schemes as an integral part of water supply. *J. Inst. Water Pollut. Contr.* **70**, 228–233.

Stander, G. J., and Funke, J. W. (1967). Conservation of water by re-use in South Africa. *Chem. Eng. Progr., Symp. Ser.* **63**, No. 78, 1–12.

Stander, G. J., and van Vuuren, L. R. J. (1969). Municipal reuse of water. *J. Water Pollut. Contr. Fed.* **41**, 355–367.

Stander, G. J., and van Vuuren, L. R. J. (1970). The reclamation of potable water from waste-water. *Water Resour. Symp.* **3**, 31–48.

van Vuuren, L. R. J., and Henzen, M. R. (1972). Process selection and cost of advanced waste-water treatment in relation to the quality of secondary effluents and quality requirements for various uses. *Conf. Appl. New Concepts Phys.-chem. Wastewater Treat.*, 1972 pp. 371–383.

van Vuuren, L. R. J., Henzen, M. R., Stander, G. J., and Clayton, A. J. (1971). The full-scale reclamation of purified sewage effluent for the augmentation of the domestic supplies of the city of Windhoek. *Proc. Int. Conf. Water Pollut. Res., 5th, 1970* pp. 1-32/1–32/9.

van Vuuren, L. R. J., Funke, J. W., and Smith, L. (1972). The full-scale refinement of purified sewage for unrestricted industrial use in the manufacture of fully bleached kraft-pulp and fine paper. *Proc. Int. Conf. Water Pollut. Res., 6th, 1972* pp. 627–636.

14

Water Reuse in the United Kingdom

G. E. Eden, D. A. Bailey, and K. Jones

I. WATER ECONOMY OF THE BRITISH ISLES

Despite their reputation for rainy and unsettled weather, the British Isles are by no means overendowed with water resources. It has been pointed out by the Central Advisory Water Committee (1971) that the average amount of water available in England and Wales is about 850 Imp gal (3.86 m^3) per head per day—an amount that is among the lowest in Europe, only Belgium, Eastern Germany, and Malta having less. The average precipitation over England and Wales is 35.59 in. (904 mm), ranging from more than 100 in. (2540 mm) over parts of North Wales and the Lake District to about 20 in. (508 mm) in parts of Southeast England. Taking into account losses by evaporation (about 50%), the runoff can be less than 5 in. (127 mm) in a large area of Southeast England: In an extremely dry year, as little as 1 in. (25 mm) of runoff has been reported (Collinge, 1967).

This situation is aggravated first by the tendency of the population to drift from the countryside to the towns (over one-third of the population lives in the six recognized conurbations) and, second, by the growth of population in the Southeast, which now contains over 40% of the population of England and Wales. At present, the volume of water used daily in England and Wales (excluding that used for cooling purposes) is about 5000 mil gal (23 mm^3), or about 95 gal (0.43 m^3) per person per day. Domestic uses account for nearly 1800 mil gal (8 mm^3), with the remainder being used by industry and agriculture; the demand is increasing by some 2.5%/year.

Though accurate figures are not available, it is estimated that about one-third of the public water supply is derived from natural upland lakes or from impounding reservoirs in the headwaters of rivers, about one-third from underground sources, and the remainder from the lower reaches of rivers. Since rivers serve also as vehicles for the disposal of sewage effluents, some one-third of the nation's water supply probably contains sewage effluent, which constitutes a widespread indirect reuse situation.

The exploitation of underground resources has probably reached, or even exceeded, the capacity of the aquifers in many areas, while the exploitation of upland waters meets with increasing public opposition if it involves the construction of further impoundments. For these reasons and also because of the geographical imbalance already mentioned, water for public supply must increasingly be drawn from lowland rivers containing sewage effluent. Furthermore, the increasing population and the increasing consumption *per capita* in the Southeast tend to increase the volume of sewage effluents discharged to rivers, while the volume of fresh water available for dilution in the river tends to decrease. Thus it seems that in the Southeast of the country the proportion of sewage effluent in public water supplies must inevitably increase.

A. Need for Direct Reuse

As a further consequence of these factors and of more general economic considerations, the cost of water seems likely to continue to rise. The cost factor in itself, together with restrictions in the actual volumes available, is likely to encourage industry to economize in water, to install alternative processes not requiring water, or to consider reuse of effluent directly on site.

Control legislation in the United Kingdom is already strict and is likely to be enforced more rigorously in the future. The greater degree of control over the industrial water economy which is implicit in a recycle or reuse situation will, of course, facilitate pollution control and may, in many cases, eliminate potentially polluting discharges.

B. Pollution Control in the United Kingdom

As from 1 April 1974 the water industry has been controlled by ten Regional Water Authorities, which together cover the whole area of England and Wales. The boundaries of these authorities have been devised with a view to producing administrative units which are to a large extent self-sufficient in water resources, though some transfers of water between authorities are still necessary. Each authority is responsible for the main sewerage systems within its area, for the treatment of sewage, for pollution control of the rivers, estuaries, and coastal waters within its area, and for the supply of potable water. Certain statutory water companies which remain independent of the Regional Water Authorities receive bulk supplies of raw water from them. Pollution control is exercised on similar lines to that existing under the former legislation; that is to say, authorization must be obtained for any industrial effluent discharged to a sewer or for any effluent discharged to a river. In giving the authorization the authority concerned specifies the conditions of flow, composition, and temperature under which the discharge may be made; in the case of a discharge to a sewer, a charge representing the cost of conveyance and treatment may be made. There are no general standards for the composition of discharges, each case being decided on its merits. This system gives great flexibility of control. Under the new administrative system, the authorization of discharges from sewage works to rivers is an internal matter for each Regional Water Authority (though there is supervision by central government), and each authority is able to plan comprehensively the management of its river systems.

C. Quality Considerations

The increasing use of rivers containing sewage effluents as sources of water supply will require a change of emphasis in the setting of quality standards for effluents and river waters. These considerations have been discussed in detail in a recent publication (World Health Organization, 1973). Some factors of particular importance in the United Kingdom situation may be mentioned here.

1. NITRATES

Certain surface and underground waters are showing an increasing nitrate concentration—a matter of significance in connection with possible cases of methemoglobinemia. The increases are by no means general and, in many cases, lack any satisfactory explanation. In some river waters, an increasing proportion of sewage effluent is clearly responsible, but, in other cases, the increase is attributable to agricultural sources (ploughing of grassland, intensive animal husbandry, and use of nitrogenous fertilizer). The situation is kept closely under review by the Department of the Environment.

2. ORGANIC MATTER

Sewage effluent contains a wide variety of organic residues derived from domestic and industrial sources. Some of these residues are readily identifiable (for example, fluorescers and surface-active substances derived from domestic detergents), while the nature of others is, as yet, unknown. In view of the undesirability of putting into public supply a potable water containing unknown substances, efforts are being made in the United Kingdom and in other European countries to identify such constituents.

The Steering Committee on Water Quality (Department of the Environment, 1971) recommended that raw water obtained from rivers and intended for the provision of potable water should undergo storage for a period of 7 days. This period was chosen as being long enough to provide a degree of dilution, an opportunity for self-purification and, at worst, sufficient time for an alternative supply to be organized, while not being so long that problems due to algal growth might be encountered.

Similar considerations apply to the virus particles present in sewage effluents and, hence, in most river waters in this country. The experience of the former Metropolitan Water Board, which drew some 80% of the London water supply from the rivers Thames and Lee, was that the normal sequence of purification processes involving storage, slow-sand filtration, and chlorination can deal effectively with this problem (Poynter, 1968).

II. INDIRECT REUSE

As has already been mentioned, about one-third of the water supply in the United Kingdom is drawn from rivers that contain sewage effluent. Since virtually all sewage discharged to such rivers (apart from occasional storm discharges) has received primary, secondary, and, in some cases, tertiary treatment, the quality of the water in these rivers can be regarded as satisfactory from the pollution control viewpoint. Nevertheless, this situation undoubtedly represents the most extensive form of reuse of effluents, even though it may be indirect and indeed involuntary.

The situation in the River Thames Basin will be discussed in some detail, partly because it is well documented and partly because of the large population supplied with water from this source.

A. Indirect Reuse in the Thames Basin

The Thames drains an area of 3810 sq miles and provides some 70% of the water supply of the London area. This is withdrawn at several points a few miles upstream of Teddington Weir*, which is the limit of tidal action. Many discharges of sewage effluent and industrial effluent are made at various points along the river and there are also many other points of abstraction of water; some of the major abstractions and discharges are shown in Fig. 1, which is based on information provided by the Thames Conservancy. These data show that under average flow conditions the proportion of effluent in the lower reaches of the river is about 14%.

The flow in the river can fall to values much lower than the average, the minimum recorded value at Teddington, allowing for abstraction, being 7.4 m^3/sec in 1934 compared with the 80-year average of 68.0 m^3/sec. The percentage of effluent in potable water derived from the river would obviously tend to be higher under these conditions, but cannot be calculated directly because of the effects of the long retention period in the river in times of low flow (the level of water in the river is maintained by a series of weirs and locks). Other factors relating specifically to the London supply are the long storage period provided for raw water after abstraction (of the order of 2 months) and the statutory requirement that the flow over Teddington Weir shall not be reduced by pumping below 90 m^3/sec.

It may be of interest to note that an independent indication of the percentage of effluent in the water at the London water works intakes is given by the boron content. Synthetic detergent formulations in general use in the United Kingdom contain a fairly high proportion of perborate added as a bleach and, as a consequence, the average boron content in sewage effluents in the United Kingdom is about 1.5 mg/liter as B. This boron appears to be unaffected by normal sewage treatment, by passage down a river, or by treatment at a water-

*Teddington Weir, incidentally, is in the western suburbs of London, the Thames as seen from Central London being the estuary, and not used for potable water supply.

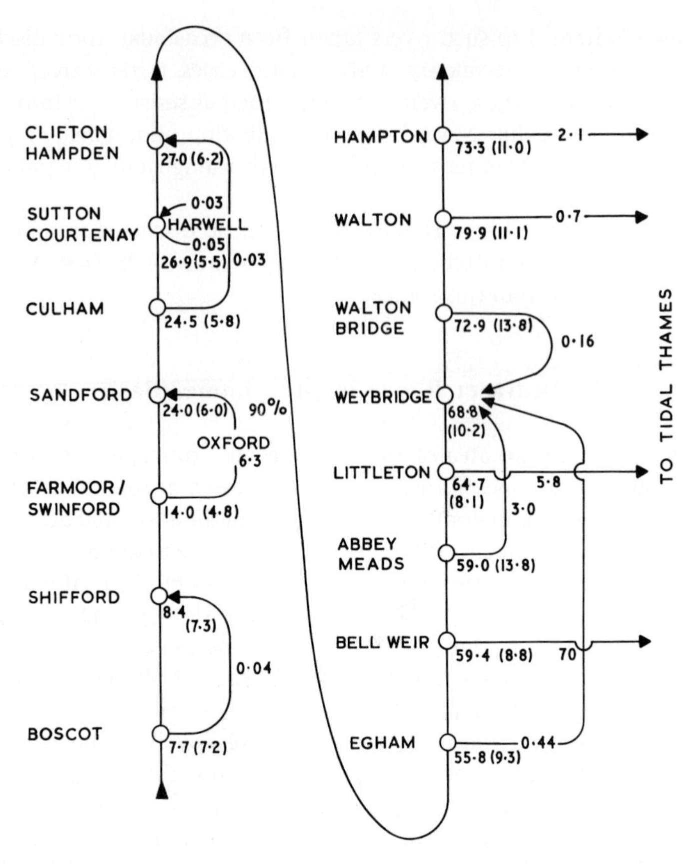

Fig. 1. Reuse of water in the River Thames Basin. Flows shown in m³/sec. River flows are averages. Figures in brackets show percentage sewage effluent under average flow conditions.

works. The average concentration in raw water withdrawn from the Thames for the London supply is about 0.2 mg/liter, indicating a proportion of sewage effluent of approximately 13%.

B. A Proposal for Intentional Indirect Reuse—The Mardyke Scheme

This imaginative scheme was devised some years ago to provide water for industrial purposes by diverting to the headwaters of a small stream a quantity of sewage effluent which otherwise would be discharged to estuarine waters. After a considerable amount of planning and pilot-plant investigation, the scheme has been put into abeyance, but the following brief description may be of some interest (Guiver and Huntingdon, 1971).

The Thamesside area of the north shore of the lower Thames Estuary has a relatively low rainfall (about 21 in.) (533 mm) and a very low useful runoff (2 in.) (51 mm). Local supplies are obtained from boreholes in the chalk, which because of overpumping in the past are increasingly subject to saline intrusion. There is little scope for further development of local sources and the requirements for potable water are being met by a bulk transfer of water from the Great Ouse river system.

A potentially cheaper source of water for industrial purposes was thought to exist in the large volumes of secondary sewage effluent discharged to the estuary from the various treatment works in the Greater London area. The conveniently placed Riverside works at Dagenham is designed to treat a dry weather flow of 0.8 m^3/sec to the Royal Commission standard (suspended solids less than 30 mg/liter, BOD less than 20 mg/liter).

The proposed scheme was that the quality of about half of this flow would be improved to a 10:10 standard by tertiary treatment, with some 0.1 m^3/sec of this being pumped to the intake of an industrial plant abstracting from the Beam river and the remaining 0.3 m^3/sec being pumped a distance of 10 km to the headwaters of the Mardyke, a small stream having an average discharge of 0.2 m^3/sec. It was planned to provide storage at the point of discharge to the Mardyke and to permit abstraction for agricultural and industrial purposes at various points downstream. The major abstraction would have been near the point at which the Mardyke flows into the estuary some 17 km below the point of discharge of the effluent.

Pilot-plant studies using various types of sand filter and nitrifying percolating filter showed that the desired improvement of the effluent could readily be obtained.

Measurements of boron ranged from 1.5 to 2.0 mg/liter in the effluent and from 0 to 0.7 mg/liter in the Mardyke water; it was considered that these levels were unlikely to have any detrimental effects on crops irrigated with the water derived from the proposed scheme, provided that the usage of boron in detergents did not increase in the future.

C. Use of Effluent to Recharge Groundwater Resources

While recharge of groundwater resources is not as yet practiced to any great extent in the United Kingdom, the Directory of Wastewater Treatment Plants (Institute of Water Pollution Control, 1972) lists over 140 works which practice some degree of recharge. These works serve a total population of $5\frac{1}{2}$ million and account for over 17 m^3/sec of sewage effluent. If all this effluent were disposed of to the ground it would represent about 10% of the total national production of sewage effluent, but the actual proportion so disposed of is not known.

III. DIRECT REUSE

The direct reuse of wastewaters in the United Kingdom has developed not so much as a matter of national water planning, but rather as a response of local management to specific local conditions. A successful industry, for example, may find that as it grows and as its water needs increase, its supplies of water from local sources may prove insufficient to meet the increasing demand. To some extent, the needs of industry can be kept within bounds by stricter internal control of water usage—for example, by economy in the use of water, by substitution of processes needing no water, and by more intensive recycling of water within the plant. Even after adopting all these measures, it may be found necessary to supplement water supplies by reuse of effluent, as in the case studies listed below.

In this section, the emphasis is placed on the reuse of effluent from sources external to the site of use rather than on internal recirculation. Nevertheless, the circumstances in which reuse of effluent rather than the use of "new" water is envisaged imply that the value of water at the site is such that internal economies, including recycling, will necessarily be employed to the maximum economic extent.

A. Water Reuse at an Integrated Steelworks (Scunthorpe)

The water system at this works has been described in considerable detail by Cook (1971), from whose paper this account has largely been abstracted. The Appleby–Frodingham works has been developed over a period of more than 100 years and, at present, extends over an area of 485 ha; steel production is over 1.5 m ingot tons. To produce 1 ton of steel in such a works requires over 200 tons of water, but, by intensive recycling, the annual consumption of water has been reduced to 5 tons/ton of steel, the greater part of this being lost as vapor to the atmosphere.

The main sources of supply, as percentages of the total flow, are enumerated in the following tabulation:

Source	%
River Ancholme	31
Effluent from the Scunthorpe sewage works (referred to as the "Ashby Ville supply")	16
Borehole water (North Lincolnshire supply)	25

The remaining 28% is obtained from plant drainage and rainfall collected on the works. The average usage of sewage effluent is 67 liters/sec, representing 79% of the total available for this source. The effluent is taken from the outlet discharge channel at the sewage works through a Venturi measuring device into a shallow lagoon of $10.5 \times 10^3 m^3$ capacity, giving a retention period of about 48 hr. The quality of this effluent in comparison with the other external supplies is indicated by the analyses given in Table I.

The Ancholme water is subject to much seasonal variation—for example, the chloride content, normally between 40 and 60 mg/liters, can rise to 1200 to 1500 mg/liter during prolonged dry weather; because of rural sewerage schemes it contains an increasing proportion of sewage effluent. The Ashby Ville water, though containing higher concentrations of polluting matter (especially detergents that tend to create foaming problems), is more consistent in quality and represents an assured supply at a shorter distance from the works. These factors are regarded as compensating very largely for the additional cost and difficulty encountered in the treatment and use of this supply.

Separate distribution systems are provided on the works, one for the Ancholme and Ashby Ville waters, and the other for the North Lincolnshire supply; at no point are the two systems interconnected even when one supply is retained as a standby for the other.

The Ashby Ville water is used without treatment as make-up to the blast furnace gas-cleaning system, and after lime-softening, to replenish losses from the turboblower and coke-oven gas-cooling and recirculating system. A flow diagram of the Ashby Ville water system is given in Fig. 2.

Table I
Analysis of Sewage Effluent (Ashby Ville Supply) Used at Appleby Frodingham Steel Works Compared with Other External Supplies

Constituent	Ancholme river water	Ashby Ville supply	North Lincoln water
Suspended solids	12	14	<3
Total dissolved solids	700	900	670
Chlorides as Cl	50	100	44
Alkaline hardness as $CaCO_3$	180	160	240
Nonalkaline hardness as $CaCO_3$	280	190	250
Total hardness as $CaCO_3$	460	350	490
Calcium hardness as $CaCO_3$	410	305	450
Ammoniacal nitrogen as N	0.1	2.0	Nil
Nitrate nitrogen as N	3.0	22.0	Trace
Permanganate value	3.0	9.0	0.2
Phosphate as PO_4	1.0	14.0	Nil
Detergent as Manoxol OT	0.2	4.0	Nil
pH value	7.8	7.6	7.7

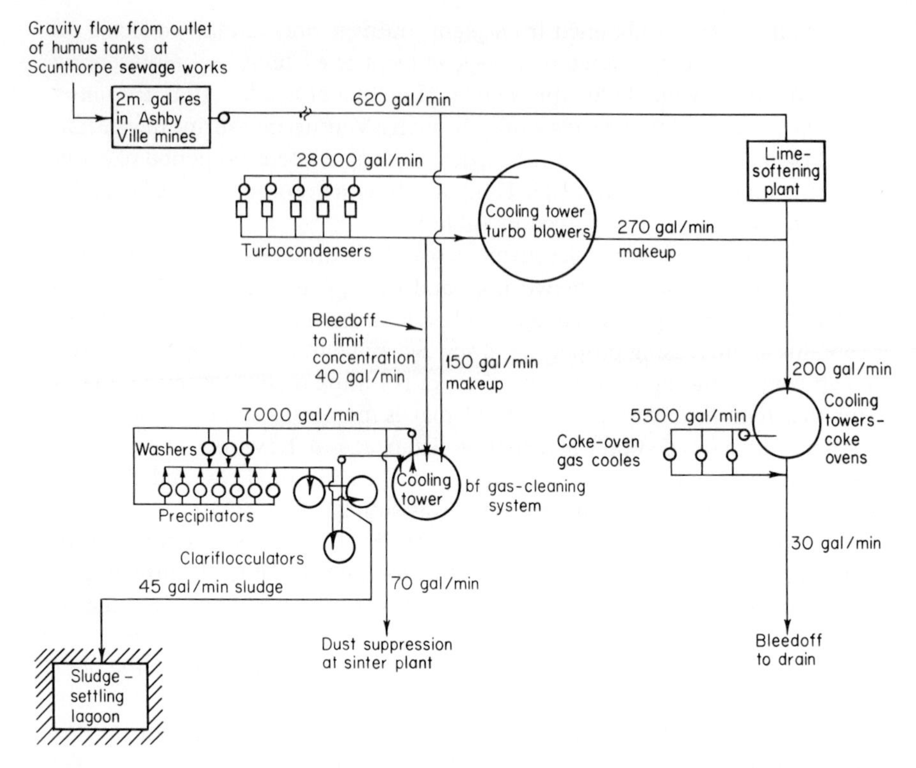

Fig. 2. Flow diagram showing reuse of sewage effluent ("Ashby Ville water") at the Appleby-Frodingham steel works (average flows for the period 1962–1969).

As the Ashby Ville water enters the delivery main to the works, it is given a shock dose of chlorine at 10 mg/liter of 90 min every 12 hr to inhibit biological growth in the mains. The lime-softening process is conventional with recirculation of sludge; typical analyses of the softened water appear in Table II.

The main disadvantage of this water and of the softened Ancholme water is that they still contain a high concentration of undesirable dissolved solids. In recirculating cooling systems appreciable concentrations must therefore be bled off; in some recent installations, the proportion may be as high as 20%. Since, at the plant, there are processes such as sinter-mixing for which waters of high salt content are quite acceptable, the high impurity level of these waters does not result in an increased demand from alternative sources. However, there are disadvantages reflected in the cost of the initial softening and in the wastage of heat in excessive blow-down.

It is considered that if the ratio of quenching to cooling and steam-raising requirements decreased considerably, the possibility of reducing the dissolved-solids contents of the incoming waters as an alternative to increasing the quantities taken in and discharged would warrant investigation.

Table II
Analyses of Softened Waters Used at Appleby Frodingham Steel Works

			North Lincoln (boreholes)	
Constituent	Ancholme river water (lime softened	Ashby Ville supply (lime softened)	Cold lime-soda softened	After subsequent base exchange softening
Suspended solids	<3	<3	<3	<3
Total dissolved solids	570	820	510	510
Chloride as Cl	50	100	44	46
Total hardness as $CaCO_3$	330	270	16	3
Nonalkaline hardness as $CaCO_3$	280	190	Nil	Nil
Alkalinity to phenolphthalein	40	50	50	50
Alkalinity to methyl orange	50	80	76	76
Phosphate as PO_4	<1	<1	Nil	Nil
pH value	10.3	10.3	10.5	10.5

Writing about the costs of water in recycling systems, Cook (1971) states that it is difficult to define the point at which expenditure on reuse of water ends and that on pollution prevention begins. The mean figures throughout the plant were 0.14 p/m^3 for straightforward cooling systems, and 0.56 p/m^3 for combined cooling and cleansing systems, excluding the cost of make-up water. About 70% of the latter figure is attributable to pollution prevention, since that expenditure would be necessary to comply with consent conditions for discharge to the river if once-through operation were to be practiced. The largest item in the cost of recycling—about half the total cost—is that of the electrical power used for pumping.

An interesting comparison is made between a recirculation system employing cooling towers and a once-through system assuming the availability of an adequate water supply. Based on certain assumptions, it is concluded that unless fresh water could be supplied to the site by gravity, with effluent disposed of similarly, recirculation becomes more economic than once-through operation if the distance between the sources of the water and the works site is more than about 2.5 km.

B. Use of Secondary Sewage Effluent at a Smelting Works (Avonmouth)

Sewage from the City of Bristol is discharged after primary sedimentation to the Severn Estuary at Avonmouth. This area is somewhat deficient in water, and existing sources were inadequate to meet the needs of an industrial complex in which zinc ore is smelted and the resulting sulfuric acid is used for the production of fertilizer-grade phosphoric acid by the wet process. Estuary water was considered but was found to be of limited utility because of its salinity and of the high costs entailed in pumping and clarification (Eynon, 1970).

The company concerned, therefore, entered into an agreement with the Bristol Corporation to make use of secondary effluent derived from the Corporation's Avonmouth sewage treatment plant. The secondary treatment plant employs the surface-aeration activated-sludge process and is now operated by the Wessex Water Authority.

The effluent is chlorinated—the normal dosage being 5 mg/liter—screened, and stored, in reservoirs providing about 1 day's storage capacity. It is used for gas scrubbing and other process requirements, and also for conveying gypsum slurry to the Severn Estuary where the gypsum dissolves in the saline water. It is also used for quenching at the Corporation's refuse incinerator and for various cooling and washing purposes at the sewage works.

The average composition (in mg/liter) of the effluent (G. R. Holden, Personal communication, 1973) for the period April 1972 to January 1973 is presented in the following tabulation:

BOD	8
COD	6.3
Ammonia as N	8.5
Nitrate as N	6.8
Suspended solids	9
Anionic surfactants	0.8

C. Use of Sewage Effluent in Wool Textile Processing (Pudsey)

The town of Pudsey, having a population of some 40,000 is situated in the woolen textile area of Yorkshire, the main industry there, until recently, being the production of cloth of the type used in uniforms.

While cloth of this kind still represents a large part of the output, large amounts of material for the fashion trade are now produced in the remaining mills. The larger mills tend to be complete units carrying out the whole processing cycle from raw material to finished cloth.

One such mill, the Troydale mill, is situated approximately 1 km upstream of the local sewage works, though the mill effluent is in fact pumped to the works (Hirst and Rock, 1973). The total water consumption at the mill is about 13 liters/sec, provided mainly from three sources, a private borehole (the yield from which is diminishing), the local stream (Pudsey Beck, which contains excessive amounts of suspended solids at times of high flow), and the town's water (an expensive source, the use of which deprives other users in the town of their supply).

Under certain circumstances, the mill has difficulty in obtaining adequate supplies of water, and this factor could restrict any future expansion. It was therefore suggested that sewage effluent might be employed, a proposal which was not unreasonable since the local stream contains a considerable proportion of effluent.

It is usually maintained that water of a high quality is needed for wet processing in the woolen textile industry, and the Woollen Industry Research Association has made an interim study of water quality requirements. Preliminary results of the study suggested that the local sewage works effluent could meet these requirements except in respect to turbidity, color, and suspended solids. Since the color of the effluent was due largely to effluent from the mill, it was thought that in this case a higher residual color could be tolerated.

Following successful laboratory trials, a plant was installed at the sewage works to treat 6 liters/sec (about half the total requirement of the mill) by sand filtration and chlorination. Table III presents comparative analyses obtained during the course of trials using this water supply at the mill. Chlorination to ensure complete absence of *E. coli* was found to need very high doses of chlorine—in excess of 50 mg/liter. It was thought, however, that in this application complete disinfection was not necessary. Residual chlorine was found not

Table III
Analyses of Representative Samples Taken during Trials at Pudsey

	Stream	Humus tank effluent	Auxiliary water supply[a]	Dye liquor[b]
Sample	1	2	3	4
Color (Hazen)	30	100	60	500
Conductivity (micro-mhos)	1000	1100	1000	1500
Turbidity (Formazin)	60	260	180	3000
Suspended solids (mg/liter)	6	50	20	85
pH	7.7	7.0	7.0	7.0
Alkalinity (mg $CaCo_3$/liter)	170	180	200	—
Total hardness (mg $CaCO_3$/liter)	350	150	200	—
Permanganate value (mg O/liter)	3	20	10	60
BOD (mg O/liter)	3	20	10	200
Chloride (mg Cl/liter)	90	160	180	200
Sulfate (mg SO_4/liter)	250	150	150	500
Chlorine dosage (mg Cl_2/liter)	Nil	Nil	20	Nil

[a] Humus tank effluent after sand filtration and chlorination.
[b] Using sample 7 before discharge to sewer.

to affect the normal process of dyeing at the mill and it was found possible to use the effluent as water supply for half the machines in the main dyehouse where several hundred pieces were dyed without complaint.

As a result of this work, a proposal was made to convert the whole of the processing at the mill to direct reuse of effluent. Because of overloading the sewage works is to be extended and will, in future, have to meet a higher standard imposed by the water authority—namely 10 mg/liter BOD, 10 mg/liter suspended matter, and 10 mg/liter ammonia. This effluent will be of higher quality than that used in the trial and will, in fact, need only chlorination to render it suitable for reuse.

The cost to the company of the reclaimed effluent will be based on the additional capital and running costs incurred between the point at which effluent leaves the normal sewage treatment cycle and the point at which it is discharged at the mill.

It is estimated that these costs will approximate to 0.45 p/m^3 for running costs out of a total cost of 1.1 p/m^3. The company will, of course, pay in trade effluent charges for the treatment of its own effluent to the standard required by the water authority.

D. Water Reuse for Bulk Transport* (Dunstable)

Until 1964, a cement company extracted limestone from a quarry near Dunstable, Beds, and transported this by rail from an adjacent siding to their works at Rugby, Warwicks, and Southam, each distant about 50 miles.

When however, the company began extracting from another site, facilities for rail transport were less conveniently situated and the two alternatives—either constructing a new rail link and sidings or road transport (involving one truck full or empty per minute)—were unacceptably expensive.

The possibility of transport as a slurry by pipeline was investigated, but since underground water at Dunstable is of the highest quality and needed for London, other sources were examined, including the canal and extractions from the rivers Ouse and Ouzel. A reliable source of relatively good quality water was found in the effluent from Dunstable sewage works.

This works treats 2.3 mgd of sewage to a 5 : 5 standard (5 mg/liter suspended solids and 5 mg/liter BOD) and the effluent, which also contains 2 mg/liter of ammonia and 70 mg/liter of chlorides is discharged via the Ouzel River to the Great Ouse River.

The local authority supply 1500 m^3/d of this effluent to the company at a fixed

*Information provided by Mr. J. R. D. Walker, Chief Engineer to The Rugby Portland Cement Company.

cost of £150 p.a. At the quarry the water is used to form a slurry with the limestone. Slurrying is achieved by mutual abrasion of pieces of limestone agitated in the water so that the slurry is of a fine consistency. The slurry, which contains 60% by weight of solid matter, is then pumped 50 miles through a 10-in. diameter pipeline to the company's works at Rugby, where about half is passed on a further 9 miles to Southam. Pumping is achieved in a single stage, with the pressure at the inlet end of the pipe being about 1600 lb/sq. in.

The cost of this method of transport amounts to some $\frac{1}{3}$p/ton-mile which compares very favorably with $2\frac{1}{2}$p/ton-mile for road transport and 3p/ton-mile for transport by rail.

E. Separate Industrial Supply (Warrington)

The provision of a separate industrial water supply which is of lower than potable quality has often been proposed as a means of conserving the increasingly scarce resources of pure water. Such systems are, however, rare, mainly because it is virtually impossible to guarantee that by negligence or by intent connections will not be made between the industrial and potable supply. Furthermore, the cost of installing a second distribution system in an existing industrial area is usually prohibitive. There is no example in the United Kingdom of a public industrial water supply employing recycled effluent, but the town of Warrington has for many years provided a separate industrial water supply derived from the River Mersey. The Mersey at Warrington carries the drainage from an extensive and highly industrialized area and is heavily polluted. Over a 2-yr period, for example, the range of suspended-solids concentrations was 24–504 mg/liter and of the permanganate value was 8–42 mg/liter (J. G. Sherratt, Personal communication, 1965); the water is often highly colored and, in general, is of poorer quality than many sewage works effluents, though as a result of tighter pollution controls the quality is tending to improve. Up to 3.6×10^3 m^3/day of this water are treated by screening, coagulation, and sand filtration, followed by chlorination; the coagulant used is a combination of alum and activated silica. The finished water, though still colored (range of Hazen values 20–232), is free of suspended matter and has an average permanganate value of 8 mg/liter. It is sold at a price considerably less than that of the domestic supply to a number of adjacent industrial establishments for use in wire drawing, condenser cooling, and board making. During the 40 yr for which the supply has been in use there have been no corrosion problems in relation to ferrous metals, though the high ammonia content (1–13 mg/liter) has been held responsible for corrosion of brass or copper fittings. The water is used apparently without complaint in a few low-pressure boilers of the Lancashire type.

A change of procedure is likely in future (W. F. Thacker, personal communication, 1973): experimentally, 8 boreholes have been constructed in the alluvium along the riverside. The water obtained in this way is of higher quality than the river water and needs a lower concentration of chemicals for treatment. If this procedure proves satisfactory, a total of 34 boreholes will be constructed, each with an output equal to the total demand, and direct abstraction from the river will cease.

F. Use of Sewage Effluents for Cooling

The greatest single use of water in the United Kingdom, more than two-thirds of the total volume, is for cooling purposes. The majority of this, about 10 to 15 $\times$ 10^9 gal/day (50–70 $\times$ 10^6 m^3/day) is abstracted from rivers and estuaries. Although the chief application of reused sewage effluents is for cooling, the 50,000 m^3/day involved represents only a very small fraction of the total quantity of water used for this purpose. Power generation accounts for most of this volume, but other major users are oil refineries, steelworks, paper manufacturers, and the chemical industry.

In instances where sewage effluent has been used for cooling, ready availability and quality were the chief factors governing its selection from alternative sources. For example, sewage effluent may contain less dissolved and sometimes less suspended matter than some estuarine sources.

Most problems associated with the use of sewage effluents for cooling can easily be overcome. Sand filtration or microstraining may sometimes be necessary to avoid deposition of organic matter in cooling systems, and chlorination (to a free chlorine residual of 0.5 mg/liter) may be practiced to control the growth of biological slimes. Adjustment of the pH to a value below 7 may also be advisable to avoid scaling.

A summary of projects in which the main purpose of reuse is cooling is given in Table IV. The data are derived largely from the Directory of Municipal Wastewater Treatment Plants (Institute of Water Pollution Control, 1972).

IV. RESEARCH PROJECTS

The following section contains brief accounts of certain pilot-plant projects. While these cannot be said to represent actual examples of reuse, they are thought worthy of inclusion as being indicative of the lines on which development may be expected to occur during the next few years.

A. Use of Ozone in Water Reclamation (Redbridge)

The object of these experiments conducted in 1966–1967 was to apply to secondary sewage effluent the so-called M.D. (Micellization-Demicellization) and Microzon processes* originally developed for the treatment of waters for potable supply (Boucher *et al.*, 1968).

The experiments were conducted at the eastern sewage works of the London Borough of Redbridge, which at the time was considered to be overloaded by 40% in terms of volume and by 60% in terms of polluting load. Sewage treatment was by primary sedimentation and biological filtration, but because of the overloading, the Royal Commission standard was rarely achieved.

The pilot plant shown in Fig. 3 provided various combinations of treatment by microstraining, chlorination, ozonation, coagulation, and rapid sand filtration at a flow of 2.5 liters/sec. The ozonation provided a maximum dose of 25 mg/liter, which was applied to the second of two contact columns, with the residue being applied to the first column in countercurrent.

"Micellization" refers to an effect by which an increase of turbidity is produced after ozonation, from the formation of negatively charged micelles. "Demicellization" refers to neutralization of the charge produced by the addition of a small dose of electrolyte and the formation of microflocs which are supposed to carry small positive charges, with these microflocs being removed by sand filtration.

In practice, there are two distinct effects of ozonation, depending upon the nature of the water being treated: (1) virtually complete removal of color, without the production of turbidity; (2) the production of colloidal turbidity due to the formation of colloidal micelles (micellization).

When the first effect is produced after prefiltration by microstraining, no subsequent sand filtration is required and the water is "finished"—clarified, decolorized, and sterilized. The M.D. process thus abridged is known as the Microzon process.

The pilot plant took effluents from the works secondary sedimentation (humus) tanks and was run for 4, 2-wk periods with the following operating conditions:

1. Microstraining, ozonation, rapid sand filtration
2. Microstraining, chlorination, ozonation, rapid sand filtration
3. Microstraining, ozonation, chemical coagulation, rapid sand filtration
4. Microstraining, chlorination, ozonation, chemical coagulation, rapid sand filtration

*Compagnie des Eaux et de l'Ozone. Brit. Pat. No. 1,052,912, 30th Dec. 1966.

Table IV
Reuse of Effluent for Cooling Purposes in the United Kingdom

Source		Quality						
Location	Sewage works	Suspended solids mg/liter	BOD mg/liter	Ammoniacal nitrogen mg/liter	Oxidized nitrogen mg/liter	Quantity m^3/day $\times 10^3$	Cost to industry £/10^3 m^3	Use made by industry
Bristol	Avonmouth	9	8	8.5	7.7	24.1	5.68	Zinc smelting fertilizer manufacture
Derby	Derby	—	—	—	—	1.1	None	Cooling gases from refuse incinerator
London	Beckton	60	57	17.0	2.5	0.7	5.50	Quenching coke
	Beddington	11	5	6.0	17.0	1.4	None	Cooling water makeup
	Deephams	12	5	3.0	—	6.8	None	Cooling gases from refuse incinerator
Nottingham	Stoke Bardolph	30	20	—	—	3.6	None	Cooling in animal processing factory
Nuneaton	Hartshill	14	10	8.4	21.7	>0.9	0.77	Cooling in offal rendering plant
Oldham	Oldham	31	21	8.0	12.9	~6.8	Based on power production	Cooling water makeup (CEGB)
Scunthorpe	Ashby	32	20	1.0	—	3.2	0.86	Cooling turbines and gases from coke ovens at steel works
Sheffield	Blackburn Meadows	30	20	—	—	~4.5	11.60	Cooling in steel industry

Stoke-on-Trent	Burslem	30	20	—	—	3.6	1.25	Cooling in steel industry
	Hanley	30	20	—	—	4.1	1.25	Cooling in steel industry
						0.1	1.25	Cooling in gas industry
						4.5	None	Cooling in tyre manufacture
	Strongford	30	20			2.2	1.41	Cooling in power production (CEGB)
Todmorden	Eastwood	—	—	—	—	Max 5.7	None	Supplies a storage reservoir for sprinkler at cotton mill
Wolverhampton	Barnhurst	11	8	4.7		20.9	Negative	To canal used by industry
		26	10	3.3		22.3	Negative	To canal used by industry
	Coven Heath	24	7	3.5		4.5	Negative	To canal used by industry

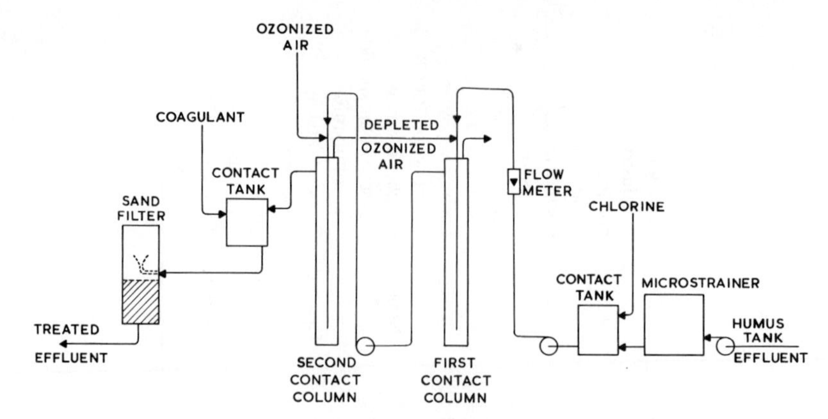

Fig. 3. Pilot plant employed in experiments on use of ozone for reclamation of water from sewage effluent.

Table V indicates the results of detailed chemical analysis of samples obtained when microstraining, ozonation, and sand filtration only were employed. A very remarkable reduction in color and turbidity were observed. Table VI shows in less detail the results obtained with other combinations of unit processes.

Microbiological examinations were carried out on samples collected at weekly intervals during the program, the results of which are shown in Table VII. With this method of presentation, the medians for some of the treatments represent much smaller numbers of samples than others, and one treatment is not directly comparable with another because they are on different samples, but the results are sufficiently clear-cut for this not to make any significant difference to their interpretation.

The table show that the microstraining made virtually no difference to the counts of any of these groups of organisms. Ozone on its own killed the majority of the organisms, but it did not produce a sterile effluent, nor did it produce such a low result as chlorine on its own. All the salmonellae and viruses naturally present were killed by ozone except, in one instance, when there was a single survivor per liter in each case. Some coliform organisms and *E. coli* survived and the numbers were generally higher when the ozone dose could not be maintained at the desired 20 mg/liter.

Chlorine produced lower counts than ozone, and although the median *E. coli* count was zero, there were two occasions out of nine chlorinated samples when counts of 13 and 25/100 ml were obtained; on one of these, a single *salmonella* per liter was isolated.

Chlorine followed by ozone was highly effective, all the coliform and *E. coli* counts being zero, the only counts obtained were in the plate counts and the colonies obtained were most likely to have originated from resistant spores.

Not much reliance can be placed on the microbiological results from the

Table V
Average Results Obtained at the Various Stages of Treatment of Humus-Tank Effluent in the M.D. Pilot Plant at Redbridge[a]

Composition	Effluent from: Humus tank	Micro-strainer	Ozonizer	Sand filter
Suspended solids	52	20	19	6
Total solids	927	—	—	896
BOD	16	10	10	8
COD	94	64	53	50
Permanganate value	17	11	8	8
Organic carbon	29	20	18	16
Surface-active matter				
Anionic (as Manoxol OT)	1.1	1.1	0.2	0.2
Nonionic (as Lissapol NX)	0.3	—	0.05	0.05
Ammonia (as N)	4.6	4.8	5.3	5.2
Nitrite (as N)	0.4	0.4	0.03	0.01
Nitrate (as N)	25.7	25.7	25.2	26.7
Total phosphorus (as P)	9.2	—	—	7.6
Orthophosphate (as P)	7.6	—	—	7.3
Total hardness (as $CaCO_3$)	—	—	—	425
Chloride	122	—	—	121
Sulfate	192	—	—	194
Color (Hazen units)	41[b]	34[b]	7.5	7
Turbidity (ATU)	70	54	26	7
Total phenol	2.8	—	—	1.0
Temperature (°C)	15.2	14.9	15.6	15.7
Dissolved oxygen (% saturation)	54	57	99	96
Conductivity (μmho/cm³)	1125	1118	1100	1150
Langelier Index	+0.10	—	—	+0.12
pH value	7.15	—	—	7.3
Pesticides (μg/liter)				
α BHC	—	Trace	Trace	—
β BHC	—	0.070	0.044	—
Dieldrin	—	0.042	0.033	—
pp DDE	—	0.083	0.072	—
pp TDE	—	<0.01	<0.01	—
pp DDT	—	<0.01	<0.01	—

[a]Results expressed as mg/liter, except where stated.
[b]Samples filtered through glass-fiber paper.

sand filter since short-term changes, particularly involving use of an effluent containing a chlorine residual, are supposed to upset the normal stabilized microflora that could be expected to develop during a long run without residual chlorine.

Table VI
Average Composition of Samples from Various Stages of Treatment during Operation of Water-Reclamation Plant at Redbridge[a]

Effluent from	Suspended solids	BOD	Anionic detergent (as Manoxol OT)	Nonionic detergent (as Lissapol OT)	Total phosphorus (as P)	Color (Hazen units)	Turbidity (ATU)
Period I (5th–19th May 1967)							
Humus tank	52	16	1.1	0.3	9.2	41	70
Microstrainer	20	10	1.1	—	—	34	54
Ozonizer	19	10	0.2	0.05	—	7.5	26
Sand filter	6	8	0.2	0.05	7.6	7	7
Period II (19th May–2nd June 1967)							
Humus tank	61	19	1.1	0.3	9.0	36	87
Microstrainer	27	12	1.1	—	—	37	61
Chlorinator	24	—	1.1	—	—	28	58
Ozonizer	19	—	0.3	0.03	—	3	37
Sand filter	5	3	0.3	0.03	8.1	3	9
Period III (1st–15th September 1967)							
Humus tank	41	15	1.3	0.3	9.1	33	41
Microstrainer	13	8	1.3	—	—	34	24
Ozonizer	8	—	0.3	0.03	—	3.5	15
Coagulant and sand filter	2	2	0.3	0.03	6.5	4	4
Period IV (15th–29th September 1967)							
Humus tank	43	14	1.2	0.2	8.1	38	58
Microstrainer	13	7	1.2	—	—	36	38
Chlorinator	12	—	1.2	—	—	28	35
Ozonizer	8	—	0.3	0.02	—	3	18
Coagulant and sand filter	1	1	0.3	0.01	5.7	2	1

[a]Flow through sand filter, 150 gal/ft^2 h (175 m^3/m^2 day); results expressed as mg/liter except where otherwise stated.

Table VIA
Addition of Reagents

Period	Reagent		
	Ozone (mg/liter)	Chlorine (mg/liter)	Coagulant[a] (mg/liter)
I	22	—	—
II	26	20	—
III	24	—	2
IV	20	20	2

[a]As $Al_2(SO_4)_3$.

All the works effluent samples contained moderate numbers of salmonellae and slightly higher numbers of virus PFU. The *Salmonella* species most frequently isolated was *S. anatum*, but other species were also isolated.

B. Reclamation of Sewage Effluent Using Powdered Activated Carbon and Coagulant (Langford)

In this work at Langford, Essex sewage effluent was treated for an experimental period in a disused waterworks—this was in fact the works at which the excess-lime or Southend process described in early books on water treatment was developed. Essex is a relatively dry part of England and experiences water shortages after prolonged dry spells. The rivers in the area are relatively small and contain a high proportion of sewage effluent. It has been found possible indeed for 3 mo to discharge sewage effluent directly above the intake to the Hanningfield reservoir (Fig. 4). Apart from raising the phosphate level of the water in the reservoir, no adverse effects could be detected.

Laboratory work by Slack (1969) had shown that the treatment of sewage effluent with alum and powdered carbon could yield a clear liquid low in phosphate and detergent and with an organic carbon content similar to that of water in the Chelmer. If such a process were applied to the effluent from the sewage works, it might yield a water that could be used to supplement low river flows and would be acceptable both in the reservoir and directly at the Langford new works intake. This process appeared likely to prove simple to operate and would need to be used only at times when the flow in the river was low.

The plant available at the waterworks comprised a stirred solids-contact reactor of the accelator type, and a hopper-bottomed upflow tank capable of flows of 9 and 4.5 m³/day, respectively, at an upflow rate of 0.5 mm/sec. Preliminary results obtained using the accelator were not entirely satisfactory,

Table VII
Microbiological Examination of Samples Collected from the Pilot Plant at Redbridge

Treatment	Coliform organisms/ 100 ml	*E. coli*/ 100 ml	37° count/ ml	22° count/ ml	*Salmonella*/liter	Virus PFU/ liter
Works effluent	2,100,000	600,000	91,000	380,000	32	62
After microstrainers	1,700,000	700,000	92,000	440,000	32	58
Ozone only	90	32	10	152	0	0
Chlorine only	1	0	46	51	0	0
Chlorine + ozone	0	0	6	2	0	0
Sand filtrate	0	0	1	3	—	—

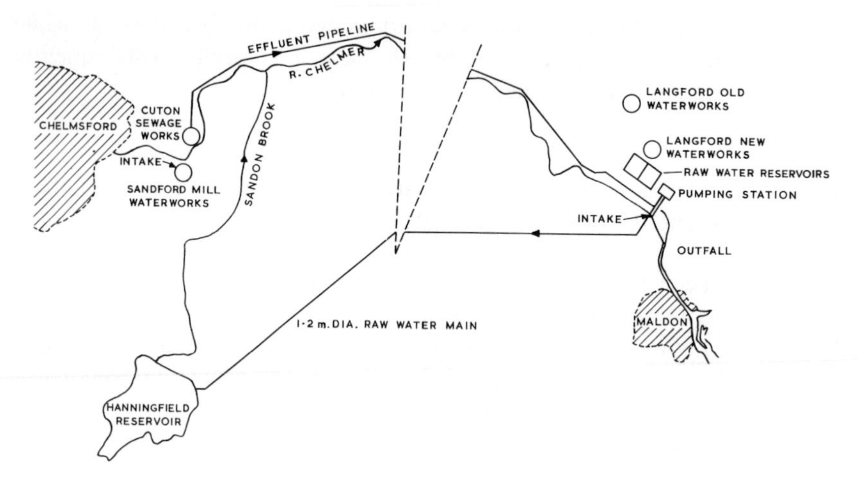

Fig. 4. Map showing parts of River Chelmer, Essex, with water intakes, waterworks, and reservoirs.

probably because the rate of agitation, intended for lime softening, was too fast and caused disintegration of the floc (Slack, 1972).

Good results were obtained, however, using the hopper-bottomed tank, with the dosage of aluminium sulfate at 150 mg/liter and carbon at 250 mg/liter; the upflow rate was 0.5 mm/sec. The plant was operated daily for about 5 wk; results for a period of 10 days are shown in Table VIII. The effluent was clear and bright, free from color, and of negligible odor; the phosphate content was low.

Table VIII
Results of Plant Trials at Langford Using the Pyramidal Upward Flow Tank (Average for a Period of 10 Days)

Composition	Before treatment	After treatment
Color °Hazen	35–45	0–5
UV absorption	0.73–0.88	0.18–0.27
PV, 4 hr, 27°C	5.5–6.5	1.4–2.1
TOC	12–21	3.5–7.0
COD	30–44	10–22
Phosphate PO_4	21–27	0.8–2.0
Turbidity	11–13	0.6–2.0
Free ammonia N	0.2–0.5	0.2–0.5
Albuminoid Ammonia N	0.5–0.6	0.2–0.4
Nitrate N	21–26	20–25
Total alkalinity $CaCO_3$	130–160	80–108
Total hardness $CaCO_3$	—	290–320
Chloride Cl	—	134–156
Sulfate SO_4	190	247
pH	7.1–7.5	6.5–7.1

Subsequent results with a sewage effluent of lower quality were unsatisfactory; the author concluded that efficient biological treatment of sewage was essential if the reclamation process was to be successful.

Further work using lime in place of alum gave good phosphate removal, but removal of organic matter was less satisfactory. It was not established to what extent this effect was due to the higher pH value (9.5–10.0) or to lower effluent quality, but it was thought that a small dose of coagulant might have been beneficial.

In view of the possibility that water treated in this way might be used for domestic supply, measurements were made of the concentrations of various metals in the sewage effluent before and after treatment with alum (150 mg/liter) and carbon; results are shown in Table IX.

Results of measurements of virus and bacteria before and after treatment are given in Table X.

C. Treatment with Activated Carbon

An appreciable proportion of the organic materials remaining in conventionally treated secondary effluent and apparently resistant to biological attack can be removed by adsorption on activated carbon. Traditionally, activated carbon has been successfully used in the water supply industry for removing color and overcoming problems of taste and odor, but hitherto its application in

Table IX

Concentrations of Trace Metals in Sewage Effluent before and after Treatment at Langford, Compared with River Water[a]

	Sewage effluent		
Trace metals	Before treatment	After treatment	River water
Arsenic	2–3	3	2–5
Selenium	0.5	< 0.5	0.5–5
Zinc	170–180	150–160	60–140
Boron	1300–1440	1360–1400	260–410
Mercury	Less than 5	Less than 5	
Lead	Less than 5	Less than 5	
Cadmium	Less than 2	Less than 2	
Molybdenum	3–4	3–12	1–2
Vanadium	15–20	10–30	3–4
Chromium	10–20	< 5–40 >	< 5
Copper	20–40	15–20	< 5–10
Nickel	15–35	20–50	< 2
Manganese	100–110	45–150	3–45

[a] Conditions of treatment: upflow rate, 0.5 mm/sec; aluminum sulfate, 150 mg/liter; carbon, 250 mg/liter.

Table X
Removal of Viruses and Bacteria from Sewage Effluent at Langford by Aluminium Sulfate and Carbon

Virus and bacteria	Before treatment	After treatment
Viruses PFU/liter	304	0.3
	750	8
Coliforms/100 ml	480,000	3600
	120,000	2200
E. coli/100 ml	27,000	130
	14,000	310
Salmonella/liter	16	1.5

sewage treatment has been inhibited by its high cost in relation to the value of the effluent.

During recent years, granular forms of activated carbon have been developed which a number of manufacturers have claimed are capable of thermal regeneration with a loss of carbon of about 5 to 10%. This has made it possible to consider the process for the treatment of sewage effluents where the required standard of purity exceeds that which can be achieved by even the most intensive application of the conventional, biological processes.

Bench-scale experiments carried out at the Water Pollution Research Laboratory showed that substantial reductions in COD, organic carbon and detergents can be achieved by a period of contact with activated carbon. A summary is given in Table XI of the results obtained from tests lasting 416 days during which secondary effluent (biological filter) was passed for 1 hr through columns of granular activated carbon derived from peat and coal—the most effective of the materials tested.

The ability of the beds of carbon to remove organic matter declined steadily throughout the tests. In the first 12 wk, the average BOD removal exceeded 75% and anionic surfactant residues were not detectable in the treated liquor. During the subsequent 48 wk, more than 50% of the BOD was removed, with the removal of surfactant decreasing from 90 to 40%.

After 60 wk, the carbons apparently had removed from the effluent between 50 and 80% of their own weight of organic matter. The carbon manufacturers have stated that adsorption alone could not account for the whole of this uptake. Subsequent microscopical examination of the used carbon revealed an abundance of bacteria, and it seems likely that biological activity was partly responsible for the high degree of removal of organic matter. Indeed it has been suggested that metabolism by bacteria of adsorbed materials releases active sites on the carbon by a form of "biological regeneration."

To permit a more practical and economic appraisal of the process, including regeneration of the exhausted carbon, a larger scale pilot plant was supplied

Table XI
Treatment of Sand-Filtered Secondary Sewage Effluent by Passage through Beds of Granular Activated Carbon

				Composition of liquor after passing through beds of granular activated carbon (mg/liter)			
Source of carbon	Nominal particle size (mm)	Surface area per unit mass (m^2/gm)	Time after start of test (wk)	BOD (8)[a]	COD (45)	Organic carbon (17)	Anionic detergent (1.0)
Peat (B)	0.8	1200	0 to 12	1	10	1	Not detectable
			12 to 24	2	16	3	0.1
			24 to 36	3	21	8	0.2–0.3
			36 to 48	3	25	11	0.4
			48 to 60	5	32	14	0.5–0.6
Coal (D)	0.8 to 1.4	1000	0 to 12	2	11	1	Not detectable
			12 to 24	2	15	4	0.1–0.2
			24 to 36	3	22	9	0.1–0.2
			36 to 48	3	25	10	0.3
			48 to 60	6	35	11	0.5–0.6

[a] Values in parentheses refer to sand-filtered effluent supplied to adsorption units.

and erected by Norit-Clydesdale Ltd. at the Rye Meads sewage works of the Middle Lee Regional Drainage Scheme (now part of the Thames Water Authority). This sewage works produces an effluent of exceptionally high quality from an activated sludge plant and might be considered as approaching the ultimate, in economic terms, in exploitation of the conventional biological treatment process.

The pilot plant consists of two columns, each containing a 6 m^3 bed of extruded activated carbon. During the first 6 mo of operation, one bed containing 0.8 mm carbon and providing a retention time of 1 hr removed on average 56% of the organic carbon and 65% of the COD remaining in the sand-filtered effluent.

The other column in the pilot plant gave less satisfactory results, presumably because the material it contained was of larger particle size (3.0 mm). It was not necessary to backwash either of the beds during the period, although this operation might have been necessary with a poorer effluent.

D. Reverse Osmosis and Ultrafiltration

The inorganic salt content of sewage is about 300 mg/liter greater than that of the water supply from which it is derived; conventional methods of sewage treatment do little to reduce the content of dissolved salts. As the proportion of sewage effluent present in rivers continues to increase, some reduction in the salinity of sewage effluents together with removal of most of the more refractory residues may eventually become necessary when the water is destined, directly or indirectly, for further reuse.

Reverse osmosis is able to remove a high proportion of the dissolved salts and residual organic matter in sewage effluents and, in many respects, is the most promising of the alternatives at present available for producing a water of very high quality. An indication of the quality of water that can be recovered from sewage effluents by reverse osmosis is given in Table XII.

Membranes having a range of permeabilities can be produced from cellulose acetate and, in many cases, it may be necessary to remove only organic matter including bacteria and viruses leaving the saline content virtually unchanged. This is the process of ultrafiltration.

A major difficulty encountered in applying membrane processes to sewage effluents is that a rapid decline in flux often occurs. This appears in two stages: an initial sharp decline during the first few hours of operation and a more gradual long-term decline. The initial loss appears to be irrecoverable and may be due to changes within the membrane. The longer term loss is accompanied by the formation of a slime layer at the membrane surface and can to some extent be recovered by flushing. As might be expected, the long-term rate of flux decline appears to be related to the quality of the effluent treated.

In pilot-scale tests conducted by the WPRL using tubular type, reverse osmosis equipment an average flux of 0.45 m/day was maintained for over 300 days in modules fed with final effluent from the Rye Meads sewage works: this effluent is well nitrified and of exceptionally good quality (BOD 5 mg/liter, suspended solids 5 mg/liter).

To assess the influence of suspended and colloidal material on the rate of flux decline, reverse osmosis units were also fed with ultrafiltered secondary effluent. This pretreatment reduced, but did not eliminate, the rate of flux decline, which indicates that an appreciable proportion of constituents causing fouling were present in solution. This observation suggested that a pretreatment with activated carbon might assist in maintaining a high flux. However, contrary to experience elsewhere, the removal from sand-filtered effluent of 65% of the COD and 56% of the remaining organic carbon by passage through beds of granular activated carbon had little effect on the performance of reverse osmosis units fed with the liquor, as compared with a unit fed directly with the sand-filtered effluent. In each case a flux of 0.45 m/day was maintained for 160 days.

The relatively high flux maintained in this plant without resort to pretreatment or flushing techniques contrasts with experience elsewhere and can possibly be attributed to the very high quality of the Rye Meads effluent which is consistently better than the 5 : 5 standard. Whether more intensive biological

Table XII
Treatment of Secondary Sand-Filtered Effluent to Various Standards by Reverse Osmosis, Ultrafiltration, and Activated Carbon Adsorption

Sample	BOD[a]	COD[a]	Org.C[a]	NO_2^- [a]	Total oxidized[a] (N)	Soluble phosphate[a] (P)	Anionic detergent[a]	Total solids[a]	pH	Conductivity (μmho/cm)	Color (hazen units)	Turbidity (ATU)
Sand-filtered secondary effluent	3	33	10	0.08	29	9.3	0.6	754	7.3[b]	1252	24	5.6
After ultrafiltration	1	14	5	0.01	27	7.2	0.4	664	6.6	1091	5.4	2.1
Less selective membrane												
More selective membrane	1	10	3	0.01	27	4.4	0.35	493	6.7	796	1.2	1
After reverse osmosis	1	10	3	0.01	4.1	0.3	0.2	55	6.7	83	0	0
Effluent from activated carbon treatment	2	13	4.4	0.06	30	9.5	0.3	720	7.4[b]	1124	1.8	1.7
After reverse osmosis	1	8	2.7	0.01	9	0.3	0.2	77	6.5	121	0	0

[a] Measured in mg/liter.
[b] The feed liquor is adjusted to pH 6.1 with sulfuric acid.

treatment to achieve a higher quality effluent was preferable to the use of flushing techniques (which reduce membrane life) or carbon pretreatment (expensive) would need to be clearly determined before the mounting of any full scale project.

Membranes are subject to biological degradation, as well as to hydrolytic damage by alkaline flushing solutions, and to general abrasion, especially when foam balls are used.

Although little can be done to reduce the latter two, if these techniques are used, biological attack can be effectively controlled by the intermittent use of biocidal agents.

Experiments at WPRL, in which pieces of membrane were suspended in activated sludge tanks, showed that bacterial attack depended on temperature, the type of membrane, and the geometry of the support system. The most rapid attack occurred in quiescent areas, which emphasized the importance of turbulence in operation and also the necessity to disinfect an installation when it is shut down even for a short period.

Weekly treatment for 1 hr with a biocidal agent increased the life of membranes severalfold; for example, weekly treatment with a 3.5 mg/liter solution of the long chain alkylbenzenesulfonate Dobane 45 trebled the membrane durability. Quaternary ammonium compounds, silver nitrate, formaldehyde, and an iodophor were also found to be effective.

Immersion in activated sludge represents extreme conditions and no attack has been observed in a plant operated for several months on sand-filtered effluents. No special steps were taken against bacterial attack, other than filling the system with 1% formaldehyde solution during periods of inoperation.

As with all separation processes, a reject stream of concentrated waste liquor (including perhaps water from flushing) would be discharged by a reverse osmosis plant, and this is not likely to be less than 15% of the initial flux. At present, it seems improbable that the whole of a works effluent would be required for full recovery. If only 25%, say, were treated, the reject—representing only 4% of the total sewage flow—could probably be returned to the works inlet. The recovery of a larger fraction would, except in coastal areas, necessitate separate final disposal, perhaps after further concentration, of the reject. A similar situation would arise if, in the future, reverse osmosis was used to control salinity in a river by removing electrolyte from a proportion of the discharged sewage effluents. (See Chapter 6 for further details on the use of pressure-driven membranes.)

V. ALTERNATIVES TO REUSE

Reuse of effluent is but one aspect of the much wider subject of water conservation. To put matters into perspective, some mention should be made of other methods of conservation employed or under consideration in the United Kingdom. Collinge (1967) has discussed them in some detail, but they may be briefly enumerated as follows.

a. Conventional Impounding and Storage Reservoirs. These have an important part to play and will no doubt be fully exploited in the future; bear in mind that proposals to construct new reservoirs meet with increasing public opposition on grounds of loss of amenity and flooding of good agricultural land.

b. Use of River Regulating Reservoirs. By releasing impounded water to rivers at times of low flow, certain advantages over direct supply reservoirs can be obtained. The stored water need be released only at times of low flow for the amenity value of the river to be enhanced, the quality of the water in it improved at times of low flow, and some degree of flood control achieved.

c. Estuarial Barrages and Impoundments. Several major schemes—for example in Morecambe Bay and in the Wash—have been the subject of major feasibility studies. The potential yields of such schemes are very high and phased development may be desirable.

d. Desalination. Such processes as distillation, freezing, and electrodialysis have been subject to considerable research and development in the United Kingdom, but mainly in view of export markets. A small 0.5 mil Imp. gal/day (2300 m^3/day) distillation unit is installed on the island of Guernsey as a standby. Plans for a desalination plant at Ipswich using a freezing process were abandoned on economic grounds at an advanced stage of design. All such processes tend to be expensive and to produce a water which is rather soft for public supply. The possibility of applying desalination processes to the renovation of effluents (which are not, like saline water, available only around the coasts) is kept under consideration.

e. Controlled Abstraction of Groundwaters. In areas where groundwater provides a major source of water for river systems the normal seasonal variation of river flows may be altered by controlled pumping of groundwater. In this way, low summer flows are augmented by groundwater, with the latter resources being made good at other times of the year. A pilot scheme of this kind in the Lambourne Valley has now been in operation for some years and supplements the flow in a tributary of the Thames. Similar schemes are under consideration elsewhere.

f. Groundwater Recharge. The use of sewage effluent for this purpose has already been mentioned, but other sources, such as surface water, mine water, and even industrial effluents, have been employed.

The broad policy issues, as summarized by Rydz (1972), are that a decision must be made between a policy of greater conservation (i.e., greater storage and use of "new" water) or one of greater reuse of effluents, possibly bringing into use rivers which, at present, are considered unsuitable as sources of public supply. The choice may well be different in different areas of the country and in reaching their decisions the new Regional Water Authorities will have a major part to play.

REFERENCES

Boucher, P. L., Truesdale, G. A., Taylor, E. Windle, Burman, N. P., and Poynter, S. B. (1968). Use of ozone in the reclamation of water from sewage effluent. *J. Instn. Pub. Health Eng.* **67**, 75–95.

Central Advisory Water Committee. (1971). "The Future Management of Water in England and Wales." HM Stationery Office, London.

Collinge, V. K. (1967). *In* "Symposium on the Conservation and Reclamation of water," Paper No. 2. Inst. Water Pollut. Contr., London.

Cook, G. W. (1971). *In* "Management of Water in the Iron and Steel Industry," pp. 11–19. Iron and Steel Institute, London.

Department of the Environment. (1971). "First Annual Report of the Steering Committee on Water Quality." HM Stationery Office, London.

Eynon, D. (1970). Waste water treatment and re-use of treated sewage as an industrial water supply. *Chem. Eng. (London)* **235**, CE6-CE7 and CE13.

Guiver, K., and Huntingdon, R. (1971). A scheme for providing industrial water supplies by the re-use of sewage effluent. *Water Pollut. Contr.* **70**, 75–85.

Hirst, G., and Rock, B. M. (1973). Proc. 1973 Pollut. Contr. Congr., Brintex Conferences, London.

Institute of Water Pollution Control. (1972). "Directory of Municipal Wastewater Treatment Plants," 4 vols. Inst. Water Pollut. Contr., Maidstone, Kent England.

Poynter, S. F. B. (1968). Problem of viruses in water. *Water Treat. Exam.* **17**, 187–204.

Rydz, B. (1972). New water—or second-hand. *Water Treat. Exam.* **21**, 182.

Slack, J. G. (1969). Sewage effluent treatment *Effluent Water & Treat. J.* **9**, 257–261.

Slack, J. G. (1972). Water reclamation from sewage effluent; experimental studies in Essex. *Water Treat. Exam.* **21**, 239.

World Health Organization. (1973). *World Health Organ.*, **517**.

15

The EPA-DC Pilot Plant at Washington, D.C.

Dolloff F. Bishop

I. INTRODUCTION

The EPA-DC Pilot Plant, located at the District of Columbia's Water Pollution Control Plant, was conceived in 1965 as a pilot facility to meet national wastewater treatment goals and for tertiary treatment of the District of Columbia's secondary effluent. The pilot facilities were constructed to permit process evaluation and to produce engineering design data for most conventional and advanced waste treatment processes, including solids handling and disposal processes. It also has been considered as a demonstration of advanced wastetreatment technology that might be applicable in future reuse systems.

Unlike early work in advanced waste treatment, the long term overall goals of the pilot facilities were not only to develop individual unit processes, but also to evaluate complete treatment systems with associated solids handling for year-round removal of carbon (C), phosphorus (P), and nitrogen (N) and to develop engineering design data for constructing these treatment systems.

In 1968, the Potomac Enforcement Conference established rigorous standards for total pollutants in effluents discharged into the Potomac River. At projected 1982 flows, the allowable concentrations of pollutants for the Washington, D.C., effluent were 4.5 mg/liter of BOD_5, 0.22 mg/liter of P, and 2.5 mg/liter of total N, respectively. With the establishment of these standards, two immediate specific goals were assigned to the pilot plant: first, to develop and evaluate on a pilot scale several treatment systems to permit the District of Columbia and its consultants to select a system for its 309 MGD advanced-treatment plant, and second, to develop detailed design data for the selected system. The water, because of the high quality required for pollution control, also has reuse potential. Simultaneous with the work for the District of Columbia, the pilot plant conducted studies related to its long-term national goals. The pilot plant is being utilized currently to continue these studies.

II. PILOT-PLANT FACILITIES

The complete EPA-DC pilot facility includes a multipurpose biological pilot plant with an oxygen activated sludge system, a physicochemical pilot plant, an ozone pilot plant, a solids handling pilot plant, and supporting analytical and research laboratories. The pilot facilities include extensive instrumentation and control systems supervised by a digital process control computer. Development studies usually at 190 to 380 m^3/day (50,000 to 100,000 gpd) may be performed with realistic diurnal flow variations of 4 : 1 minimum to peak flow.

These pilot facilities permit the evaluation of two basic treatment approaches, physicochemical treatment and advanced biochemical treatment. Flexible piping arrangement permits operation of various complete treatment systems which may include combinations of physicochemical and biochemical processes. Flow rates in the plant are automatically controlled from central analog control panels or from the digital computer for any desired flow pattern. All equipment is above ground with complete accessibility.

A. Multipurpose Pilot Plant

The multipurpose pilot plant (Fig. 1) consists of two parallel treatment systems with each system containing primary sedimentation, diffused and/or mechanical aeration, and secondary sedimentation. Grit removal, air-flotation clarification, and air-flotation thickening, are shared between the two parallel treatment systems. In one system, the biological reactors are covered for oxygen activated sludge and for anaerobic denitrification.

The primary clarifiers in each system may function as conventional settlers or as chemical flocculator-clarifiers and, at 380 m^3/day (100,000 gpd), provide nominal overflow rates of 22 m/day (550 gpd/ft^2). The processes after primary treatment may receive either pilot plant primary or the District's main plant primary effluents.

The two open aeration tanks in one parallel system permit operation of essentially all types of air-activated sludge processes including conventional plug flow, modified aeration, step aeration, contact stabilization, nitrification, and totally mixed aeration. Air may be supplied by porous diffusers or by new static mechanical mixers with up to 23 liters of air per liter of wastewater (3 ft^3/gal) at the maximum capacity of 750 m^3/day (200,000 gpd) through each of the treatment systems.

Two 380 m^3/day (100,000 gpd) oxygen-activated sludge pilot plants designed by the Linde Division of Union Carbide (Fig. 2) were installed in the covered reactors within the multipurpose pilot plant. Each oxygen reactor includes four contacting stages with a total detention time of about 2 hr at a nominal 380

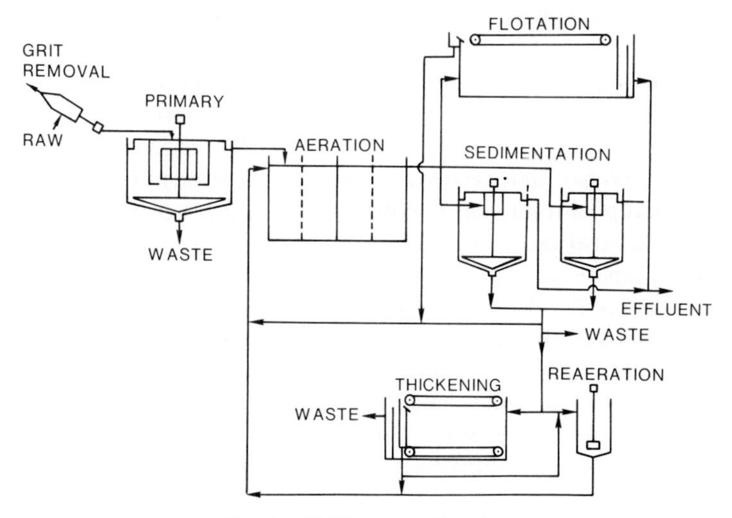

Fig. 1. Multipurpose pilot plant.

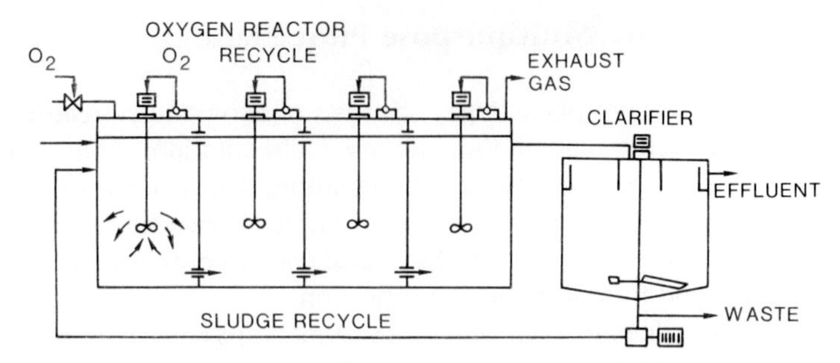

Fig. 2. Oxygen-activated sludge pilot plant.

m³/day (100,000 gpd) flow. The four-stage reactor permits air tight cocurrent contacting of the mixed liquor and the oxygen stream for efficient oxygen use. The cocurrent contacting between stages with oxygen recirculation within each stage produces parallel decreases in oxygen concentration in the overhead gas and in the BOD_5 (5-day biological oxygen demand) of the wastewater, and the pure oxygen permits satisfactory oxygen transfer to mixed liquors of high solids concentration. The oxygen reactors may be used for organic carbon removal or for nitrification.

Another covered reactor is designed for denitrification (biological conversion of nitrate to nitrogen gas) of nitrified effluents at nominal flow of 190 m³/day (50,000 gpd). The reactor contains mechanical mixers and four separate stages to simulate "plug" flow.

The secondary sedimentation tanks for the biological reactors include six conventional center feed clarifiers. Two of the clarifiers are nominally sized for up to 380 m³/day (100,000 gpd), two for up to 280 m³/day (75,000 gpd), and two for 190 m³/day (50,000 gpd) of effluent. A single air-flotation clarifier may be employed in any of the parallel treatment systems to separate the activated sludge from the water under aerobic conditions and, as an alternate, may function as a backup conventional gravity clarifier which can handle a flow of 380 m³/day (100,000 gpd).

The multipurpose pilot plant also includes a small laboratory scale [0.75 m³/day (2000 gpd)] pilot system with two activated sludge reactors with appropriate clarifiers (for parallel or series operation), multimedia filters, and carbon columns. One of the activated sludge reactors may be covered for denitrification studies. The small pilot systems are used to evaluate process feasibility before full pilot studies or as alternate systems for evaluating operating modes in direct comparison to the large pilot studies. The small pilot systems markedly increase the efficiency and flexibility of the engineering development work.

B. Physicochemical Pilot Plant

The highly automated (analog and digital automation) physicochemical pilot plant at a nominal capacity of 190 m^3/day (50,000 gpd) (Fig. 3) includes chemical clarification; either air stripping, selective ion exchange, or breakpoint chlorination for ammonia removal; filtration; pH control; carbon adsorption; disinfection with Cl_2; and a sludge processing and chemical recovery system.

The chemical clarification equipment includes two systems both with two clarifiers of nominal 40 m/day (1,000 gpd/ft^2) overflow rates for a 190 m^3/day (50,000 gpd) flow. One system is designed for tertiary treatment of secondary effluent; the other for treatment of raw wastewater. The clarification systems remove particulate organics and the chemically precipitated phosphorus from the wastewater. These systems, designed for the two-stage excess lime process, include various types of reactor-clarifiers using external sludge recirculation and are operated in series with an intermediate recarbonation tank. Each clarifier system may also be operated in a parallel mode providing single-stage chemical clarification of wastewater by $Ca(OH)_2$–Na_2CO_3 or $Ca(OH)_2$–$FeCl_3$ precipitation.

If high lime pH precipitation is employed, the high pH of the effluent from either the first or second lime clarifier permits the removal of ammonia from wastewaters by air stripping. The multistage air-stripping system consists of five cross-flow cooling towers in countercurrent operation. Air stripping is not used in cold weather due to low efficiencies or freezing.

The filtration equipment is divided into two parallel systems of two filters in each system. Each of the two filtration systems is designed for treating any clarified wastewater at a nominal capacity of 190 m^3/day (50,000 gpd). The

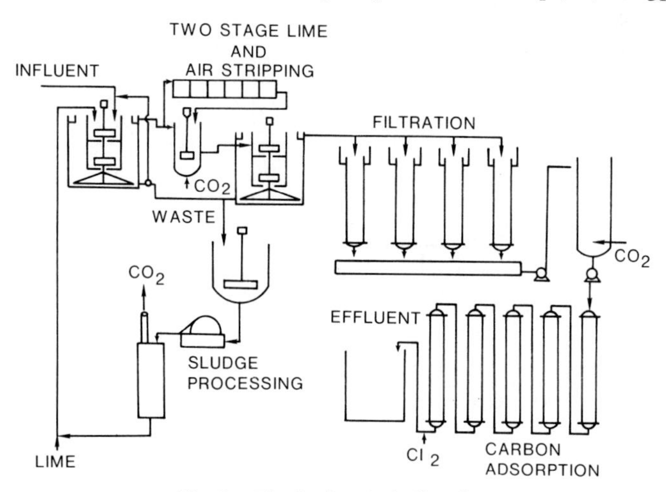

Fig. 3. Physicochemical pilot plant.

filters may be operated as either dual or multimedia filters. Each filter processes 95 m^3/day (25,000 gpd) at a surface-loading rate of 0.12 m/min (3 gpm/ft^2) and can receive a maximum of 380 m^3/day (100,000 gpd) at a surface-loading rate of 0.49 m/min (12 gpm/ft^2). An automatic backwash with air scrubbing permits a maximum backwash rate of 0.815 m/min (20 gpm/ft^2) and produces 50% expansion of the media.

After clarification or filtration, the water (clarified secondary or clarified raw wastewater) may be neutralized with carbon dioxide in a stabilization tank prior to further treatment. The neutralization prevents calcium carbonate scaling and is needed for the subsequent carbon adsorption or selective ion exchange processes. Alternately, the water may be breakpoint chlorinated to oxidize ammonia to nitrogen gas prior to carbon adsorption.

The breakpoint chlorination system permits either two-stage or single-stage chlorination in in-line static or mechanical mixers (1.6-sec. nominal detention time). The system includes automatic control of chlorine dosage, residual chlorine, and pH in the breakpoint reactor and a 1-min detention time after chlorination.

The carbon columns provide a maximum empty-bed contact time of 35 min at a surface loading of 0.285 m/min (7 gpm/ft^2). The carbon system includes

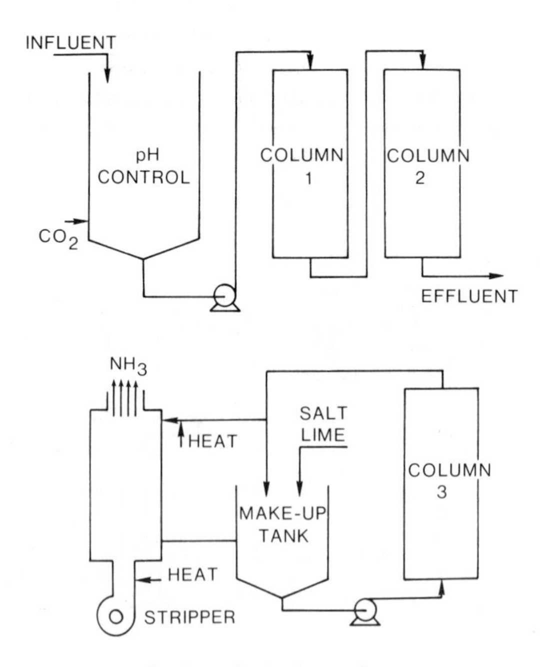

Fig. 4. Selective ion exchange.

automatic backwash and surface-wash systems. A piping manifold permits taking loaded columns off the influent end of the train and placing virgin carbon columns on the effluent in countercurrent operation.

An alternate ammonia removal process, selective ion exchange (Fig. 4), may be located either ahead or after carbon adsorption and consists of two ion exchange columns in service and one under regeneration. The 1.0-m (3-ft) diameter, 3.3-m (10-ft) high butyl-rubber-lined exchange columns, each filled with 955 kg (2100 lb) of 20 by 50 U.S. mesh clinoptilolite, operate under 3.38 atm (35 psig) at a design surface loading of 0.200 m/min (4.95 gpm/ft^2). The ammonium ions in the wastewater are exchanged on the clinoptilolite for sodium ions and are later concentrated in the regenerant brine.

The 1.0-m (3-ft) diameter, 3.5-m (11-ft) high fiberglass stripping column for brine regeneration is packed with 2.3 m (7 ft) of 2.5-cm (1-in.) polypropylene Intalox saddles. Ammonia enriched limesalt brine is recycled through the stripping column from the regenerant holding tank. Up to 14 m^3 (500 cfm) of electrically heated air may be used to strip the ammonia from the regenerant brine.

C. Ozone Pilot Plant

The ozone pilot plant designed, constructed, and operated for the EPA by the Air Reduction Company (Wynn *et al.*, 1973) consists of six reactors (Fig. 5) each with 10-min detention time when operated in series 190 m^3/day (50,000 gpd) and uses pure oxygen to feed the ozone generators. Ozone-oxygen gas may be contacted cocurrently or countercurrently with the wastewater flow.

The oxygen-residual-ozone stream is then recycled through a compressor and

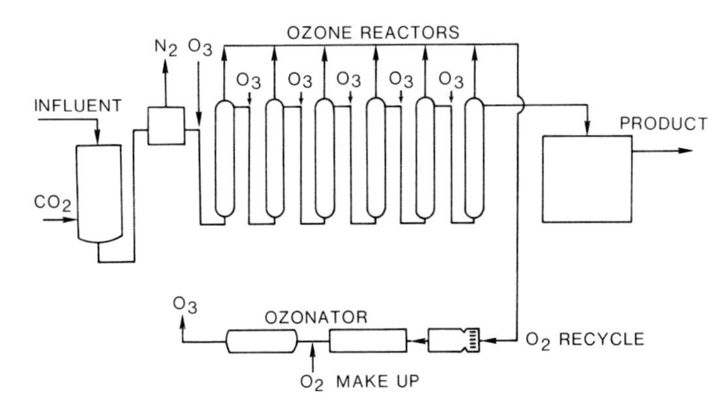

Fig. 5. Ozone pilot plant.

drier to the ozonator. The recycled oxygen loop provides efficient use of oxygen and prevent ozone leakage to the environment. Ozone generation with oxygen is nearly twice as efficient as with air. The wastewater entering the process is vacuum degassed to prevent buildup of nitrogen in the recycled oxygen stream. Control of the influent water pH is also available ahead of the reactors.

D. Solids Handling Pilot Plant

The solids handling pilot plant (Fig. 6) includes gravity or air-flotation thickening, parallel dewatering processes of vacuum filtration or centrifugation, and incineration or recalcination in a fluid bed furnace or in a multiple hearth furnace. Each furnace has associated thickeners, a 1.0-m (3-ft) diameter vacuum filter and 15-cm (6-in.) solid-bowl centrifuge.

The multiple hearth system is physically located in the physicochemical pilot plant. The fluid bed system is in a separate facility. The furnaces are designed to permit study of incineration of organic sludges and recalcination of lime sludges. A special solids injector mounted in the top of the fluid bed reactor properly distributes organic solids and lime or other chemical sludges into the furnace.

III. EARLY TERTIARY TREATMENT

In late 1967, a physicochemical pilot system (see Fig. 3) for carbon and phosphorus removal from the District of Columbia modified secondary effluent (75% removal of BOD_5 and SS across primary and secondary treatment) was placed on stream. The initial system consisted of two-stage high pH lime treatment, filtration, pH control, and carbon adsorption. High pH $Ca(OH)_2$–

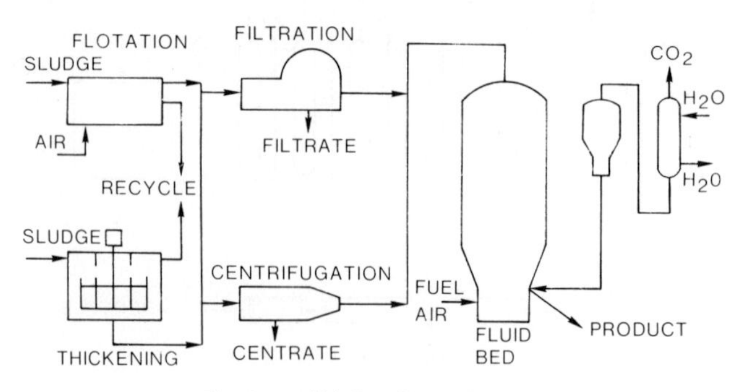

Fig. 6. Solids handling pilot plant.

Na_2CO_3 clarification was also evaluated. The work continued through the fall of 1969 with NH_3 air stripping added in early 1969.

The study (O'Farrell *et al.*, 1969) revealed the importance of secondary effluent quality on physicochemical treatment as a tertiary process. The poorly bioflocculated solids in modified secondary effluent included colloidal bacterial solids that were difficult to remove by chemical clarification and filtration. The colloidal materials containing organic carbon, phosphorus, and nitrogen required supplemental flocculants (15 mg/liter of $FeCl_3$) to produce relatively efficient carbon and phosphorus removals (residual TOC, $\sim$ 3 mg/liter; residual total P, 0.2–0.4 mg/liter).

Studies of air stripping of ammonia (O'Farrell *et al.*, 1972) in 1969 revealed that a high scale rate of 125 mg/liter of $CaCO_3$ precipitated from the Washington, D.C., water onto the stripping towers during the high pH operation (pH 11.5). The hard scale within 70 days of summer operation reduced the air flow rates from 2.6 to 1.9 m^3 of air per liter of water (350 to 250 ft^3 of air per gallon) and thus reduced the ammonia stripping efficiency from 85% to less than 60%.

Laboratory studies (Stamberg *et al.*, 1970) at the pilot plant further revealed that chemical clarification of the raw wastewater produced more effective clarification and lower phosphorus residuals than clarification of low-quality secondary effluent. These studies led to the consideration of physicochemical treatment of raw wastewater.

IV. PILOT STUDIES FOR THE DISTRICT OF COLUMBIA

With the availability of the biological pilot plant in late 1969, the work at the pilot plant evolved into a detailed evaluation of the two basic treatment approaches for C, P, and N removal: physicochemical treatment and advanced biochemical treatment. In cooperation with the District of Columbia, the pilot plant began evaluating several parallel treatment systems: three-stage activated sludge treatment (Fig. 7), high pH physicochemical treatment of raw wastewater (Fig. 8), and conventional-tertiary treatment with an efficient secondary process (Fig. 9). The pure oxygen activated sludge process for BOD_5 removal was evaluated as an alternate to the first-stage (air-activated sludge) in the three-stage activated sludge system. Step aeration with mineral addition and filtration (see Fig. 10) was also studied on a small pilot scale.

The basic three-stage activated sludge treatment system (Fig. 7) consisted of (1) primary sedimentation, (2) activated sludge for BOD_5 removal, and (3) nitrification and denitrification for nitrogen removal, with filtration of the denitrified effluent for residual solids removal. Phosphorus removal was

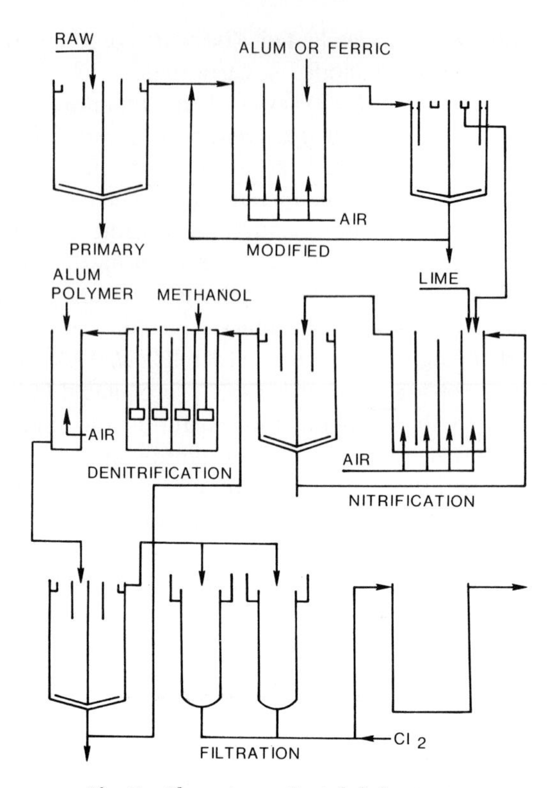

Fig. 7. Three-stage activated sludge treatment.

achieved by mineral addition within the first and third activated sludge processes. Step aeration, pure oxygen, and modified activated sludge were considered as unit processes for carbonaceous BOD_5 removal. Evaluation of the processes in the pilot plant revealed that modified aeration (the present District system) was more appropriate for the complete treatment system.

The physicochemical treatment system (Fig. 8) consisted of (1) two-stage lime clarification on the District of Columbia's raw wastewater for solids and phosphorus removal, (2) filtration for residual turbidity removal, (3) selective ion exchange or breakpoint chlorination for nitrogen removal, and (4) granular carbon adsorption for soluble organic removal. Ozonation was tested intermittently to remove residual organics from the carbon effluent.

The conventional tertiary treatment system (Fig. 7) consisted of (1) primary treatment and step aeration for solids and BOD_5 removal, (2) two-stage lime clarification for phosphorus removal, and (3) filtration for turbidity removal. Breakpoint chlorination (Cassel *et al.*, 1972) and ammonia stripping (O'Farrell *et al.*, 1972) were intermittently evaluated in the system for nitrogen removal.

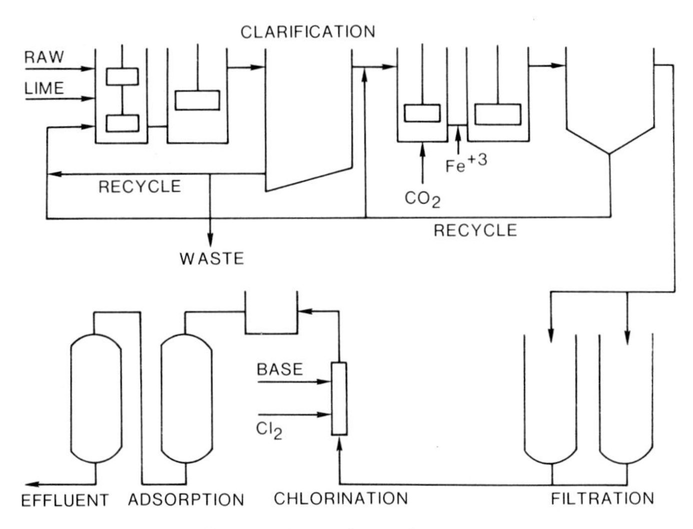

Fig. 8. Physicochemical treatment.

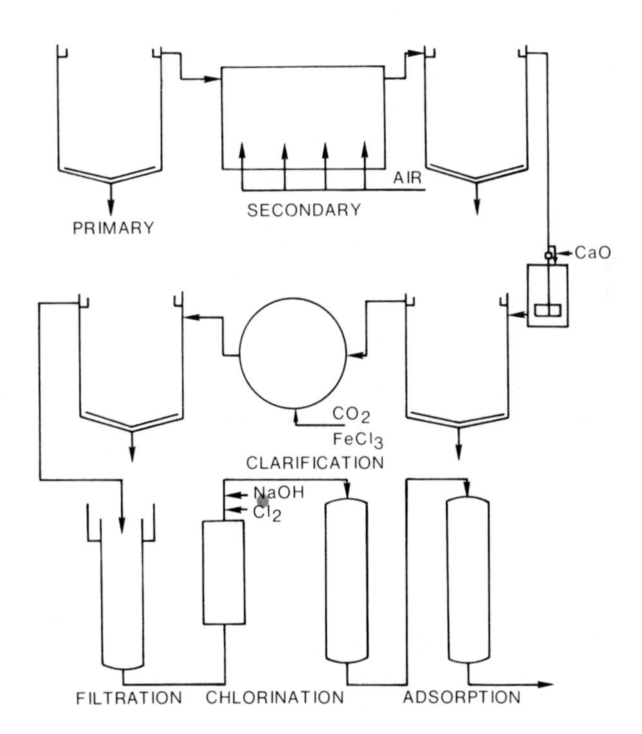

Fig. 9. Conventional-tertiary treatment.

Selective ion exchange was tested in the physicochemical system (Bishop *et al.*, 1972a). Carbon adsorption for removal of residual organics has been evaluated on chemical clarified secondary wastewater in the District of Columbia and was thus only tested on a small pilot scale.

Though the early studies indicated that each of the evaluated systems (conventional tertiary treatment required the selective ion exchange process) had the potential for meeting the discharge standard, the District of Columbia selected the three-stage activated sludge system with mineral addition as the most compatible and the most economical with the existing district facilities. The EPA-DC Pilot Plant then began and completed the development of the detailed design data for the advanced wastewater treatment plant that is planned for the District of Columbia. Cocurrently with the work on the three-stage activated sludge system, research and development for national application continued on various alternatives for physicochemical treatment, including automation of the system with both analog and digital control systems. Work also continued on oxygen activated sludge treatment.

A. Three-Stage Activated Sludge System

The process design data (Heidman *et al.*, 1975) developed for the District of Columbia's three-stage activated sludge treatment system (Fig. 7) included the following:

Chemical requirements (alum, $FeCl_3$, lime, methanol)

The BOD_5 removal rates and the kinetics of nitrification and denitrification for sizing the biological reactors

The settling and thickening characteristics and solids production for design of clarifiers and sludge wasting and recycle systems

The sludge handling characteristics

Automation (Yarrington *et al.*, 1974) of the chemical feeds (alum, powdered CaO in pH control, and methanol) and dissolved oxygen control were also completed. Work is currently underway to develop F/M ratio control for the BOD_5 removal stage.

The most important system design and process interactions are briefly described below. The first activated sludge stage, with 2.5 hr of detention time, was operated with 1100 to 1800 mg/liter of mixed liquor solids to provide a controlled effluent BOD_5 level of 15 to 35 mg/liter to the subsequent nitrification process. The residual BOD_5 provided for reasonable bioflocculation in the nitrification stage. Control of the BOD_5 residual was necessary because NH_3 breakthrough occurred if the BOD_5 entering the nitrification stage exceeded 35 mg/liter during the winter, especially during peak hydraulic flows. In the first stage, alum or $FeCl_3$ was required at a mole ratio of approximately 1.1 : 1 Al (or Fe) to P to insolubilize most of the phosphorus.

In the second-stage activated sludge unit (nitrification), the pH of the influent was adjusted with powdered quick lime (CaO) to produce an effluent pH of about 7. The powdered quick lime (CaO), fed directly into the reactor without slurring, significantly simplified the feeding and pH control systems. The lime neutralized the nitric acid produced during nitrification and thus enhanced the nitrification kinetic rates. The kinetic rates for nitrification during the winter determined the reactor size. For District wastewater with a 2 : 1 peak-to-average flow and approximately 2500 mg/liter of MLSS, the detention time in the reactor was approximately 4 hr for an average flow and 2 hr at sustained peak flows. The design was for a minimum water temperature of 13°C (55°F).

In the third stage (dentrification), methanol was added at about a 4 : 1 weight ratio of CH_3OH to NO_3^- (nitrate expressed as nitrogen) to serve as a carbon source in the anaerobic reactor. The anaerobic reactions reduce the nitrate to N_2. The winter denitrification kinetic rates in Washington required a 4-hr reactor similar to the nitrification reactor. A 30-min aeration stage immediately after the denitrification reactor and before the denitrification settler served three functions: to remove nitrogen gas from the mixed liquor to improve sedimentation, to aerobically remove excess methanol, and to serve as a mixer during the addition of alum to precipitate residual phosphorus. The second alum addition of approximately 35 mg/liter increased the total Al : P molar dosage ratio to about 1.7 : 1.

The settlers for modified aeration with mineral addition and for nitrification with average overflow rates of 0.88 m/hr (520 gpd/ft^2) had conservative detention times of 3 hr. The settler for denitrification with an average overflow rate of 1.35 m/hr (790 gpd/ft^2) had a detention time of 2.5 hr.

The three-stage activated sludge system with mineral addition reliably provided high-quality water (Table I). The residuals of BOD_5, COD, P, and N in all the effluents are averages of data from 10 months of continuous operation. The

Table I
Effluent Qualities of Three-Stage Activated Sludge Treatment

	Amount of effluent[a] (mg/liter)			
Treatment	BOD	COD	Total P	Total N
Primary	99.2	231	6.8	23.6
Aeration	23.6	64.3	1.5	16.7
Nitrification	15.9	22.4	0.87	13.5
Denitrification	9.0	26.7	0.46	2.1
Filtration	2.7	15.9	0.15	1.4

[a]Effluent qualities through the filtration stage are averages from 10 mo of continuous operation.

addition of carbon adsorption, although not required for the discharge standard, reduced the residual COD to 5 mg/liter and thus produced final effluent qualities of 5 mg/liter of COD, 0.15 mg/liter of P, and 1.4 mg/liter of N.

B. High pH Physicochemical Treatment

The physicochemical treatment system, representative of the second basic approach to advanced wastewater treatment, was applied to the District of Columbia's raw wastewater with the 4 : 1 diurnal flow. In the first stage of the precipitation process (Bishop *et al.*, 1972a), raw wastewater, a lime slurry at a dosage of approximately 350 mg/liter as CaO, and recycled solids were rapid mixed and then flocculated usually for about 30 min in a turbine-mixed flocculator. Later, powdered CaO as an alternate to the lime slurry was fed directly into the rapid-mix tank without loss of liming efficiency. The lime increased the wastewater pH to above 11.5 and precipitated and removed bicarbonate, phosphate, and magnesium ions from the water. The magnesium hydroxide flocculated the organic solids that were removed with the mineral precipitates in the settler.

The limed water, after sedimentation, flowed through an open channel to recarbonation. Here carbon dioxide was added (at an approximate average dose of 145 mg/liter for District-of-Columbia wastewater) in a turbine-mixed recarbonation tank with a nominal 15-min detention. The CO_2 reduced the wastewater pH from 11.5 to 9.5 and precipitated the excess calcium ions added in the liming stage. Fifteen mg/liter of ferric chloride were also added in the recarbonation tank to form the flocculant $FeOH_3$. The water was flocculated in a turbine-mixed flocculation basin for 15 min.

In each stage of the two-stage clarification process, settled solids were recycled from the bottom of the settler, at a flow rate equal to 10% of the average influent flow to the reactor, to provide nuclei for chemical precipitation. Wasting of solids at flows equal to 1.5% of the influent maintained the solids balances in the slurry pools of the settlers. A successful clarifier operation was obtained with peak flow rates in each clarifier of 1.35 m/hr (2000 gpd/ft^2).

The overflow from the second settler was pumped to a distribution box ahead of two filters. It then flowed by gravity at a nominal loading of 0.12 m/min (3 gpm/ft^2) through dual-media filters packed with coal and sand to remove residual particulates.

In the early work, ion exchange (clinoptilolite) was employed for ammonia removal instead of chlorination and thus avoided mineralization (addition of NaCl or $CaCl_2$) of the water. The wastewater was neutralized with CO_2 usually to pH 7 before application of the ion exchange process. The ion exchange bed, located either before or after carbon adsorption, replaced the NH_4^+ with a sodium ion and with efficient operation reduced the residual ammonia nitrogen to about 1 mg/liter.

Breakpoint chlorination could be applied to remove the residual ammonia in the ion exchange effluent. If the breakpoint process is located ahead of carbon adsorption, the residual chlorine minimizes biological activity on the carbon columns. In the actual pilot study, however, chlorination was not employed after ion exchange and heavy anaerobic biological activity occurred on the carbon columns. The activity produced H_2S and significantly increased carbon losses during backwash. (Similar work at the Pomona Pilot Plant corrected this problem by the introduction of 5 mg/liter as N of $NaNO_3$.)

Another approach evaluated at the EPA-DC Pilot Plant was breakpoint chlorination of the ammonia to nitrogen gas (Pressley *et al.*, 1972). In the reaction, the chlorine formed hypochlorous acid which oxidized the ammonia first to monochloramine and then to N_2 gas. Base was required to neutralize the HCl produced by the chlorination reactions. The "breakpoint" occurred at the point where NH_3—N was reduced to zero and free available chlorine was detected.

The amount of chlorine required depended upon the ammonia and non-ammonia chlorine demand and the amount of residual free chlorine in the wastewater. In the District of Colombia's lime-clarified raw wastewater, the Cl : NH_3—N dosage weight ratio usually was 9 : 1. A static mixer provided rapid and complete mixing for 1.6 sec average detention time. Control of pH at approximately 7 and free residual chlorine dosage control provided conditions favorable for the oxidation of NH_3 to N_2 and, thus, minimized undesirable production of nitrogen trichloride and nitrate.

After a 1-min holding time, the flow was pumped through downflow granular carbon columns with a detention time of approximately 30 min to remove most of the soluble residual organics. The carbon also removed the residual total chlorine from the breakpoint process. In two-stage carbon treatment, one-half of the carbon was replaced with virgin carbon at each replacement in a countercurrent fashion. The chlorination ahead of adsorption minimized the H_2S production on the carbon columns and the columns required backwashing only once every 2 days without excessive carbon losses. The waste solids from the physicochemical treatment system were thickened and classified with a centrifuge into carbonate (centrifuge cake) and noncarbonate solids (centrate). The centrifuge cake was recalcined to produce CaO and CO_2 for reuse in the clarification process. The solids in the centrate were thickened and then dewatered by pressure filtration. In full-scale practice, the filter cake would be disposed of either by incineration or on the land. The physicochemical system excluding solids handling has been fully automated (Bishop *et al.*, 1973) with a digital process control computer.

Table II
Effluent Qualities in High pH Physicochemical Treatment

	Amount of effluent (mg/liter)[a]				
				Total N[b]	
Treatment	BOD	COD	Total P	Cl_2	Ion exchange
Raw	129	307	8.4	22	23
Clarification	24	55	0.27	15	17
Filtration	20	49	0.18	14	16
N removal	—	—	—	2.6	3
Carbon adsorption	6	15	0.13	2.5	2

[a]Effluent qualities for BOD, COD, and total P represent 10 mo of continuous operation.
[b]Effluent concentrations for N represent typical efficient operating performance for continuous operation of chlorination and selective ion exchange.

The physicochemical system produced water of excellent quality (Table II). The residuals of BOD_5, COD, and P in all the physicochemical effluents are averages of data from 10 mo of continuous operation (but not always with a nitrogen removal process). The nitrogen residuals represent typical operation of the physicochemical system with breakpoint chlorination or with selective ion exchange. Ozonation of the carbon effluent in laboratory (Roan *et al.*, 1973) and intermittent pilot tests indicated about 60 to 70% removal of the COD from the carbon effluent. Thus, the physicochemical system with O_3 as the final process has produced excellent final effluent qualities of 5 mg/liter of COD, 0.13 mg/liter of P, and 2.5 mg/liter of total N on the District of Columbia's wastewater.

C. Conventional Tertiary Treatment

Conventional tertiary treatment was continuously operated with the exception of nitrogen removal for 10 months. During the study, air stripping of ammonia (O'Farrell *et al.*, 1972) was operated for 4 months; short pilot studies on breakpoint chlorination of filtered secondary effluent (Cassel *et al.*, 1972) without chemical clarification were performed later. In the system, the primary sedimentation and the step aeration process usually provided 90% or better removal of solids and BOD_5. However, near the end of the study, filamentous growth developed in the aeration process and reduced the system's treatment efficiency and further demonstrated the requirement of efficient biological treatment in conventional tertiary systems.

The effluent from the step aeration process was fed, usually with an average flow of 220 m^3/day (58,000 gpd) and a diurnal variation of 125 to 380 m^3/day

(33,000 to 100,000 gpd), to a physicochemical treatment system consisting of either one or two lime-treatment solids contactors with a recarbonation tank for two-stage operation, an air-cooling tower (for ammonia stripping), two dual-media filters, a neutralization tank, and a small, activated carbon column. Each clarifier with an internal reactor consisted of four zones; a primary mixing zone (rapid mix), a secondary zone (flocculation), a clarification zone, and a sludge densification zone. Solids from the densification zone from each clarifier were externally recycled at 10% of average flow to either the primary mixing zone or to the recarbonation tank. The clarification system was operated with a maximum overflow rate of 3.4 m/hr (2000 gpd/ft^2).

During the two-stage clarification operation, the pH was reduced to approximately 10.5 in the recarbonation tank where the average contact time was 15 min. Fifteen mg/liter of $FeCl_3$ were added to the primary mixing zone of the second clarifier to improve the flocculation of the precipitated calcium carbonate. Single-stage operation consisted of pumping sodium carbonate solution to either the primary or secondary mixing zone of the first clarifier to precipitate the excess calcium ions and to eliminate the need for a recarbonation tank and second-stage clarifier.

Because of the high scaling rates (Section II) above pH 11.5, air stripping of ammonia was applied to the second-stage clarifier effluent at pH 10.5. The stripping system included five crossflow cooling towers each packed with polypropylene grids. The water was pumped to a distribution box located on the top of each tower and flowed downward over the grids. The air was drawn countercurrent to the flow of water between towers and crossflow within the packing of each tower. The wastewater was then filtered through two gravity fed dual-media filters.

A side stream of the filtered effluent was neutralized to pH 7–8 and pumped through four 7.6-cm (3-in.) activated carbon adsorption columns. At a constant rate of 1.2 liters/min (0.32 gpm), the loading to the column was 0.285 m/min (7 gpm/ft^2) with a total empty bed contact time of 30 min. The columns were manually backwashed everyday at a rate of 0.61 m/min (15 gpm/ft^2).

While the tertiary treatment demonstrated excellent removals of phosphorus and organic carbon (O'Farrell and Bishop, 1972), the nitrogen removal by ammonia stripping was not satisfactory for the District of Columbia's treatment requirements, especially because of air temperature effects on the stripping process. Specifically, two-stage lime precipitation of a good quality step aeration effluent in the solids contactor units produced low residuals of 0.13 mg/liter of phosphorus and 2.1 mg/liter of BOD_5. Dual-media filtration of the chemically clarified effluent reduced phosphorus (as P) and BOD_5 concentrations to 0.09 mg/liter and 1.4 mg/liter, respectively.

Single-stage clarification (pH 11.5 and sodium carbonate addition) with a good quality secondary effluent produced a phosphorus residual of 0.53

mg/liter as P in the clarified effluent. The precipitated phosphorus, in the clarified effluent, however, was filterable as filtration produced an effluent with an average residual phosphorus concentration of 0.10 mg/liter as P. Operational difficulties in the pumping of sodium carbonate solution intermittently caused excess soluble calcium ion concentrations in the water entering the filter. The postprecipitation of these ions as $CaCO_3$ in the filters caused cementation and channeling of the filter.

Recalcined lime, with recoveries up to 75% of the total solids slurry and 60% of the CaO, was recycled to the clarification system without a loss in efficiency of the tertiary system.

In the ammonia-stripping studies (O'Farrell *et al.*, 1972) with inlet water and air temperatures of 25° and 24.5°C (77° and 76°F), respectively, and an air to liquid ratio of 2.45 m^3 of air/liter of liquid (327 ft^3 air/gal liquid), 80% of the ammonia was air stripped at pH 10.5 from the effluent of the two-stage lime system. A reduction in inlet air temperature to 6°C (43°F) reduced the removal efficiency by 30%. At pH 10.5, the calcium carbonate scale rate was 16 mg/liter $CaCO_3$. Activated carbon adsorption removed 55% of the total organic carbon from the neutralized filter effluent with an average TOC residual of 3.7 mg/liter.

The effluent qualities (Table III) for BOD_5, COD, and P in conventional tertiary treatment also represent averages of data from 10 months of continuous operation. Nitrogen removals and the effluent qualities after chlorination represent typical operation. Nitrogen removal with breakpoint chlorination requires a complex control system, but efficiency is basically independent of seasonal and wastewater variations except for ammonia content.

D. Selected Process Studies

The most important process studies related to the work of the District of Columbia were on step aeration with alum addition (Fig. 10) and effluent

Table III
Effluent Qualities of Conventional Tertiary Treatment

	Amount of effluent[a] (mg/liter)			
Treatment	BOD	COD	Total P	Total N
Raw	129	307	8.4	22
Secondary	19.7	59	6.7	17
Clarified	4.2	22.7	0.34	15
Filtered	3.3	21.7	0.14	14
Chlorinated	—	—	0.14	3

[a]The filtered effluents for BOD, COD, and P represent 10 mo of continuous operation. Nitrogen removals represent typical results of short pilot tests.

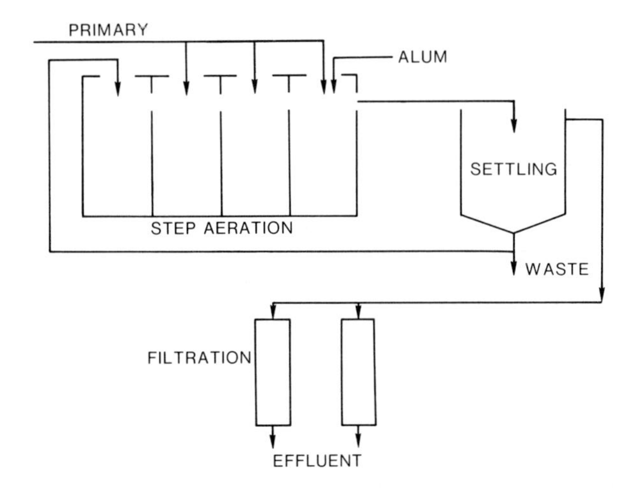

Fig. 10. Step aeration with mineral addition and filtration.

filtration (Hais *et al.*, 1972), on the oxygen activated sludge process (Stamberg *et al.*, 1972), and on control of filamentous growth in the aeration process by H_2O_2 addition (Cole *et al.*, 1973).

1. STEP AERATION WITH MINERAL ADDITION

The small [7500 liter/day (2000 gpd)] step aeration process was operated with a 4-hr aeration time (based on the influent flow) and an overflow rate in the secondary settler of 0.72 m/hr (425 gpd/ft^2). A solution of alum was added to the last aeration pass to supply a 2 : 1 Al/P weight ratio. Lime (25 mg/liter) was added as CaO to the mixed liquor to stabilize the wastewater pH during periods of low alkalinity. The dual-media filter [62-cm (24-in.) bed depth] and the tri-media filter [100-cm (40-in.) bed depth], each loaded at 0.098 m/min (2.4 gpm/ft^2), regularly exhibited filter runs between 24 and 32 hr.

Dual and trimedia filtration both provided effective tertiary solids separation. The trimedia filter removed between 5 and 10% more of the pollutant residuals from the secondary effluent than did the dual-media filter.

The system with trimedia filtration and with the fourth pass mixed liquor pH typically between 6.7 and 6.9 produced filtered effluent residuals of 5 mg/liter of BOD_5, 27 mg/liter of COD, 0.6 mg/liter of total phosphorus (as P) for 3 mo of stable operation. These residuals correspond to removals of 96% of the BOD_5, 90% of the COD, and 93% of the phosphorus from the influent primary wastewater.

Operation of the pilot system with a fourth pass pH between 6.3 and 6.6 improved the removals of BOD_5 and total phosphorus. The pollutant residuals after trimedia filtration for the 20 days of operation in this pH range were 3.6 mg/liter of BOD_5 and 0.23 mg/liter of total phosphorus.

The results indicated that pH control was essential to consistently produce low phosphorus residuals (~ 0.3 mg/liter of P) from the process. In the District-of-Columbia wastewater, the control of effluent pH in the relatively narrow range between 6.3 and 6.6 necessitated alternate dosing of acid and base. The alternate dosing depended upon the prevailing pH and alkalinity of the influent wastewater.

2. OXYGEN-ACTIVATED SLUDGE PROCESS

The oxygen-activated sludge process was originally evaluated at the EPA-DC Pilot Plant as an alternate BOD_5 removal process to reduce area requirements of the three-stage activated sludge system. Later, the studies were continued to further evaluate sludge production and sedimentation rates.

Briefly, the process in the oxygen-activated sludge pilot plant (Section II, A) was operated on District-of-Columbia primary wastewater over a wide range of F/M ratios (0.26–1.0) and with reactor detention times between 1.5 and 3 hr. With the short reactor detention times, the mixed liquor suspended solids concentrations varied from 2700 mg/liter to 8100 mg/liter. The dissolved oxygen concentration was typically in the 4–8 mg/liter range. The microorganisms in the oxygen reactor were visually similar to those in a parallel step aeration process. However, based upon solids production at various operating conditions, the volatile solids in the oxygen process were more active than those in the parallel step-aeration process.

Over the wide range of operating conditions, the oxygen aeration system produced a good quality secondary effluent. The residual soluble 5-day BOD_5 averaged less than 5 mg/liter and indicated essentially complete insolubilization of BOD_5 with average aeration times of between 1.5 and 3.0 hr. Thus, the final product quality depended upon the clarification efficiency. With good clarification, the process produced typical BOD_5 concentrations of 10–25 mg/liter in the District-of-Columbia wastewater (80–92% removal from the districts primary effluent).

The oxygen requirements depended upon the COD consumed (the total influent COD minus the total effluent and waste sludge COD) and the nitrification demand. The study confirmed that each unit amount of COD consumed required a stoichiometric amount of 0_2 input. As expected, the total amount of oxygen required increased with decreasing sludge production. In warm temperatures with low F/M ratios, nitrification occurred and also stiochiometrically further increased the oxygen requirements. During the study, the process consistently used over 90% of the supplied oxygen.

During the summer, satisfactory liquid solids separation with mixed liquor solids concentrations as high as 7000 mg/liter was achieved at sustained peak overflow rate of 3.3 m/hr (1960 gpd/ft^2). In winter, the maximum sustained peak rate was 1.65 m/hr (975 gpd/ft^2) for a mixed liquor concentration of approximately 5500 mg/liter. The study also revealed that the sludge in the underflow from the clarifiers achieved concentrations of 1.0 to 1.4% solids in a clarifier with 1.9 hr of hydraulic detention time and 2.0 to 2.4% in a clarifier with 2.8 hr of hydraulic detention time.

With the high rate of O_2 transfer available in the oxygen process, selected F/M ratios were achieved with smaller reactors than in air-activated sludge by increasing the mixed liquor concentration. For the mixed liquor concentrations usually employed in the study (4000–7000 mg/liter), the suspended solids entering the clarifier exhibited hindered (zone) settling and revealed marked decreases in initial settling velocity as the mixed liquor solids concentration increased. Initial settling velocities varied from 2.3 m/hr (8 ft/hr) to 5.3 m/hr (18 ft/hr) at 4000 mg/liter and from 0.9 m/hr (3 ft/hr) to 2.9 m/hr (10 ft/hr) at 7000 mg/liter.

Solids production varied from about 1.0 to 0.35 kg of solids produced per kg of BOD_5 applied at respective F/M ratios of 1.0 and 0.33 kg of BOD_5 applied per kg of solids.

3. CONTROL OF FILAMENTOUS GROWTH BY H_2O_2

As indicated by previous pilot studies (Section IV, C), poor performance of biological processes caused by filamentous growth reduced the product quality from advanced wastewater treatment systems. The wastewaters in the District of Columbia developed sphaerotilus filamentous organisms when the activated sludge processes were operated with a 3–5 day sludge-retention time (SRT). Laboratory tests at the Dupont Experimental Station and pilot tests at the EPA-DC Pilot Plant revealed that H_2O_2 addition to the recycled solids provided a rapid solution to the bulking activated sludge caused by sphaerotilus growth.

In the study, the heavy filamentous growth in the bulking pilot process was eliminated several times by applying from 20–200 mg/liter of H_2O_2 (based on the influent plant flow) to the recycled sludge for periods from one to several days. The sludge volume index decreased from over 400 to less than 150. The high dosage levels produced the improvement in 1 day. An effective dose of H_2O_2 was a function of time and dosage concentration. The optimum dosage and time were not determined and may vary from plant to plant.

While the H_2O_2 eliminated the bulking condition, the activated sludge process redeveloped filamentous growth if operated under conditions (3–5 day SRT) where the filamentous metabolism was competitive with that of the normal zoogloeal aerobic bacteria. For economic application of the H_2O_2 technique, the biological process must have the flexibility of operation under conditions which suppress the filamentous growth.

V. RECENT WORK

In the early work at the pilot plant, the high pH physicochemical treatment and the three-stage activated sludge systems were considered as chief examples of the two basic advanced treatment approaches. Each system removed C, P, and N and provided both intermediate and final effluents with water qualities suitable for many reuse applications. Other systems at the pilot plant represented combinations of processes from these two systems. The design approach on the two systems briefly described in Sections III, A and III, B of this chapter, while modified by interactions between the different process combinations in other treatment systems, generally may be employed when these processes are arranged into other treatment systems.

The two treatment systems, high pH physicochemical treatment and three-stage activated sludge treatment, reliably produce high quality water on a year-round basis. The high pH physicochemical system produces substantial quantities of sludge. Lime recovery and CO_2 production are generally required even in small scale plants. The three-stage activated sludge system requires relatively long detention times and uses methanol or other organic additives. These system characteristics represent appreciable energy and operating costs (Bishop *et al.*, 1972b) and provided the incentive for additional studies to reduce costs. These recent studies, therefore, evaluated an important combination system and an alternative basic treatment system. Each of these two systems reduces the energy requirements and the costs of advanced wastewater treatment.

The combination system (Fig. 11) consisted of lime precipitation, nitrification, denitrification with alum addition, and, if desired, filtration and carbon adsorption. The physicochemical system (Fig. 12) consisted of low pH single-stage $CaO-FeCl_3$ clarification, breakpoint chlorination, carbon adsorption, and filtration with alum addition ahead of the filters.

A. Lime Precipitation–Nitrification–Denitrification Treatment

The lime precipitation, nitrification-denitrification system has been operated with both low pH single-stage and two-stage lime clarification. All effluent qualities (Table IV) represent typical operation of the system with single-stage $CaO-FeCl_3$ precipitation. Final phosphorus removal by filtration of the denitrification effluent was not demonstrated during this pilot plant study. The moderately high COD in the denitrification effluent represented a small overdose of methanol; however, correction of the methanol overdose and addition of filtration and carbon adsorption should easily produce a final water quality

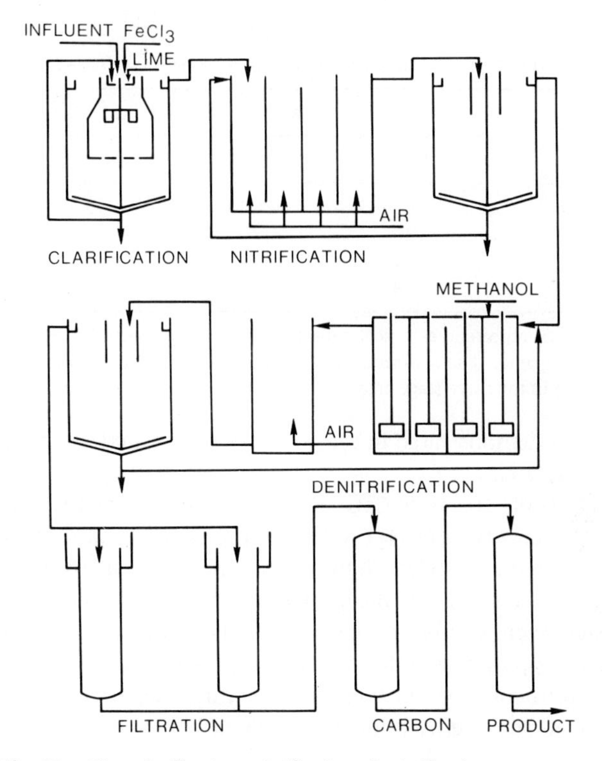

Fig. 11. Lime clarification–nitrification–denitrification treatment.

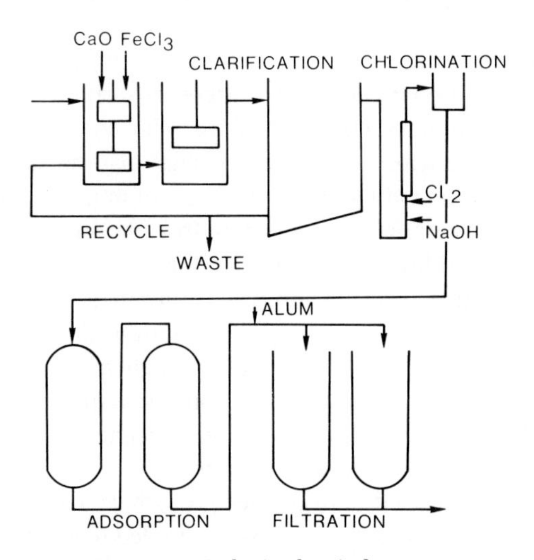

Fig. 12. Low pH physicochemical treatment.

Table IV
Effluent Qualities of Lime Precipitation–Nitrification–Denitrification

	Amount of effluent[a] (mg/liter)			
Treatment	BOD	COD	Total P	Total N
Raw	100	251	6.2	21.8
Clarified	30	66	1.3	16.8
Nitrified	15	26	1.2	15.9
Denitrified	18	42	0.7	1.9

[a]Effluent qualities represent a typical month of continuous operation. Filtration was not employed in the sequence. The performance of filtration after denitrification should produce results similar to those in the three-stage activated sludge system in Table I.

similar to that demonstrated in the three-stage, activated sludge system, Section III, A (see Table I).

The low-pH (pH 10–10.5) $CaO-FeCl_3$ clarification system did not necessarily require lime recovery nor CO_2 production (no recarbonation stage). The wastewater pH was reduced to less than pH 8 by the CO_2 and HNO_3 produced in the nitrification process. The nitrification and denitrification kinetic rates observed on the system were similar to those observed in the three-stage activated sludge system. The appropriate pilot plant design approach on low pH physicochemical treatment (Section V, B) and the three-stage activated sludge system (Section III, A) may be used for this combination system.

B. Low pH Physicochemical Treatment

In the early work on advanced waste treatment, residual COD ($\sim$20 mg/liter) was found in the effluents from carbon adsorption columns which had been applied to secondary effluents. Small pilot studies (Bishop *et al.*, 1967) revealed that this refractory COD was colloidal material that was too small to be removed by filtration through the carbon column and too large to diffuse into the pores of the carbon for adsorption. In addition, filtration studies on the mineralized (alum addition) denitrified effluent in the three-stage activated sludge system (O'Farrell and Bishop, 1974) revealed that dual or multimedia filtration of a wastewater treated with alum removed much more phosphorus than filtration of that same wastewater through a 0.45 μm Millipore filter. Finally, since the breakpoint chlorination process requires base for pH control, that base may be supplied by the lime addition in chemical clarification, as well as by direct addition in the breakpoint process.

The above facts combined with the need to reduce sludge production and the costs of the high pH physicochemical treatment led to low pH physicochemical treatment. The pilot equipment was rearranged to provide low-pH $CaO-FeCl_3$

single-stage clarification ($CaO \sim 200$ mg/liter, $FeCl_3 \sim 45$ mg/liter), followed by breakpoint chlorination, carbon adsorption, and finally by filtration with 20 mg/liter of alum $[Al_2(SO_4)_3 \cdot 14\ H_2O]$ added directly ahead of the dual-media filters.

The low-pH clarification system consisted of a 1-min rapid mix, a 15-min flocculation period, and a 2-hr settling period with a 1.75 m/hr (1040 gpd/ft^2) overflow rate. External sludge recirculation from the clarifier to the rapid mix tank at 10% of the influent flow was also employed in the process. Part of the base required for the breakpoint was supplied by the single-stage clarification process. Design, operating, and control variables for the breakpoint, carbon adsorption and filtration processes were similar to the high pH physicochemical system.

Typical results (Table V) revealed that the low pH system (total detention time ~ 4 hr) provided similar final water quality as the larger two-stage high pH system (total detention time ~ 7 hr) with significantly less lime requirement and lower sludge production. In addition, lime recovery was not required for CO_2 production. The brief studies also revealed that the solids were satisfactorily handled by the vacuum filters and centrifuges available in the pilot plant.

Table V
Effluent Qualities from Low pH Physicochemical Treatment

	Amount of effluent (mg/liter)				
Treatment	BOD	COD	SS	P	Total N
Raw	107	271	133	7.2	23.3
Clarified	30	69	19	1.75	18.5
Chlorinated	—	—	—	—	4.9
Carbon	13	31	6	1.32	3.8
Filtered	4.9	14	3	0.16	2.2

REFERENCES

Bishop, D. F., O'Farrell, T. P., and Marshall, L. S. (1967). Studies on activated carbon treatment. *J. Water Pollut. Contr. Fed.* **39**, 188.

Bishop, D. F., O'Farrell, T. P., and Stamberg, J. B. (1972a). Physical-chemical treatment of municipal wastewater. *J. Water Pollut. Contr. Fed.* **44**, 361–371.

Bishop, D. F., O'Farrell, T. P., Stamberg, J. B., and Porter, J. W. (1972b). Advanced waste treatment systems at the EPA-DC Pilot Plant. *AIChE Symp. Ser.* **68**, 11–22.

Bishop, D. F., Schuk, W. W., Samworth, R. B., Bernstein, R., and Fein, E. D. (1973). *In* "Pollution Engineering and Scientific Solutions" (E. S. Barrekette, ed.), pp. 522–547. Plenum, New York.

Cassell, A. F., Pressley, T. A., Schuk, W. W., and Bishop, D. F. (1972). Physical-chemical nitrogen removal from municipal wastewater. *AIChE Symp. Ser.* **68**, 56–64.

Cole, C., Stamberg, J. B., and Bishop, D. F. (1973). Hydrogen peroxide cures filamentous growth in activated sludge. *J. Water Pollut. Contr. Fed.* **45**, 829–836.

Hais, A. B., Stamberg, J. B., and Bishop, D. F. (1972). Alum addition to activated sludge with tertiary solids removal. *AIChE Symp. Ser.* **68**, 35–42.

Heidman, J. A., Bishop, D. F., and Stamberg, J. B. (1975) Carbon, nitrogen, and phosphorus removal in staged nitrification–denitrification activated sludge treatment. *AIChE Symp. Ser.* **71**, 264–277.

O'Farrell, T. P., and Bishop, D. F. (1972). Lime precipitation in raw primary and secondary Wastewater. *AIChE Symp. Ser.* **68**, 43–55.

O'Farrell, T. P., and Bishop, S. (1974). Filtration of effluents from staged nitrification–denitrification treatment. *76th Nat. Meet. AIChE, 1974.* (Unpublished.)

O'Farrell, T. P., Bishop, D. F., and Bennett, S. M. (1969). Advanced waste treatment at Washington, D.C. *Chem. Eng. Progr., Symp. Ser.* **65**, 251–255.

O'Farrell, T. P., Frauson, F. P., Cassel, A. F., and Bishop, D. F. (1972). Nitrogen removal by ammonia stripping. *J. Water Pollut. Contr. Fed.* **44**, 1527–1535.

Pressley, T. A., Bishop, D. F., and Roan, S. G. (1972). Ammonia nitrogen removal by breakpoint chlorination. *Environ. Sci. Technol.* **6**, 622–628.

Roan, S. G., Bishop, D. F., and Pressley, T. A. (1973). Laboratory ozonation of municipal wastewaters. *Environ. Protect. Tech. Ser.* 670/2-73-075 U.S. Environmental Protection Agency, Washington, D.C.

Stamberg, J. B., Bishop, D. F., Warner, H. P., and Griggs, S. H. (1970). Lime precipitation in municipal wastewaters. *AIChE Symp. Ser.* **67**, 310–320.

Stamberg, J. B., Bishop, D. F., and Kumke, G. (1972). Activated sludge treatment with Oxygen. *AIChE Symp. Ser.* **68**, 25–31.

Wynn, C. S., Kirk, B. S., and McNabney, R. (1973). "Pilot Plant for Tertiary Treatment of Wastewater with Ozone," Environ. Protect. Tech. Ser. EPA-R2-73-146 U.S. Environmental Protection Agency, Washington, D. C.

Yarrington, R. B., Schuk, W. W., Bishop, D. F., and Fein, E. D. (1974). "Computer Operations at the EPA-DC Pilot Plant," *76th Nat. Meet. AIChE, 1974.* (Unpublished.)

SUBJECT INDEX

D